MICROBIC DISSOCIATION

Philip Hadley

From the Hygenic Laboratory at the University of Michigan Ann Arbor, Michigan

as originally published iin the Journal of Infectious Diseases Volume 40, No. 1 (1927)

FOREWORD
S. H. SHAKMAN

InstituteOfScience.com
Santa Monica, CA USA
2012

Published in the USA
by the Institute Of Science

FOREWORD
S. H. Shakman

INNATE INSTABILITY OF MICROBIAL TYPES

Although the popular contemporary view holds that the realm of bacteria is distinctly separate from that of viruses, and within each realm numerous separately distinct entities exist, the reality of the situation is not so clear cut. The fundamental line between what is considered a bacterium and a virus is apparently essentially determined by the capabilities of filters, while true size is actually a continuum from larger organisms which get caught in a filter to smaller ones that pass through. Microscopic and submicroscopic organisms are commonly observed to grow smaller or larger depending on environmental conditions.

In this work Philip Hadley provides a comprehensive discussion of the immense body of prior work on the occurrence of dissociation of bacterial forms, including much accomplished even prior to the start of the Twentieth Century. Hadley's work has apparently, unfortunately, not previously been published independently in book form, which may help explain why this exhaustive treatise has thus far evaded a more-deserved and widespread audience.

For even the general reader, Hadley's introduction and review can be mind-opening, offering a compelling contrast to the rather rigid conceptualizations of microbial species that have persisted up to the current time. The bulk of the remaining work is admittedly more technical, but nonetheless must be considered essential for any and all professionals who wish to explore the intimate and pervasive significance of microbial variation relative to any and all biological processes.

"The chief point … is that there has been slowly developing in bacteriological literature a large body of scattered, and up to the present time uncorrelated, data regarding peculiar culture changes and species instability which, although little appreciated by the majority of bacteriologists and lacking adequate explanation and interpretation, promises to serve as an entering wedge into some of the inscrutable problems that today confront us. And it may be prophesied that, when more of the facts are known, it may be shown that many aspects of observed bacterial instability, far from being a sign of chaos, are in reality indicative of the action of certain biologic laws which are orderly in their operation and widespread in their application among unicellular microorganisms." Hadley herein seems to provide a worthy framework for what he referred to as this "new but definite branch of bacteriological study which is just beginning to emerge from the state of scattered observation."

Hadley discusses in detail "an ever-increasing mass of evidence pointing to the instability of bacterial species [which] ... may have a more significant bearing on problems of virulence, infection and immunity than many have supposed." [p, 5]

In reaction to the advocacy of extreme variability by Nägeli in 1877, Hadley contrasts Cohn's "equal vehemence on the conservatism of bacterial types", which position was joined by Robert Koch and his disciples who "succeeded in forcing upon the bacteriological world a saner - perchance too sane - view definitely opposed to that of Nägeli and his colleagues. ... . the Cohn views gradually became established as a sort of dogma...". [p.8] Hadley laments that "The dogma of the absolute constancy of specific bacterial types was strongly entrenched." [p. 9]

In his concluding observations (p. 286), Hadley asks the bottom line question: "What is 'the normal' bacterial type? .. is there such a thing ...? I believe that a careful consideration of the data assembled in the preceding pages tends to establish the view that 'normal culture' or 'normal type' in the absolute meaning of these terms and as commonly employed, is something of a myth. ... Shall we regard as 'normal' the disease form, the convalescent form or the old laboratory form? Also what shall we say regarding the intermediates and the filtrable forms? ... bacteriologists for the most part have been chiefly interested in their culture tube collections. It is these organisms - often tame, domesticated things - that have been set up as 'types' and as standards of normality. It would seem to me much more accurate ... to regard a culture as normal relative to a given condition of environment... ."

Hadley cites "evidence that microbic dissociation, as an adaptive reaction, stands in close relation to a type of reproduction about which we as yet know little. ... it is becoming increasingly clear that we shall never know what a bacterial species really is until we acquaint ourselves with the outermost limits of its variability ... ." [p. 288]

"The truth we shall eventually come to .. is that the free-living microorganism is potentially a kaleidoscopic thing, in which the power of responding successfully to a changing environment by alterations in the body state, both morphologic and biochemical - and even by self-destruction, if need be, in order to generate another and more stable type - stands as its one most important attribute. ... [288-9]

"...most cultures, when first secured from their natural habitat and placed upon the usual cultural mediums, possess great potential variability. Each apparent species is surrounded by its small group of satellites to each of which we unwisely attempt to assign a classificatory niche." Hadley suggests we concentrate not on methods of classification but rather "on the one thing that is most essential - the problem of the nature and origin of variations. ... there is no class of organisms more favorable than bacteria for studying the possible influence of environment in determining the trend of hereditary variation." [p.290]

> "...it may eventually be demonstrated, not that a foreign filtrable virus gives rise to dissociation and to autolysis in the d'Herelle sense; but on the contrary, that the fundamental physiologic reaction, of which both microbic dissociation and transmissible autolysis are only different modes of expression, gives rise to the filtrable virus." [p.296]

A Brief Chronology of Subsequent Works on the Subject of Dissociation

This chronology incorporates some early works on dissociation, plus selected updates through the late 20th century on related subjects, as compiled and previously presented in *Reference Manual Rosenow Etal* published by InstituteOfScience.com.

Arthur Kendall, *Science* 1931: "It is postulated that a majority, if not all, known bacteria can and do exist in a filterable and in a non-filterable state. ... The belief that bacteria may have a filterable state is a very old one." Kendall goes on to list types: B. typhosus, coccus from influenzal cases, Rosenow's poliomyelitis streptococcus, Dochez's scarlet fever streptococcus, B. paratyphosus alpha, Noguchi's leptospira icteroides, Staphylococcus aureus, B. typhosus and coccus from the "flu" cases, have thus been ... made filtrable, filtered and recovered. ... both staphylococcus 'phage' filtrates, and Bedreska 'staphylococcus antivirus" have yielded perfectly typical cultures of Staphylococcus aureus upon cultivation in the proper manner." [134] "... bacteria within the intestinal canal are in an environment that should encourage their continued existence in the non-filterable state. Rather the contrary condition would appear to prevail in the respiratory tract." [137]

"the problem of immunity, in light of these observations, takes on a new aspect. On the one hand, the beneficial effects reported during the use of phage and Bedreska Antivirus for therapeutic purposes would seem to be related, not to enzymes or toxins, but rather to the presence of viable, filterable stages of bacteria, although, of course, enzymatic and toxic effects are not disproven by any means." [138]
[In suggesting that the beneficial action of a suspected antitoxin may actually be that of an antibacterial vaccine (filterable stage of the bacteria), Kendall allows for the removal of the only exception that Wright had possibly allowed for the action of serum therapy to be other than that of vaccine therapy!]

> H.H. McKinney HH, J. Heredity 28 (1937), 51-57, "Virus Mutation and the Gene Concept:
> "It is possible that the virus represents a filterable form of some larger organism, or it may represent a degenerated organism which has retrogressed by a series of mutations to a stage where a few genes or perhaps a single gene remains to perpetrate as a virus. However, it does not seem necessary to assume that the virus represents a stage in a degeneration process, since it may represent a stage in a progressively complex evolutionary development from molecules, and it may possess a simple metabolism. In any case, the primary virus and its mutants doubtless reflect a series of closely related compounds which function essentially as genes. The several characters of a given virus may reflect properties of a single compound and changes in any of these characters - mutations - may reflect changes in this compound."

Tulasne R, Nature 164 (Nov. 19, 1949), "Existence of L-Forms in Common Bacteria and their Possible Importance" discusses the dissociation of Proteus vulgaris in the presence of high concentration of penicillin into "pleuro-pneumonia-like", "dwarf, sub-microscopical, 'filtrable' forms", "easily identified ... as colonies of the type L" and shown by ultra-microscopical and histo-chemical study to comprise "desoxyribonucleic granulations ... having the aspect and the dimensions of a normal Proteus 'nucleus' (about 0.2-0.3u). ... In the case of a transfer to a medium without penicillin, quite normal Proteus cells soon appear, each of which seems to originate from one of the pleuropneumonia-like bodies." ...

Reversible dissociation of Proteus

Hence a bacterium as common as *Proteus* gives, under the action of penicillin, dwarf, submicroscopical, 'filtrable' forms with possible reversion from the dwarf to the normal form (see diagram).

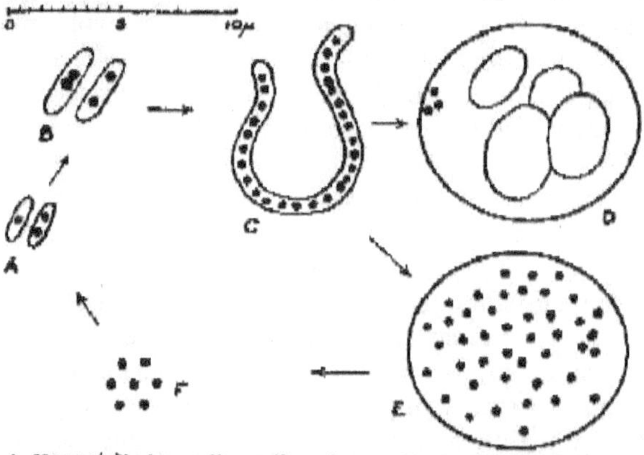

A, Normal *Proteus* cells, resting phase; *B*, normal *Proteus* cells, lag phase; *C*, swarming filament; *D*, large body in lysis; *E*, large body with desoxyribonucleical granulations; *F*, pleuropneumonia-like body (desoxyribonucleical granulations)

Observations concerning such transformations ... are not exceptional"; probably in the near future they will become very common." Tulasne goes on to cite reports of such transformations in cultures of Streptobacillus moniliformis; Bacteroides spp., spontaneously or under penicillin action, with possible reversion to "normal" form; some Gram-negative and Gram positive sporulated bacteria; typhoid bacilli; Pasteurella pestis and Salmonella enterididis. "... Thus it seems to be clear that the appearance of such submicroscopical forms is a general phenomenon .... .

"We are of opinion that these submicroscopical forms of bacteria may be considered as normal resistance forms which the micro-organisms adopt against various noxious agents. They are selected by those agents but not produced by them. They are formed by bacteria which are nearly reduced to nuclei.

"The existence of such submicroscopical, filtrable forms of bacteria may have great importance for the pathology of infection. It is probable that they can be selected under various influences or appear spontaneously 'in vivo', and it is quite possible that their pathological potentialities are different from those of corresponding visible forms. Indeed it may be that the whole problem of the 'filtrable forms' of bacteria (expecially those of Mycobacterium tuberculosis and of Treponema pallidum) should be entirely re-investigated. This may well lead to the solution of some outstanding general problems of pathology and epidemiology".

Louis Dienes and Howard J. Weinberger, Bacteriological Reviews 15 (1951), "The L Forms of Bacteria", 245-288, provide a review of literature on L Forms from 1935:

"In 1935 Klieneberger isolated from cultures of Streptobacillus moniliformis a strange organism which she designated as L1. This organism differed in many respects from bacteria and resembled the organisms of bovine pleuropneumonia. ... " The authors relate that all investigators including Klieneberger have now concluded that L1 "was a growth form of the bacillus. ... Apparently the bacteria undergo a strange transformation in response to various influences, and they survive and multiply as tiny soft forms often unrecognizable with the usual bacteriological methods. The study of these forms is in a preliminary stage and their significance in the life and activity of bacteria is not known. ... [245]

The authors note that Heilman and others were able to regain the bacilli after making transfers from single L1 colonies, demonstrating clearly that the bacilli grow from L forms. [250] Heilman also studied growth requirements, finding that "The bacillus grew on certain media in the absence of native animal proteins. The L1 grew only if such proteins were present. ... The authors further note "Strains isolated from pathological processes are usually highly pleomorphic [with a greater tendency to produce L1 colonies], while strains isolated from the pharynx of rats are often only slightly pleomorphic and produce few L1 colonies.

"The bacillus is very sensitive to penicillin; the growth of some strains is inhibited by the presence of 0.02 unit per ml of the medium. L1 grows even in the presence of 10,000 units of penicillin per ml., and strains of L1 occurring spontaneously in a culture are just as resistant as those developing in the presence of penicillin." In the case of Bacteroides, the authors note "Both the bacillary and the L forms of Bacteroides grow only under anerobic conditions and the L forms only in the presence of animal serum." [251]

"After the discovery of the L1, the next L type culture observed was obtained by Dienes from a bacillus cultivated from a human dog-bite wound." [262]
"... it is of interest that the initial stages of [L-form] development were observed in several genera and families. Thus, this ability to transform into such forms may well be a general property of bacteria." [265]

"It is apparent that the small forms are reproductive [and] that they multiply by division like bacteria. ... Dienes' impression is that the small forms are often elongated and distinctly bcillary. They may grow into short curved filaments and sometimes have the appearance of tiny bipolar stained bacilli. [269] "The small forms gradually swell into the large forms and in most cultures the majority of the organisms are in various stages of this process. ... All transitional forms present from the smallest to the largest in cultures and organisms of varying size may be seen in short chains." [Note similarity to E.C. Rosenow , JAMA 67, 1916, 1203), regarding poliomyelitis "In all the liquid mediums during the early days of growth, chains are often found in which there are single members of all sizes and shapes - large diplococci, large coccus forms, small diplococci and small coccus forms."]

"The change of structure [to L-forms] and slowing down of the metabolism may secure the survival of the bacteria in conditions in which the usual bacterial form cannot survive." [272]

"Although the rigid bacterial membrane is lacking in the pleuropneumonia-like organisms and L forms, a definite cell boundary is clearly seen in stained preparations and with electron microscope, both in the small organisms and the large bodies. The boundary between the individual organisms inside the large bodies, regardless of whether they develop into bacilli or into L forms, only becomes visible when the smaller forms have segmented from each other. The L forms once separated do not coalesce with each other to form larger ones. The large body is derived from a single small organism which probably undergoes multiplication remaining inside of a common envelope. [272-3]
"Pleuropneumonia-like organisms cultivated from urinary infections often are indistinguishable morphologically from L type cultures. ..." [276]
[L forms] belong to the smallest living forms capable of independent life."
"... it is difficult to believe that the similarities between the L type cultures and the pleuropneumonia organisms extending to so many points are accidental. ... [283]

"These characteristics represent a simplification of the bacterial structure, and when bacteria are transformed into L forms, they assume these properties. It is very likely that the origin of a similar complex group of properties is the same in the pleuropneumonia group. This group probably descended at some time from the bacteria and became stabilized to live in this form. If this supposition is correct ... the pleuropneumonia group ... represents a growth form which various bacteria can take up. ...

"The speculations just discussed suggest further speculations concerning the development of the viruses. Two aspects of the problem should be distinguished. One is the phylogenetic developmentof the viruses developed from other microorganisms by gradual loss of properties as a consequence of parasitism. The discovery of the L forms and their properties gives support to this idea. The other aspect of the question is whether viruses develop from bacteria at the present time. The observations made thus far with the L forms offer no evidence for this supposition. We have no information on the role of L forms in the pathogenic action of bacteria. Well-controlled, positive observations on this point would be of great interest. Until they are made, speculations have little value.

"It is outside the scope of this review to discuss the possible relationship of the L forms to the so-called filtrable forms of bacteria, to bacterial gonidia and to various pleomorphic forms described in the literature. Klieneberger suggested in a recent paper [Klieneberger-Nobel, E., 1951, "Filterable forms of bacteria.", Bact. Rev. 15, 77-103] that some of these are probably analogous to the L forms. She feels especially justified to identify the filtrable G forms of Hadley [Hadley P, Delves E and Klimek, J, 1931, "The filterable forms of bacteria: I. A filterable stage in the life history of the Shiga dysentery bacillus. J Infect Dis., 48, 1-159] with the L forms, and she regards the discovery of the latter as confirmation of the older observations.

The reviewers do not agree with Klieneberger's conclusions. ... Unless a technic is devised by which the filtrable forms can be produced regularly so that their properties may be studied, their very existence remains questionable."

De Robertis EDP, Nowinski WW, and Saez FA, CELL BIOLOGY, Fifth Edition, W.P. Saunders, Philadelphia 1970, p. 9: "Among agents that have the smallest living mass, the best suited for study are microbes of the so-called pleuro-pneumonia group (PPLO) ... . These agents range in diameter from 0.25*u* (the limit of resolution of the optical microscope) to 0.1 *u*; thus their size corresponds to that of some of the large viruses."

De Long, R. [J. theor. Biol. (1977) 64, 761-764, "On Bacterial Dissociation"]: "One of the most common phenomena found among bacteria is dissociation... .

"Bacterial dissociation is characterized by many changes occurring in a population simultaneously. These changes are readily reversible. The changes occur in the absence of any known mutagens. Dissociated bacteria have been designated as smooth (S) and rough (R) forms depending on their colonial morphology.

Some of the simultaneous, associated changes occurring along with colonial morphology are as follows: (1) cellular morphology, (2) antigenicity, (3) synthesis, (4) virulence and (5) cellular division. The percentage of bacteria in a population manifesting such changes is large. Populations of bacteria shift from S to R and R to S quite readily along with the associated changes and at high percentages (50% to 100% is not uncommon)."

De Long argues that the frequency of dissociation is far too great to be explained by the hypothesis of mutation, and that "the back mutational rate's frequency makes the mutational hypothesis seem almost ludicrous". DeLong calculates that the probability of both direct and reverse mutation occurring would be less than 1 x 10 (-16), and thus that dissociation relates to a "more encompassing mechanism". The author suggests dissociation may relate to bacterial reproduction, and points out that "Such characteristics as virulence, antigens, syntheses and diagnoses are dependent on whether the bacteria are of the S form or the R form."

In Madoff, Sarabelle, ed., The Bacterial L-Forms, 1986, Lewis Thomas, in the Forward, declares: "L-forms are charming creatures ... absolutely irresistible. They are ... beautiful, nothing less. ... There is a shared hunch that they may turn out to be significant pathogens in certain chronic human diseases, perhaps including genitourinary disorders, even rheumatoid arthritis (my own hunch)."

Madoff [p.2]cites the mid-1930s work of Klieneberger and Dienes as having "ushered in a new era in the world of microbiology" after the former isolated pleuropneumonia-like organisms from a culture of streptobacillus moniliformis. Dienes demonstrated their reversion to the bacillary form [termed "revertant bacteria"] and also "established that bacteria could continue to multiply in the absence of their rigid cell wall". Madoff provides a listing of some 24 bacterial genera from which L-forms have been derived, illustrating that this quality is quite general: Agrobacterium, Bacillus, Bacteriodes, Bartonella, Bordetella, Brucella, Clostidium, Corynebacterium, Erysipelothrix, Escherichia, Flavobacterium, Haemophilus, Listeria, Neisseria, Proteus, Pseudomonas, Salmonella, Sarcina, Serratia, Shigella, Staphylococcus, Streptobacillus, Streptococcus, Vibrio.

Madoff offers that "L-forms probably represent a polygot mixture of considerable variability." As for the induction of L-forms, "The penicillins are the most effective inducing agents. Other antibiotics that interfere with cell-wall synthesis ... have also been used." It is noted that some species "require increased osmotic protection for L-form growth." Among other agents, the combination of sucrose and sodium chloride has been reported to be successful. As for size, the author notes "Although 'elementary bodies' as small as 200-300 nm are seen, the size of the smallest units capable of cell division remains uncertain. Mechanisms of cell replication, as in mycoplasma, appear to vary from binary or asymmetric division to budding or segmentation of the small dense bodies from large spherical forms. ... In summary, L-form processes indicate that if there are orderly mechanisms in the life cycle of L-forms they are not entirely clear. It is not yet known how the genome segregates in the formation of the new and viable L-form elements."

Mil'ko & Egorov, ibid, 1986(4):6-19, "Role of temperate phage in bacterial dissociation"; Medline abstract: "... dissociants may appear in bacteria population from spontaneous mutations and transfer of genetic material (conjugation, transformation, transduction). ... The role of temperate phage has been shown in splitting of bacteria into variants in the genera Mycobacterium, Corynebacterium, some Bacillus, Clostridium, Staphylococcus, some enterobacteria, Yersinia, Vibrio Pseudomonas, Rhizobium, Nostoc..."

Bergh etal 1989, per Sherr 1989, found 3 to 7 orders of magnitude more femtoplankton (viruses of <0.2 um in size) in natural waters than expected. Bergh etal referred to these as bacteriophages. As per other items in this section, it might be suggested that these may be viral analogs to larger identifiable bacterial types, rather than "bacteriophages".

Rosengarten R and Wise KS, Science, 1990 Jan 19, 247(4940):315-8, "Phenotypic switching in mycoplasmas: phase variation of diverse surface lipoproteins". Medline abstract: "The ability of some microorganisms to rapidly alter the expression and structure of surface components reflects an important strategy for adaptation to changing environments, including those encountered by infectious agents within respective host organisms. Mycoplasma hyorhinis, a wall-less prokaryotic pathogen of the class Mollicutes) is shown to undergo high-frequency phase transitions in colony morphology, opacity, and expression of cell-surface-protein-antigens which spontaneously vary in size ...". This is seen as "part of a complex system that controls interactions of these organisms with their hosts."

Pavolva IB, etal [Zhurnal Mikrobiologii, Epidemiologii i Immunobiologii, 1990 Dec.(12):12-5], present an electron microscopy study of bacterial development, asserting that "heteromorphous growth of cells is inherent in the normal cycle of development of bacteria in the population and that this process is reversible. It has certain regularities, common for different bacteria, in the variability of the natural L-transformation of bacteria."

Mil'ko ES & Egorov NS, Biologicheskie Nauki, 1992(5):89-96, "The effect of physicochemical environmental factors on the growth of gram-positive bacterial dissociants". Medline abstract: "The growth of R- [rough], S-[smooth] and M (g)-dissociants of Streptococcus lactis, Bacillus coagulans, Rhodococcus rubropertinctus under the action of some physico-chemical factors: temperature, pH, ultraviolet (UV) rays, high concentration of NaCl and storage have been compared. R-variants gain selective advantage under the influence of UV-irradiation, high temperature and storage; S-variants-- at decreasing of active pH of medium; M (g)-variants -- at decreasing of growth temperature, high values of pH, increased NaCl concentration. The dissociation has been concluded to enlarge the limits of the species survival."

Bove JM, Clin. Infect. Dis., 1993 Aug, 17 Suppl 1:S10-31, "Molecular features of mollicutes", states "It is now firmly established that the mollicutes are true eubacteria. They have evolved regressively (i.e. by genome reduction) from gram-positive bacterial ancestors ... specifically, from certain clostridia. Many of their properties, such as small genome size, small number of rRNA operons and tRNA genes, lack of a cell wall, fastidious growth, and limited metabolic activities, are seen as the result of this evolution. Other properties, such as the anerobiosis of their earliest evolving members ... have been inherited from their eubacterial ancestors. However the mollicutes are not simply wall-less gram-positive bacteria. They have properties of their own. ... [including] peculiar systems for pathogenicity, cell adhesion, antigenic variation ..."

Hurley JC, Lancet, 1993 May 1, 341(8853):1133-5, "Reappraisal of the role of endotoxin in the sepsis syndrome", discusses evidence that release of endotoxin from gram negative bacteria "is not directly responsible for complications of sepsis syndrome", but "rather ... is a marker for transition of gram-negative organisms to cell-wall-deficient forms (L-forms) that may persist undetected despite antibiotic therapy directed against the parental form. This transition has two consequences in compromised patients: L-forms cause organ failure and the serve as a sanctuary from which cell-wall-intact revertants may arise."

L. H. Mattman, Cell Wall Deficient Forms - Stealth Pathogens, 1993
Mattman's "dual thrust" is "to describe the unrecognized omnipresent role of wall-deficient-organisms..." and "to note that the majority of unexplained negative cultures concern infection with these variants." The author notes that shapes "are almost endlessly variegated: and that "Binary fission ceases; budding is one of the common forms of reproduction. ...These variants are critically important in septicemia, menengitis, urinary tract infection, heart valve infection, arthrides, blinding ocular inflammation, and a host of other maladies."
The author relates that "L-Forms circulating in the blood were demonstrated in several general classes of thromboembolic disease", with specific mention of post-operative pulmonary emboli, thrombophlebitis and blood clots. It is noted that "Altemeier noticed that Enovid (Norethnodrel and Mestranol) is a growth factor for

certain wall-deficient bacteria. Thus an explanation exists for thrombi which have been reported in relatively young women taken birth control pills." [123]
The author notes that presumed causative agents for arthritis were cultured as early as 1893 and "observed in direct stains and cultures in 1896"; that streptococci produce rodent arthritis, and "that both L-Forms and parent Group A streptococci produce arthritis when injected intraarticularly into rabbits." [166]

The authors refer to Kendall's work in the 1920s which had "found that filtrable units of the Staphylococcus, Streptococcus and Salmonella typhi could be coaxed back to their parent forms. ... Although William H. Welch regarded Kendall's work as a distinct advance, great skepticism was expressed by most microbiologists. Unfortunately, this was just prior to the demonstration by Klieneberger and by Dienes that filtrable organisms could be grown on solid medium and their sequential reversion steps followed." [209]

The work of P. Hadley and co-workers, with filtrable units of Shigella shiga, and works of F.R. Heilman of Mayo is also noted. For the record, Heilman had co-authored five articles between 1934 and 1939 with E.C. Rosenow, and the cited Hadley work refers to a Rosenow and Towne, J.M. Res. 36, 175, 1917].

Weiser JN, etal., J. Infect. Dis., 1993 Sep, 168(3);672-80, "Relationship between colony morphology and the life cycle of Haemophilus influenzae: the contribution of lipopolysaccharide phase variation to pathogenesis", discusses variants with lipopolysacchaide (LPS) of differing molecular weights, appearance either transparent (heavier LPS) or opaque (lighter LPS), colonization abilities and virulence: "Colonies of Haemophilus influenzae are heterogenous in appearance because of phase variation in opacity. ... The more transparent variants expressing a higher-molecular-weight LPS were serum sensitive and could efficiently colonize the infant rat nasopharynx after intranasal inoculation. In contrast, the fully opaque variant expressing a smaller-molecular-weight LPS was serum resistant, unable to colonize the nasopharynx, and more virulent when intraperitoneally administered. Organisms disseminating into the blood-stream from the nasopharynx changed phenotype from transparent to opaque. ..."

Weiser JN, etal., Infection and Immunity, 1994 Jun, 62(6):2582-9, "Phase variation in pneumococcal opacity: relationship between colonial morphology and nasopharyngeal colonization.", relates that electron microscopy "suggests that autolysis occurs earlier in the growth of the transparent variant". The authors note an apparent "selective advantage" for this variant in its "efficient and stable colonization" of the nasopharynx; and conclude "that phase variation which is marked by differences in colonial morphology may provide insight into the interaction of the pneumococcus with its host.

S. H. Shakman
InstituteOfScience.com
Santa Monica, CA, USA
October 3, 2012

J. Infect. Dis., Vol. 40, No. 1.

MICROBIC DISSOCIATION

THE INSTABILITY OF BACTERIAL SPECIES WITH SPECIAL REFERENCE TO ACTIVE DISSOCIATION AND TRANSMISSIBLE AUTOLYSIS

SIX PLATES

Philip Hadley

From the Hygienic Laboratory of the University of Michigan, Ann Arbor

OUTLINE

Received for publication, Oct. 20, 1926.

1. INTRODUCTION

For the past three decades there has been accumulating an ever increasing mass of evidence pointing to the instability of bacterial species. For the systematic bacteriologist the situation has already become somewhat alarming; while, for those who have never taken the problems of classification too seriously, it at least has inconvenient aspects. But the present state of affairs, as reflected in both publications and discussions, may not be entirely free from elements of danger, since the growing chaos tends to discourage sincere attempts to alleviate the difficulties by searching for the causes that underlie them. While it may be true that greater knowledge of systematic relationships may not contribute significantly to useful knowledge of the bacteria in this or that relation, the inherent problem involved concerns closely microbic variation; and this, in turn, may have a more significant bearing on problems of virulence, infection and immunity than many have supposed.

When confronted on every hand with such pictures of bacterial instability, certainly of a pattern too extensive and intricate to admit of clear exposition, but the general nature of which most bacteriologists realize, it is logical, first to inquire whether the confusion we observe is pure chaos, or whether there exists any trace of orderliness amidst the general disorder, the extreme possibilities of which were first pointed out by Nägeli.[361] Cocci become rods and rods cocci or spirals; forms of growth change overnight; motility is lost and regained; fermentation reactions are modified by time and opportunity; spore formers become sporeless; hemolytic activity comes and goes; capsulated bacteria lose their capsules, and capsules are gained by noncapsulated forms; antigenic power vanishes and reappears; cultures become spontaneously agglutinative or fail of agglutination; virulent cultures become harmless and harmless cultures virulent. It may be many years before we learn the cause of all these apparent incongruities; but we are bound to attack the problem as best we may and to grasp at any possible thread of orderly change running through the tangled web of bacterial variation.

I believe that sporadic citations in the literature of the past thirty years, and more frequent references within the past ten years, indicate that there exists such a thread, common to at least some of the changing bacterial patterns. Indeed, it is becoming clear that one possible approach to the problem may be established through the medium of a new but definite branch of bacteriological study which is just beginning to emerge from the state of scattered observation, and to assume organized form. It is one, moreover, which has attracted less attention than it deserves; and which, it may be predicted, will soon be found to possess unlooked-for significance in its bearing on several important problems relating to bacterial variation and "mutation," serologic reactions, virulence, immunology and serum therapy. I refer particularly to a study of certain more or less orderly mutation-like changes in bacterial cultures, often marked by autolytic processes and occurring either more or less spontaneously, or under the influence of the bacteriophage or lytic principle. Certain aspects of this new study thus concern the phenomenon which has been termed "microbic dissociation," while others are related to transmissible bacterial autolysis. Still others are more clearly related to culture and cell changes that have been reported by a small group of investigators in connection with the so-called "life cycles of bacteria." The chief point, however, is that there has been slowly developing in bacteriological literature a large body of scattered, and up to the present time uncorrelated, data regarding peculiar culture changes and species instability which, although little appreciated by the majority of bacteriologists and lacking adequate explanation and interpretation, promises to serve as an entering wedge into some of the inscrutable problems that today confront us. And it may be prophesied that, when more of the facts are known, it may be shown that many aspects of observed bacterial instability, far from being a sign of chaos, are in reality indicative of the action of certain biologic laws which are orderly in their operation and widespread in their application among unicellular microorganisms.

It is the chief aim of this paper: to expand this view regarding microbic dissociation; to examine in some detail its causes, modes of expression and effects; to make certain comparisons between microbic dissociation and transmissible autolysis with special reference to the mode of action and effects; to suggest forms of reference for some of the apparently more important "types" of bacterial cultures involved—by means of which thinking may be kept clearer and the interchange of ideas made more effective; to reexamine the nature of bacterial "species" and of bacterial "mutations" in the light of the dissociative process; and to consider the probable biologic significance of dissociation in the life of the bacterial culture and of the bacterial species.

To all this it may be added that this work has one further purpose which, though not announced in every section, it is hoped will be kept in the mind of the reader throughout. It relates to the problem of the bacteriophage (transmissible bacterial autolysis), about which a further word may be said at this point.* In the field of bacterial autolysis, as we view it today, there are two classes of phenomena demanding special attention. The first of these is microbic dissociation; the second is the classical phenomenon of d'Herelle, termed by him "bacteriophagy," but by Bordet [62] "transmissible bacterial autolysis." Regarding the first, which has received little attention up to the recent time, there is small question regarding its normal, and apparently spontaneous, nature, although the reaction may easily be forced. Regarding the latter, however, there exists a division of opinion regarding the causal agency—one school (that of d'Herelle) maintaining the bacteriophage (Protobios bacteriophagus, syn. Bacteriophagum intestinale d'Herelle, 1918) to be of a living, virus-like nature; the other school (that of Bordet) maintaining the phenomenon to be due to an inherited, vitiated metabolic state of the organism concerned. Whatever may be the relative merits of these respective views, it is perhaps of importance to note that both bacteriophagic action and microbic dissociation (as this subject will be developed in the following pages) have a common meeting ground in a single aspect of physiologic behavior—namely the instability of the bacterial culture in its reproductive function. In view of this fact it has seemed worth while to conduct a detailed study of phenomena bearing on this subject so far as rendered possible by the literature; and this with a view to ascertaining to what extent, if at all, we can detect any relation in cause, nature or mode of expression, obtaining between these apparently divergent phenomena, proceeding out from a presumably common physiologic state. I believe it is only by a study of such a comprehensive sort that we can derive a sound support for logical and effective methods of investigating the nature and meaning of bac-

* If the reader is not already familiar with the phenomenon of the bacteriophage (the d'Herelle phenomenon), much that appertains to this phase of the subject will be to him meaningless. The reading of d'Herelle's most recent (1926) and fully comprehensive work [348] is recommended. For an understanding of the most widely recognized alternative view relating to the theory of the bacteriophage, I would recommend a fairly recent paper by Bordet.[61]

teriophagic action, about which, notwithstanding the most recent and elaborate researches of d'Herelle and his supporters, one must believe the last word has not yet been said.

2. BRIEF CHRONOLOGICAL REVIEW OF THE DEVELOPMENT OF OUR KNOWLEDGE REGARDING DISSOCIATIVE PHENOMENA

In surveying the present state of our knowledge of microbic instability, and particularly that phase of it which has become known as microbic dissociation, as this will be presented in detail in later pages of this review, it will prove most convenient to regard it as related to certain special fields of bacteriological thought and practice. In order, however, to present a clear notion of the course of development of our knowledge in these fields, it may be of advantage to preface this consideration with a brief résumé by means of which there may be brought to light a few of the more important events in their chronological order.

Our present knowledge of microbic dissociation has its roots in numerous older observations on microbic variability and assumed transmutations. The early and common notions of transmutability among the fission fungi, current through the middle of the nineteenth century, were crystallized in the views of Nägeli,[361] published in his Lehrbuch of 1877. For him the fission fungi represented only a single type of cell, highly variable and capable of passing from one morphological state to another, as well as capable of undergoing profound alterations in biochemical behavior and fermentative capacity. Thus, a "generic and specific" differentiation of the bacteria on the strength of their morphological and biochemical characters was not justified. By this time, however, the botanist, Cohn,[103] had begun to insist with equal vehemence on the conservatism of bacterial types and had presented a system of genera and species which was based solely upon morphological and biochemical grounds. This conception of constancy of species formed a part of the inheritance of Robert Koch, whose discoveries and observations, together with those of his disciples, succeeded in forcing upon the bacteriological world a saner—perchance too sane—view definitely opposed to that of Nägeli and his colleagues—namely, a conviction of the constancy of bacterial types. In bacteriological thought the pendulum of opinion regarding the fixity of species, in these early days, was swinging wide, and the Cohn-Koch views gradually became established as a sort of dogma, in that there was recognized for each bacterial species only one limited colony and culture type, all the cells of which were characterized by the same predetermined morphology.

As Baerthlein[30] pointed out years later, there arose in this way conceptions of normal colonies and normal bacterial morphology, which have never ceased to influence—and in later years, unfavorably—both bacteriological thought and practice. What did not conform to the "normal" was relegated to the field either of degeneration or of "involution" forms—a mischance in which in later years Loeffler and others of his school did much to assist. These degeneration and involution forms have served, even to the present day, as the final repository for all forms of all bacteria which depart from expected "normality."

Although, during the eighties and after, the trend of bacteriological thought turned strongly to the side of Koch's laboratory and here found strong support in the monomorphic views of Migula[353] (1897), other writers, notably Hueppe, Kruse, and Gruber, held tenaciously to the conception of bacterial variability, although perhaps not to the extreme view earlier held by Nägeli. The persistence of these views found expression particularly in the characteristic work of K. B. Lehmann and Neumann,[301] the "Systematik und Handatlas," familiar to all students of bacteriology. In the micrococcus group, in Bact. pneumoniae, Bact. acidi-lactici, B. lactis-sporogenes and in the pyocyaneus-fluorescens group, they pointed out that many accepted species were often only the expression of variability in a single species. These as well as succeeding intimations of bacterial variability, such as those of Neumann for M. aureus, of Kossel and Kolle and of Kruse for the vibrio of cholera ("Stamm VI"), of Gruber and Firtsch for the vibrio of Finkler-Prior, of Schottelius and Dieudonné for B. prodigiosus—all fell on deaf ears. The results obtained were commonly regarded as due to contamination; or, if they were accepted as facts, it was believed that they possessed nothing of practical significance. The dogma of the absolute constancy of specific bacterial types was strongly entrenched.

It probably can be said with truth that slight general interest in, or acceptance of, the fact of bacterial variability even in a relatively minor degree (to say nothing of the "transmutability of species" as upheld in extreme form by Nägeli, the Buchner brothers and others of their school), was apparent until about 1906 when Neisser (1906), and Massini (1907) injected into the somewhat torpid literature of the day their observations on "mutations" in a strain of B. coli, termed B. coli mutabile—observations too clearly significant to be brushed aside. Although at once registered as contaminations by Kolle, they were fortunately quickly confirmed (at Kolle's instigation) by Kowalenko

(1910) who took the precaution to work with single cell cultures. These results, apparently remarkable enough at the time, were quickly extended into other fields. The result was to bring into the field of study of bacterial variation the de Vriesian term, "mutation," imported from the botanical literature. The work of Neisser and of Massini served as a stimulus to many similar observations and the literature from 1906 to 1914, especially in the English school, is rich in allusions to mutation-like phenomena among the bacteria. Many of these observations, particularly from German sources, are summed up (to 1914) in the splendid monographs of Eisenberg. Eisenberg attempted to analyze the various types of variation and to classify them according to their significance—modifications, fluctuations and permanent changes, the last referred to by the terms, mutation, clone-formation, etc.

Regarding the justification of the terms, modification or mutation, much has been written in subsequent times, but without leading to fruitful results. Some investigators have observed cultural alterations in countless numbers believed to be of mutational significance, while others, as Ernst Lehmann, have regarded such variations as observed by Neisser and Massini as not entitled to the term, mutation, which according to their view should be reserved for organisms possessing a sexual form of reproduction. Lehmann held the same view for the "pure lines" or "biotypes" of Johannesen. As we shall see later the question of the justification of the use of the term, "mutation," depends entirely upon the significance which we shall attribute to the phenomenon of microbic dissociation as the subject will be evolved in the following pages.

But perhaps the greatest significance attaching to the controversy over mutations lies in the circumstance that many bacteriologists were admitting the existence of some sort of bacterial variation; were, indeed, recognizing wide, and sometimes apparently permanent, departures from the long assumed, constant "normal type."

The expression of these views, from the time of Neisser's and Massini's work on B. coli mutabile in 1906 and 1907, up to about 1921, which marks the beginning of another stage in the study of microbic instability, took various forms. Indeed, it is seldom that we see the problem attacked consciously and directly. Practically all the earlier observations of present significance are byproducts of some other work; and in many instances were not recognized even by the authors themselves as being related to the general problem of mutation. Some of the observations were concerned with morphology or cultural type, some with biochemical reactions, some with serological reactions and others with virulence. These aspects of the general problem will be considered independently in due time. But it is of interest to point out in passing that, subsequent to the first observations on the agglutination phenomenon by Grubler and Durham in 1896, many studies on the discrepancies in this reaction played indirectly into the hands of the "mutation problem." These studies began with the work of Nicolle (1898) on self-agglutinating cultures of the typhoid bacillus. They were carried forward rapidly by the English school, to a considerable extent by Savage and his followers, and culminated in the valuable work of Arkwright in 1921 dealing with dissociation as a recognized phenomenon among various intestinal bacteria.

Extending more or less parallel with this study of variation in its relation to serologic reactions was a study of the variants in relation to fermentation power. These studies were initiated soon after the introduction of B. coli mutabile by Neisser (1906) and were prosecuted particularly in England by Ledingham, Penfold and their associates as well as by Reiner Müller and others in Germany. They culminated in the significant work of Müller and of Penfold on daughter colony mutants in members of the colon-typhoid-dysentery group. In Germany the study of variation was carried further by Eisenberg in his splendid series of papers on bacterial variability; also by Bernhardt who in 1915 introduced several new and significant points bearing upon the dissociation problem as we have come to know it today. In 1917 appeared the work of Weil and Felix on B. proteus X19, in which unknowingly they approached the dissociative reaction from a different viewpoint; but in such a manner as to make their results converge with other data towards the central problem.

Although these early studies were largely concerned with variability related to serologic and fermentative reactions, the field of morphological variation in colonies and cultures, as a point in itself, had not been abandoned. In 1918 appeared the classical work of Baerthlein representing a study of colony variation and correlated features in fourteen common bacterial species, each analyzed with a wealth of detail. This important work not only correlated many previous observations regarding purely morphological aspects of colonial and cultural variation, but extended the observations in such a manner as should have cleared away, once and for all, any doubt regarding the absurdity of the old dogma of "normal bacterial types"—at least in the extreme form in which it was maintained during the Cohn-Koch regime. At the same time, any far reaching conception of variability that could have been based on Baerthlein's observations must have fallen far short of the extremes of bacterial transmutability pictured by Nägeli and others of his school.

While we may well hesitate in attaching significance to all of the various colony types of each organism described by Baerthlein as actual variations or mutations, there can be no doubt that, among these types, there exist two or three significant forms which we shall see play an important part in microbic instability and dissociation. The most important fact clearly established by Baerthlein's exhaustive and detailed treatment is that colony-variability may always serve as a criterion of cultural variability; moreover, that with the various colony types are closely correlated other highly important characteristics of the organism, including individual morphology of the cells, pigment production, slime production, fermentation reactions, serologic reactions, and virulence. With reference to these matters, Baerthlein appears to have been the first to grasp in a broad way a view of some of the important consequences of bacterial instability; and to place on record a mass of valuable data much of which we are able today easily to interpret into the phenomena of microbic dissociation.

Probably the first investigator to recognize the fundamental significance of Baerthlein's contribution to the study of bacterial variation was Arkwright, already mentioned, whose studies possessed the merit of getting at the kernel of truth which, in Baerthlein's study, was somewhat overshadowed by a wealth of detail often including some inconsequential aspects of variation. Arkwright apparently recognized among the many colony variations reported by Baerthlein two which were of special significance; and these he was able to identify in the case of many organisms of the colon-typhoid-dysentery group. They are the socalled rough (R) and smooth (S) types. He pointed out clearly the trend of the dissociative process (which had already been observed by Baerthlein) from S to R, but not commonly nor easily from R to S. He pointed out again the characteristics associated with the R and S types, their cultural and biochemical features and their serologic interreactions; and discussed the significance of these variations from the viewpoint of the mutation hypothesis.

Although the fundamental work of Baerthlein attracted but little notice at the time of its publication, the significance of Arkwright's contribution was seized upon at once, first and foremost by the English school, while in this country the same line of experiment was developed independently in the hands of de Kruif. In England Arkwright's observations were at once carried over to the streptococci by Cowan (1922), to the pneumococcus by Griffith (1923), to B. enteritidis by Topley and Ayrton (1924) and by Goyle (1926); also to Salmonella types by P. B. White in 1925. They also underlie the chief significance of the dissociative features of the studies by Atkin on the meningococcus (1923) and on the gonococcus (1926). They were carried over to the work of Arkwright and Goyle in 1924 on a comparison of the S and R strains of B. typhosus and B. dysenteriae with the H and O forms of Weil and Felix, and continued in this direction by Goyle in a report of exceptional value appearing in 1926; also by Balteanu in his thorough antigenic analysis of the cholera vibrio in 1926.

In 1921 de Kruif also had reported on the dissociation of Bact. lepisepticum of rabbit septicemia, clearly demonstrating the S and R types, the former virulent, the latter not. De Kruif's results were extended into interesting fields by Webster. The work of Griffith (1923) on the pneumococcus was confirmed and extended by the studies of Reimann and of Amoss in 1925. All of these last studies have served to confirm in a somewhat startling manner the relation of the dissociative reaction to the problem of virulence as well as to many other peculiar aspects of bacterial behavior.

Suffice it to say in concluding this brief chronological review of the earlier studies bearing directly or indirectly upon the dissociation phenomenon, that the data now accumulating, although not meeting by any means Nägeli's surmise regarding the extent of transmutability of bacterial species, can unquestionably be accepted as indicating the existence of a highly unstable state of the average bacterial culture when placed under conditions of changing environment; and as indicating likewise a common trend in the variational processes in all bacterial species so far observed.

As we proceed with the following sections of this review we shall find our inquiry leading into many fields of bacteriological thought and practice. And as we do so we may remind ourselves of a phrase from a work of Laurent.[297] Referring to the old question of bacterial variation he said: "Assurément le transformisme indéfini des Bactéries rèvé par l'ancienne école polymorphiste n'est qu'une chimêre. Mais quel vaste champ de recherches reste à explorer dans la voie des variations physiologiques des microbes!"

So little has the science of bacteriology progressed in the past three or four decades, so far as our intimate knowledge of the bacteria is concerned, that these words, uttered with reference to the old "Ruber of Kiel," come to us today almost as freshly significant as when they were written more than thirty-six years ago.

3. THE NATURE OF THE REACTIONS INVOLVED IN MICROBIC DISSOCIATION

In the earlier development of bacteriology, as has been shown, the predominating conceptions of the constancy of specific bacterial types, taken in conjunction with the fear of unwarranted generalizations, has undoubtedly done much to perpetuate the view that each species was a "law unto itself" with respect to its kind of variation and physiologic behavior. But we are now beginning to see that, as in microbic respiration for example, as this subject has been illuminated by Novy and Soule,[375] certain fundamental aspects of physiologic behavior are not peculiar to this or that species, but are the property of many, and probably all, bacterial forms; and perhaps of all living protoplasm. In a somewhat similar way it is becoming clear that there may also exist fundamental trends of variation in bacteria; and that these are observable, not only in different genera, but also in different families and orders, and probably in higher forms such as Cladothrix, Ascomycetes and the higher fungi as well.

In the majority of scattered instances in which significant dissociations have been noted the study of the variations themselves, as I have already pointed out, has seldom been the primary aim of the inquiry. They have been noted "in passing," so to speak; and our present knowledge of them has been built up largely through resort to reports on widely different lines of research, as for example: the effect of dyes on intestinal bacteria, or of antiseptic substances on typhoid or anthrax bacteria; the "spontaneous agglutination phenomenon," variations in virulence of many species, varying reactions to phagocytosis, the "double antigen" problem of Proteus X19, the body and flagellar antigen of Smith and Reagh,[445] secondary colony formation, and various other bacterial adaptations to unfavorable environments. In other instances they have involved merely casual observations on colony form or other physical aspects of growth. Moreover, it is seldom, at least up to recent times, that an investigator has seen, in his own results, a phenomenon analogous to other and similar phenomena described earlier or elsewhere. Few attempts have been made to study dissociation in itself; and, so far as I am aware, no attempt has been made to correlate these interesting observations or to indicate the extent to which many of them contribute to the formulation of a general scheme of variation and adaptation underlying bacterial instability. First, however, we must understand the chief features of the dissociative process as it occurs on solid and in liquid culture mediums. The observations immediately following have been made by so many laboratory bacteriologists that it is unnecessary, except in more exceptional cases, to cite original references. I wish to present here merely a general picture of dissociation phenomena as they occur on common culture mediums and with ordinary cultures such as B. coli, B. typhosus, Friedländer's pneumobacillus, B. anthracis, B. subtilis, B. pyocyaneus, etc. Next we shall turn to some less obvious and more unfamiliar instances in which it is apparent that dissociative reactions also play the predominant rôle, especially in biochemical and serological reactions.

Dissociative Phenomena on Solid Culture Medium.—Superficially at least, microbic dissociation involves the partial or total transformation of a pure line culture of normal type into one or more subtypes often differing in cultural, morphological, serological and biochemical characters from the original. The phenomenon may be accompanied by the slow or rapid disappearance of the mother culture. If it occurs slowly no macroscopic culture changes simulating lysis appear; if rapidly, a sort of disintegration or lysis of the old culture mass may be seen to occur over broad or limited areas.

Dissociation may occur naturally (spontaneously) or it may be forced by appropriate measures. It may occur in colonies or in broad surface growths on solid medium, or in liquid medium. Naturally that occurring in colonies is the most significant if we are led to believe that the colony has been founded by a single cell. Much of the later work involving dissociation (de Kruif,[118, 119] Webster,[481] Reimann,[405] Amoss,[8] Mellon,[325, 344] Jordan,[273, 274] has involved this precaution. These results, however, are in no way different from the earlier results involving colony isolation only. It is highly important that certain recent work involving single cell isolations seems to validate many earlier observations in which only colony isolations were employed; and even some in which apparently only mass inoculations were practiced, as in the early but important observations of Bordet and Sleeswyk[64] on B. pertussis in 1910.

The phenomenon reveals itself on solid culture medium in many ways. These include primarily: secondary colony formation without erosive action; secondary colony formation with erosive action; lysis and transformation over limited or broad areas without erosive features and with delayed secondary colony formation. These reactions, involving a greater or less disappearance of the old, normal culture mass and the generation of new bacterial forms, may occur either in mass cultivations on agar or in single colonies.

Formation of Secondary Colonies: In the simplest and probably most commonly observed cases dissociation occurs in the form of colonies within colonies without any striking signs of lysis. The old colony, having attained fair growth, undergoes a sort of degeneration, becoming glassy and translucent, not in patches, but over the entire surface. Within it arise daughter colonies or papillae (Knöpfe), from two or three to 100 or more, depending on the colony size. When such a colony, especially if old, is cultured in mass it is often only the daughter colonies that yield growth, although the culture in the glassy areas may sometimes perpetuate itself, if not too old. If the subculture is made early in the life of the primary colony both types of organisms will appear in varying numbers. Sometimes the old culture disappears only slowly and well formed papillae are present after three to seven days.

Although Günther [218] had seen and described secondary or daughter colonies as early as 1895, Preisz [393] (1904) was one of the first to mention typical and atypical strains of B. anthracis in relation to formation of secondary colonies (Knötchen) and spore formation. He made use of slightly alkaline agar and on such medium showed how variations in the form of the rods went hand in hand with cultural variations. The colonies observed when one day old might be either whitish or bluish, smooth or rough, sharply delimited or indistinct, rich in spore production or weak. These characteristics usually appeared divided between the two chief types of colony in the relations mentioned, the former perhaps representing the S type, the latter the R. In some of the larger and more translucent colonies there arose, sometimes after a day or two, sometimes only after a week or more, granular, half-moon shaped "Knötchen" which slowly transformed themselves into "secondary colonies." The translucent races which formed spores most quickly and richly always gave primary colonies with the most numerous secondaries. After a time the centers of these secondaries became pale and translucent, sometimes star-shaped and surrounded with a slightly raised, whitish ring, as if the transparent center had been drawn back to the periphery. Sometimes similar clear areas arose in the primary culture mass itself, without the presence of obvious secondary colonies, a point of extreme significance as we shall see later. The secondaries sometimes showed a "ring-structure," with the clear center still intact.

Preisz attributed the formation of the secondary colonies to the germination and subsequent growth of spores, forming a new race of young bacilli of quite different form from that of the original colony. The consistency of these colonies was often tough and tenacious, so that the entire colony could sometimes be removed from the medium. This phenomenon is very characteristic of many "extreme" R forms. As sporelessness increased, the primary colonies became bluer and the rods longer and more filamentous. Such races, Preisz held, were "headed for destruction"; the cultures would "run out." In conclusion, he noted similar secondary colonies in colonies of B. diphtheriae, Vib. cholerae and Sp. Finkler-Prior. The secondary strains always possessed the greatest longevity. In closing he makes the following pertinent (in relation to our problem of dissociation) observation: That asporogenous as well as sporogenous bacterial species can show exceptional individuals whose vital power is attributable to a splitting of the race; and that it is perhaps by reason of this diversity of type that many statements regarding tenacity of life of nonsporeformers are so little in accord.

The phenomenon observed by Preisz involved important aspects of dissociation and is of unusual interest, not only for its bearing upon variation, but also because of its apparent relation to bacteriophagic phenomena. In this aspect it has been treated further by Pesch and by Katzu whose work will be reviewed shortly. Preisz showed: first the origin of secondary colonies within primary colonies of the blue translucent R type; second, the lysis of the centers of these colonies, with the production of erosions; and third, the occasional appearance of tertiary colonies on the sites of lysis. Here, it may be noted in passing, we have in a rough way, appearances suggesting the production of the lytic areas by the bacteriophage. The only striking difference is that, although Preisz could clearly recognize the source and origin of his lytic spots (Löcher), we do not know the origin of lytic areas. Both Pesch and Katzu noted the resemblance but were unable to detect the presence of a filtrable lytic agent.

The formation of tertiary colonies, as reported by Preisz is of special interest. These were observed to form in the bare areas produced by the lysis of the larger secondaries, and were attributed to the germination of still other spores remaining after the dissolution of the first spore colony lying within the primary. The same sort of phenomenon was reported later for B. anthracis by Pesch,[387] who extended the original results by the following observation. The smaller, whiter and more compact anthrax colonies after some days sent out from their borders a thin, transparent and expansive growth simulating the structure of the blue or translucent colonies. If, however, the plates were held longer at 37 C. there arose colonies of the opaque, whitish type in the blue, marginal growth. Seeding from these gave a mixture of both colony types. He also noted that the blue or translucent form of growth appeared to "repress" the growth of the whitish, opaque type. As suggested later, we may provisionally regard the whitish colony as the S form and the translucent as the R.

Daughter colonies were later observed in many bacterial species: In the intestinal group by Neisser,[362] Massini,[320] Kowalenko,[286] Burri,[88] R. Müller,[356] Thaysen,[460] Penfold,[385] Baerthlein[27-30] and many others, including Eisenberg[148] who in 1906 described daughter colonies in a considerable number of species. Similar observations were made by Beijerinck[42] for B. nitroxus and by Engelland (according to Enderlein) for M. albus and M. aureus growing on agar containing tartar emetic. Here the secondaries, which were much larger and more strongly colored, overgrew the smaller primary colonies. Atkin[21,22] has recently studied daughter colonies of Diplococcus gonorrheae and the meningococcus. Enderlein[159] reported brilliant red daughter colonies of B. prodigiosus in colorless primary colonies grown on agar containing from 6.5 to 8% of salt. It is undoubtedly true, as Enderlein[160] states, that daughter colonies may be observed at times in cultures of all species of bacteria. In some they may appear within a few days, while in other species they may not be seen until the cultures have grown for several weeks, or even months.

On the other hand, as Enderlein also notes, it may sometimes be difficult or even impossible to ascertain whether a primary colony produces daughter colonies. This can result either from the circumstance that they may be so small as to escape notice, or that they may be so numerous and develop so quickly, that they fully obscure the primary growth. Often minute areas of granulation represent the beginning of secondary growth; and these may occur either deep in the colony mass, superficially, on the surface, or at the rim. All colonies should naturally be examined with a hand lens as well as with the No. 3 objective. If a colony contains only one type of organism (i. e., that of the mother culture itself) it is, according to Enderlein, an isomorphic colony; while if it contains two or more types of culture (cyclostages), it is a heteromorphic colony. Aside from the statements of Enderlein, and the older records of Preisz[393] (1904) regarding tertiary colonies in the case of B. anthracis, I am not aware that more than one kind of secondary colony in a single mother colony has been reported. This phenomenon has, however, been confirmed with respect to giant colonies of Streptococcus fecalis growing on Gordon deep "trypagar" plates by Faith Hadley.[499]

We may now ask ourselves—what is the nature of the daughter colonies arising as secondaries in the mother colony? One might be inclined to consider them as representing the R culture formed within the S form. This may be the case in certain instances, and particularly in those cultures in which secondary colonies form in the culture background of old and desiccated agar slants; but there are certainly exceptions; and these are found chiefly in colonies having an age of three to seven days and observed before the effects of desiccation can have become prominent. Under such circumstances, the chief difficulty in the way of interpreting the daughter colony "mutants" as "rough" forms is that, neither in the papillar state nor in subcultivations from papillae, are the centers of modified growth rough. Indeed, they are much more likely to be smooth, as anyone can ascertain by simple examination with a hand lens in good reflected light. This fact is shown particularly in daughter colonies of Streptococcus fecalis, Streptococcus mitis, and Streptococcus salivarius (Faith Hadley [499]). Moreover, Reiner Müller [357] has indicated that when one subcultures a papilla occurring in a typhoid colony the resultant colony growth is often a large, mucoid colony which, in turn, sends out after some days a delicate film (regeneration fringe) representing the normal culture form. These mucoid colonies therefore appear to be "intermediate" and can regenerate the original culture. They are, however, very unlike the R type colonies. Often, as shown by Krumwiede and his collaborators,[288] the mucoid form (B. paratyphosus) may generate the true R culture, rather than the normal S. The same point has been shown by several others for other bacterial species and will be considered in detail in section 6 of this paper. That the true R type colony may, however, develop in the mother culture mass after many days or weeks has been shown for Streptococcus fecalis by Faith Hadley.[499] Here, in giant colonies measuring from 12 to 18 mm. in diameter (on Gordon's trypagar plates grown for several weeks) daughter colonies accompanied by distinct papillae made their appearance in a few days. It appeared that, if the secondary colonies occurred near the surface of the mother colony, they would form papillae; if the colonies were deeper no papillae resulted. The first colonies (and papillae) to arise were coarsely granular by transmitted light, but at the same time smooth by reflected light. After two to three weeks, however, a second type of daughter colony might arise, lying much deeper and sometimes imbedded in the agar itself. These were extremely rhizoid and gave no resemblance to a streptococcus colony. They might be easily mistaken for crystalline deposits unless one removed such a colony in toto, crushed it on a slide and examined stained preparations. Here there appeared a streptococcus in long chains, quite different from the typical fecalis which shows only a slight tendency to chain formation. When transplanted to fresh medium such colonies produced a certain number of "extreme" rough colonies having a thin, coarse, dry and coarsely granular structure and an irregular, jagged margin with many filaments extending out into the medium. Among these roughs were several other very peculiar colony forms. When, on the other hand, the smooth secondary colonies were transplanted to fresh medium many reverted to the normal type, while others began to take on R characteristics and with further repeated transfer to approach more closely the true R type. Streptococcus fecalis colonies were thus found to give rise to at least two forms of daughter colonies, one arising early, the other appearing only after a longer interval and typifying the R form. The subject of daughter colonies in this organism is, however, very complex and requires much further study.

Regarding the general question of daughter colonies a few other points may be mentioned. Regarding their location in the colony, they may appear either in the central part or at the rim. There may be only a single one, or there may be so many that they quickly become confluent and give such an appearance as to suggest that the whole colony is made up of the daughter colony type of culture. If they appear only at the rim of fresh growth, they present the appearance of a rosette. If, in such a structure, the colonies fuse together they produce the effect of a rim-wall about the colony which then manifests a depressed central area. If the original culture regenerates circumferentially outside the rim, then after a time another rim of secondaries may be formed, thus giving concentric zones of growth such as have been often observed by bacteriologists. Enderlein [160] described in detail some of these colony pictures, and they have been reproduced in photographic form by Eisenberg [148] in his interesting study of secondary colonies in 1906.

In the cases mentioned in the preceding pages dissociation usually occurs without striking signs of lysis in the primary colony mass other than a gradual "melting away," and increasing transparency, of the old culture mass, such as is often seen to accompany the process of aging. There exist, however, two other types of dissociation characterized by greater speed of reaction ("suicide cultures"), and sometimes by marked erosive disturbances in the colony or culture mass. The suicide culture has no doubt often been seen by bacteriologists though seldom described. Collins [106] has reported a culture of this sort from a rabbit abscess. I have studied one coming from the gastric mucosa of a rabbit dying in course of immunization to B. typhosus; also from air of the laboratory. Probably it was such a culture of slow dissociative reaction described by Moto [350] as his "creeping culture" (B. helicoides). These cultures are characterized by extremely rapid transformation and often leave in their wake a secondary growth. This type will receive further consideration on a later page.

Erosive Phenomena: Dissociative reactions characterized by greater speed and observable either in colonies or in broad surface growths also occur in the form of erosive phenomena. This was first described for colonies of B. anthracis by Preisz [393] in 1904, and has been rediscovered at intervals since that date. It also appears in the dissociation of B. pyocyaneus as this has been described by Canzik [89] and others including myself.[224] The same phenomenon is undoubtedly concerned in the peculiar erosive ("bacteriophageähnliche") reactions seen by Sonnenschien [448] in a Monilia culture from the throat; in B. cereus (Andervont and Simon [9]); also in cultures of Saccharomyces.

The phenomenon of dissociation in B. pyocaneous shows strong resemblance to the Preisz phenomenon in B. anthracis; but with this exception: It has been impossible to determine the origin of the erosions. They appear to develop from areas of granulation, as in the Preisz phenomenon, but it has not been demonstrated that they develop from colonies, although this seems highly probable. In the case of pyocyaneus the granulations which subsequently may be centers of erosive action make their appearance first on the surface of the colonies (on agar slant cultures) as metallic flecks. These gradually enlarge and deepen to produce erosions of large size and with perpendicular walls. Over the floor of the pockets, as was also true in B. anthrax, lies a thin film of culture.

The phenomenon observed and pictured by Sonnenschien in Monilia shows striking resemblance to the case of pyocyaneus. Unfortunately, however, he made no reference to different types of colony or variations in cell form. He found no evidence of transmissible lytic action such as I have observed in pyocyaneus.

Transformation and Delayed Formation of Secondary Colonies: Still another manifestation of the dissociative reaction involves the appearance of a fringe of new and modified growth at the border of a colony or streak, from which it extends in a delicate film like a halo (regeneration fringe). It was pictured by Sanarelli [423] in 1897 for the cultures of B. icteroides obtained in his yellow fever researches. It is probably the same reaction as the "fringe growth" on colonies of B. anthracis described by Pesch, as already mentioned. It has been described moreover by Neisser and by Massini for B. coli mutabile and can be seen in almost any plating of sewage polluted water on Endo plates, as I have often observed when they are kept for a week or more at room temperature. It has also been reported by Bernhardt,[49] Baerthlein [30] and others. I have observed the phenomenon in colonies of B. diphtheriae, B. malleus, staphylococcus and in a proteus-like culture. Possibly a similar growth has been mentioned by Braun and Schaeffer [66] for the O form of proteus X19 of Weil and Felix, which is described later. It does not ordinarily issue from a normal culture, but from an intermediate form (see section 6).

Dissociative reactions may show themselves also in another manner. In cultures of many intestinal bacteria, Friedländer's pneumobacillus, etc., grown upon alkaline (P_H 7.8 to 8.0) agar the borders often show curious bluish or translucent, wedge shaped invaginations which may unite to produce a slowly growing translucent fringe. The lysis of staphylococcus colonies observed by Twort [468] was probably a closely related phenomenon. This has been mentioned by Gratia for B. coli, and I have seen it in nearly all members of the intestinal group. It likewise may appear as blue sectors in colonies, when they are from a few days to a week old. The translucent fringes and invaginations in such cases are made up almost exclusively of R type culture, while the S form has largely disappeared from these areas. If such areas are plated they yield almost exclusively thin, irregular, translucent colonies. It is important to note that in cases of this sort, there exists not only an extension of the fresh growths to new agar surface, but also an extension of the dissociation area backward and into the original culture mass which may eventually be wholly consumed, leaving only a thin film of transparent growth (Hadley [226]). This reaction I have described and pictured in another place for some unknown air and water bacteria. The colonies may become so transparent as to make easily possible the reading of printed type lying beneath the plate. Such cultures are in reality, "slow suicides." They have been clearly pictured, but not adequately described, by Gildemeister [198] in 1916 for a culture of B. paratyphosus B, isolated from a carrier.

In connection with growth fringes, regeneration fringes, or "halos," it is important to observe that cultures of the R or of the intermediate type (O) may arise (though indirectly, as we shall see) from the S; or that cultures of the S type may arise (perhaps directly) from the R. The former is certainly the more common. The growth fringes of B. proteus, as described particularly by Braun and Schaeffer,[66] are of interest in this respect. Here, however, the pure R type is perhaps not yet definitely known, but we have two clearly marked forms, the spreading (H) and the restricted (O) to be considered later in detail. In this case the intermediate O sends out the fringe of H. In proteus cultures many bacteriologists of laboratory experience have observed the "ring-growth," in which, starting from a single, small colony on a plate, eight to ten or more concentric rings of alternating transparent and whitish growth may be found spreading from the common center. I have studied several of these cases and have concluded that they represent alternations of S and O culture indicating alternate dissociations and recoveries. The true R form, however, was not present in the cultures which I have examined. The characteristic S or O cultures can be isolated from the respective ring areas. It seems likely that "ring-growth" in some other bacterial species may have a similar explanation.

Certain observations, however, particularly on B. proteus, make it appear that the question of regeneration fringes may not always be so simple as the regeneration of typical R (or O) from S or of typical S from R culture. The point in question has been made clear particularly in a study by Braun and Schaeffer on the double antigenic proteus HX19 and the single antigenic OX19, as well as on X2 strains. In view of some later deductions it is important to present this matter in some detail.

Braun and Schaeffer attempted to ascertain whether the HX19 of Weil and Felix [483] could be transformed into OX19 by disinfectants or by "starvation" (for methods see section 11). By growing on phenol agar and on agar impoverished in beef-tea content they succeeded in obtaining the O type which was fairly constant in its changed (non-spreading) character in contrast to the H type (spreading), even when returned to common mediums. It sometimes happened, however, that these O type cultures (and particularly a natural OX19 obtained from Weil) gave a regenerated growth quite different from the original H though possessing the H spreading character. This new growth, which emanated from a single streak of OH19 on an agar plate, eventually spread over the entire plate. Unlike the normal HX19 (also spreading), however, this new growth was extremely delicate and quite invisible except in certain plate areas and in favorable light. It appeared more like "angehauchten Glas" and developed much more slowly than the normal spreading HX19 culture. Other important aspects in the study of Braun and Schaeffer will be presented later in the appropriate sections, but we shall find it convenient to bear in mind this almost "invisible colony growth" emanating from the O type culture.*

The extent of dissociative reactions in a culture is important, whether in single colonies or in broad surface growths. The beginnings may show merely as areas of surface depression, sometimes central, sometimes marginal (Hadley [225, 226]), in which otherwise opaque (S) growth becomes translucent or even transparent over a sharply demarked area. There may be one or many such dissociation centers. Sometimes they are shallow and temporary; but more frequently they are progressive with increasing age of the culture; and in a few instances as in B. anthracis (Preisz,[393] Katzu [276]), in B. pyocyaneus (Canzik,[89] Hadley [225]) and in Monilia (Sonnenschien [448]), develop into well marked erosions which may within a week's time consume nearly the entire colony or culture mass. Particularly in the case of B. anthracis and Monilia the normal culture is replaced in the eroded areas by a dull, wrinkled and highly tenacious growth. The case of B. pyocyaneus I have described in detail elsewhere.

Suicide Cultures: There remains to be mentioned a striking mode of dissociation no doubt observed by many bacteriologists who have had laboratory experience, and this relates to the socalled suicide cultures, which have received mention by Collins [106] but have not obtained a significant place in the literature. Such cultures, after attaining within the first 18 to 20 hours a luxuriant growth, then seem literally to melt away within the space of another 12 to 18 hours until nothing is left but a thin, transparent film covering their former site. Such cultures are manifestly lytic and there is some evidence that they may be lysogenic. Because of the usual lack of corresponding S type cultures one is handi-

* Referring to the terms H (double antigen type) and O (single antigen type) of Weil and Felix, we shall see subsequently that, despite the antithetical views of Arkwright and Goyle,[18] we are justified in regarding them identical with what will be termed the S (H) and the O (O) type cultures; R was presumably not observed by Weil and Felix, but may have been by Fejgin.[169]

capped in the study of the action of the filtrates. These cultures often fail to grow in beef broth. On agar, however, they may leave secondary colonies on the old culture site. Cultures from these colonies undergo no further lytic changes; they are resistant to the suicidal action. Regarding the position of these suicide cultures, they seem to occupy a place intermediate between those cultures manifesting the more common form of active dissociation and those which manifest frank, transmissible autolysis. Like both, their reaction is characterized by the elimination of S organisms and a generation of the R. Such cultures are of great interest and demand further study with reference particularly to their lysogenic abilities. Certain aspects of this matter will be considered later.

To the above a few incidental points may be added. When dissociation occurs in a culture the growth energy always is markedly enhanced; we obtain, as I have pointed [226] out for B. pyocyaneus and some other cultures, a veritable proliferative growth accompanying or preceding the reaction. In connection with lysis accomplished by the bacteriophage in liquid mediums it may also be recalled that Bordet [61] has drawn attention to the necessity of "growth before lysis." D'Herelle [247] also has pointed out this reaction as a characteristic of all cells when attacked by a filtrable virus. Another minor point is the circumstance that, when the erosive type of dissociation occurs in agar slant cultures, it appears largely on the bottom half or two-thirds of the slanted surface. This also is true of the lytic areas caused by the bacteriophage, as shown by d'Herelle. We shall have reason later to return to these simple facts.

Dissociation in Liquid Medium.—It is also true that dissociation takes place in liquid mediums, and here even more actively than on solid surfaces (de Kruif,[120] Webster,[481] Mellon,[326, 327] Soule [450]). Sometimes the dissociation, accompanied by a greater or less loss of the S type organisms, registers itself by partial clearing. In other cases it can be detected only by enumeration of colonies when the broth cultures are plated or streaked on a solid medium. Here, however, the results may be misleading owing to the repressive action of the R type colonies on the S (de Kruif,[120] Pesch; [387] see also section 11). On such plates one may observe the presence or absence of R type colonies, or the relative proportion to those of the S type. Dissociation in liquid medium may proceed rapidly or slowly depending on its composition, the nature of the organism and its stage of development. According to Feiler [167] it proceeds more rapidly at 37 C. than at 22 C. Certain substances added to broth may hasten the process, while others may delay it (Webster [481]).

Rapid passage through an unfavorable medium increases the R type, while passage through a favorable medium increases the S type (de Kruif [120]). These aspects are considered more fully in section 11 on incitants to dissociation. Cultures which contain mixtures of S and R types have no doubt often been discarded because of suspected contaminations (Eisenberg,[150] Bernhardt,[49] Preisz [394]). In reality they may be mixed, but they are not on that account contaminated. By the time the dissociation has progressed sufficiently far to give a predominance of the R form, it usually manifests an agglutinative form of growth in broth, and will not remain suspended evenly in salt solution. Under these conditions it may still be kept in suspension by decreasing the salt concentration (Arkwright,[16] Shibley,[440] and others) to 0.42 or even to 0.21%. Although the R type agglutinates spontaneously, the precipitate is not always granular but may be flocculent. The granular precipitates are more commonly produced, as we shall see later, by another form of culture (P. B. White,[487] Goyle [206]). These variations in the form of growth, it may be noted, are not unlike what is often observed in the top and bottom growth of certain yeasts, among which I have little doubt it will be shown that a somewhat similar dissociation is also present.

So much insistence on the chief types of dissociates, S and R, is likely to leave the impression that these forms comprise all the significant participants in the reaction. Such a view, however, would be a great misapprehension of the actual situation. While it is indeed true that the "smooth" and "rough" forms are the most common and striking types, and seem to represent respectively the beginning and end of the reaction, a careful reading of Baerthlein and a consultation of many other works (Firtsch,[178] Eisenberg,[150] Bernhardt,[49] Arkwright,[17] Goyle [206]) impresses us with the fact that there are other forms of consequence. Baerthlein [30] often describes from three to five different colony types for the many organisms considered in his classical study of colony variation. Among these types it is always the S and R forms which stand out most clearly; but what of the others? Are they chance variations or are they of deeper significance? At present we cannot state, but the latter seems more probable. One type in particular appears in the literature with peculiar constancy, though usually mentioned as existing in small numbers. This type, which has been mentioned for many bacterial species, shows colonies that are large, round, regular, convex, fleshy and mucoid; they often resemble colonies of Bact. aerogenes. Microscopic examination reveals long filaments, coccus forms and often giant cocci. It is fairly stable in propagation and some evidence suggests that it is the mother form of the R type; at least it is different from the true S and R forms. I mention this and the other variations only for the purpose of indicating before we proceed further that the problem of dissociation is by no means limited to the S → R "mutation," although this is at present its clearest aspect and one that will occupy our attention sufficiently. Sometime the relation of these other colony types (which it is scarcely feasible to discuss further at present) to the complete dissociative reaction will undoubtedly become clearer. At the present time all that we see clearly is that some of them are intermediate forms lying between S and R, and that a few of them present an extremely curious behavior when we attempt to cultivate them. To a more detailed consideration of this point we shall return in sections 5 and 12 of this paper.

This brief review of the general nature of the dissociative reaction is sufficient to indicate its main lines of action, and the sort of phenomena concerned. It will be the object of the forthcoming pages to expand into greater detail some of these issues which thus far have been only briefly mentioned. And, as we progress, we shall see that these apparently simple changes enter into many fields of bacteriological theory and practice.

4. RÉSUMÉ INDICATING THE EXTENT OF THE DISSOCIATIVE PHENOMENA, TOGETHER WITH THEIR PARALLEL TRENDS AMONG BACTERIA AT LARGE; SUGGESTIONS FOR TERMS OF REFERENCE TO PRIMARY DISSOCIATES

Extent of Phenomenon.—In the previous section I have dealt with some of the manifestations of dissociation but only in such a way as to indicate roughly its superficial mechanism and effects in ordinary cultures, without any attempt to look behind these surface phenomena to the deeper significance of the reactions. Since it would be impossible in the space available to give a summary of even a small part of the numerous references bearing upon the problem of dissociation, it is my intention to present in the pages immediately following, only a skeleton outline showing the implication of this phenomenon in a variety of bacterial reactions, but without in any case attempting to exhaust the subject. Subsequently I shall give a more detailed review of its bearing on certain aspects of bacterial behavior that are of greater practical significance in laboratory bacteriology. The following citations, arranged roughly according to subject, thus make no pretense of completeness, but are taken as representative of instances in which it seems fairly certain that microbic dissociation was at work, although often not recognized as such at the time when the observations were made. As will be observed, the majority of the citations deal with microbic variations occurring either spontaneously or under the stimulus of modified environment. References to the source of the observations will be found in the literature list at the end of this paper.

(1) Microbic dissociation is concerned with the production of daughter colonies of a character manifestly different from the mother type on a background of old or dying culture; also when daughter colonies appear within mother colonies. The following examples may be presented:

B. anthracis (Preisz,[393, 394] Eisenberg,[149, 151] Pesch,[387] Katzu[276]).
B. coli mutabile (Neisser,[362] Massini,[330] Beneke,[45] Kowalenko,[286] Hubener,[260] Sobernheim and Seligmann,[447] Eisenberg,[154] Baerthlein[29]).
B. cyanogenes-lactis (Eisenberg[148]).
B. diphtheriae (Preisz,[393] Bernhardt,[49] Trautmann and Dale,[467] Enderlein[159]).
B. enteritidis (Eisenberg[148]).
B. fluorescens (Eisenberg[148, 153]).
B. herbicola (Beijerinck[42]).
B. indicus (Eisenberg[148]).
B. kielense (Eisenberg[148]).
B. malleus (Eisenberg[148]).
B. megatherium (Eisenberg[148]).
B. mycoides (Eisenberg[148]).
B. nitroxus (Beijerinck and Minckmann[43]).
B. prodigiosus (Eisenberg,[148] Baerthlein,[28] Enderlein[160]).
B. pyocyaneus (Eisenberg[148]).
B. ramosus (Eisenberg[148]).
B. septicemiae hemorrhagicae (Eisenberg[148]).
B. sporiferus (Eisenberg[148]).
B. subtilis (Eisenberg,[148] Soule[450]).
B. typhi-murium (Eisenberg[148]).
B. typhosus, paratyphosus, dysenteriae (Eisenberg,[154] R. Müller,[356] Baerthlein,[30] Burri,[88] Ledingham,[300] Thaysen,[460] Jacobsen,[265] Penfold,[383] Schröter and Gutjahr,[435] Gildermeister,[198] Morishima[352]).
Bact. pneumoniae (Eisenberg,[153] personal observation).
Corynebacterium diffidens (Enderlein[160]).
Diplococcus gonorrheae (Atkin[22]).
Diplococcus meningitidis (Bernhardt,[49] Atkin[21]).
Micrococcus albus and aureus (Lehmann,[301] Engelland, according to Enderlein;[160] Eisenberg[148]).
Sarcina aurantiaca (Eisenberg[148]).
Sarcina lutea (Eisenberg[148]).
Sarcina pulmonum (Eisenberg[148]).
Spirillum albensis (Eisenberg[148]).
Spirillum Finkler-Prior (Firtsch,[178] Preisz.)
Streptococcus fecalis, mitis, salivarius, hemolyticus (Faith Hadley[498]).

Vibrio cholerae (Preisz,[393] Eisenberg,[150] Enderlein,[160] Balteanu[38]).
Vibrio metchnikovi (Eisenberg[148]).
Vibrio rumpell (Eisenberg[148]).

(2) Microbic dissociation is probably at work in the production of the often observed regular or irregular colonies in the following cases:

B. acidophilus (personal observation).
B. anthracis (Bongert,[55a] Preisz,[394] Wagner,[476] Baerthlein[28]).
B. avisepticus (Manniger[317]).
B. coli (Neisser,[362] Massini,[320] Baerthlein,[29] Eisenberg,[154] Bernhardt,[49] Prell,[396] Arkwright[16] and others).
B. cholerae suis (Baerthlein,[27] Orcutt,[376] White[487]).
B. enteritidis (Baerthlein,[27] Topley and Aryton[464]).
B. diphtheriae (Corbett and Phillips,[110] Zupnik,[495] Schick and Ersettig,[430] Slawyk and Manacatide,[443] Bernhardt[49] and others).
B. dysenteriae (Steinhardt,[453] Baerthlein,[30] Arkwright[16]).
B. herbicola (Beijerinck[42]).
B. lactis-erythrogenes (Dyar[142]).
B. paratyphosus (Baerthlein,[27, 30] Savage,[425] Breinl and Fischer,[71] Jordan[273]).
B. proteus (with special reference to spreading or restricted growth, Baerthlein,[30] Braun and Schaeffer[66] and others).
B. pyocyaneus (Baerthlein[28] and others).
B. tetanus (Hilda Heller[241]).
B. whitmori (Stanton and Fletcher[451]).
Bact. pneumoniae (Eisenberg,[153] Baerthlein[28]).

Also in the production of rough or smooth, soft or wrinkled colonies in many of the instances mentioned above; and in the following:

B. anthracis (Preisz[393] Wagner,[476] Gratia,[212] Nungester[490]).
B. avisepticus (Bernhardt[49]).
B. lepisepticum (de Kruif[119, 120]).
B. proteus (Felix[172]).
B. of Schweinerotlauf (Wychelessky[498]).
B. subtilis (Soule[450]).
B. (thermophilic) (Koser[285]).
Streptococcus fecalis (Faith Hadley[409]).
Streptococcus hemolyticus, viridans (Cowan,[111] possibly Macchiati[314]).
Diplococcus pneumoniae (Baerthlein,[28] Blake and Trask,[52] Griffith,[215] Reimann,[405] Amoss[8]).
Vibrio cholerae (Celli and Santori,[93] Kolle,[282] Kruse,[292] Eisenberg,[150] Balteanu[38]).
Vibro proteus (Firtsch[178]).

Also in the peculiar colony form or culture growths of:

Actinomyces annulatus (Sector formation, Beijerinck[42]).
B. alkaligenes (Mellon and Yost[332]).
B. botulinus (Reddish,[402] probably).
B. butyricus (Schattenfroh and Grassberger,[427] Bredeman[68]).
B. diphtheriae (Kurth[294]).
B. gasterophilus (Sandberg[424]).

B. helicoides—"creeping colonies,"—(Muto[350]).
B. influenzae (Grassberger[207]).
B. lactis (Sandberg[434]).
B. pertussis (Bordet and Sleeswyk[64]).
B. pestis (Gotschlich,[204, 205] Dudschenko,[137] Shibayma,[439] Klein[279]).
B. (proteus) fluorescens (Jager[267]).
B. of Rauschbrand (Schattenfroh and Grassberger,[427] Meiszner[330 a]).
B. tuberculosis (Karwacki[507]).
Diplococcus gonorrheae (Atkin,[22] perhaps Lavrinowicz[299]).
Diplococcus meningitidis (Atkin[21]).
Micrococcus tetragenus (Eisenberg,[153] Wreschner[492]).
Sarcina lutea and other chromogens from air and water (Eisenberg[148]).
Staphylococcus albus, aureus, citreus (Lehmann,[301] Baerthlein,[28] Eisenberg[148]).
Streptococcus hemolyticus and viridans (Cowan,[111] Enderlein[160]).
Streptococcus fecalis (Faith Hadley[420]).
Vibrio Finkler-Prior (Firtsch[178]).
Oïdium albicans (Draper[502]).

(3) Microbic dissociation is present in the curious erosive phenomena observed in slant agar cultures or in plate colonies of the following organisms:

B. anthracis (Preisz,[393] Pesch,[387], Katzu[270]).
B. cereus (Andervont and Simon[9]).
B. paratyphosus (Gildermeister[198]).
B. pyocyaneus (Canzik,[89] Blanc,[58] Quiroga,[399] Sonnenschien,[448] Hadley[234] and others).
Micrococcus aureus, probably (Seiffert, on authority of Sonnenschien[448]).
Monilia (Sonnenschien[448]).
Unknown bacteria from air and water (Hadley[234]).
Also probably in all "suicide" cultures.

(4) The results of the dissociative reaction are shown in the non-agglutinative, or the spontaneously agglutinating growth of many bacteria in broth or serum, as in the following cases:

B. anthracis (Markoff,[319] Wagner,[478] Gratia[212]).
B. cholerae-suis (Orcutt,[376] White[487]).
B. diphtheriae (Corbett and Phillips,[110] Zupnic,[498] Slawyk and Manicatide,[442] Bernhardt[49]).
B. (diphtheroids) (Mellon[327, 328]).
B. dysenteriae (Steinhardt,[453] Benians,[46] Arkwright,[18] Arkwright and Goyle,[18] Zoeller[497]).
B. enteritidis (Topley and Aryton[464]).
B. lepisepticum (de Kruif[119, 120]).
B. melitensis, paramelitensis (Bassett-Smith,[30] Et. Burnet[83]).
B. paratyphosus (Weil and Felix,[484] Baerthlein,[27] Arkwright[16]).
B. subtilis (Soule[450]).
B. typhosus (Nicolle,[366, 367] Savage,[425] Steinhardt,[453] Moon,[351] Teague and MacWilliams,[458] Ishii,[262] Arkwright,[16] Arkwright and Goyle,[18] Feiler,[167] Burnet,[87] Gardiner and Walker,[196] Krumwiede,[288] Goyle[206]).

Bact. pneumoniae (Baerthlein,[28] Hadley[225]).
Diplococcus pneumoniae (Reimann,[405] Amoss,[8] Takami[457]).
Streptococcus viridans and hemolyticus (Mary Cowan[111]).
Streptococcus fecalis (Faith Hadley[499]).

Also in the phenomenon of the double and single antigen:

B. aertrycke (Furth[190]).
B. cholerae-suis (Smith and Reagh,[441] Furth,[190] White[487]).
B. dysenteriae (Arkwright,[14] Breinl,[69] Arkwright and Goyle[18]).
B. enteritidis (Gruschka,[217] Arkwright and Goyle,[18] Goyle[206]).
B. paratyphosus (Weil and Felix,[484] Furth,[188] Breinl and Fischer[71]).
B. pertussis (Bordet and Sleeswyk,[64] Krumwiede, Mishulow and Oldenbusch[290]).
B. proteus X19 (Weil and Felix,[483] Braun and Salomon,[67] Braun and Schaeffer,[66] and many others).
B. typhosus (Joos,[272] Furth,[191] Arkwright and Goyle,[18] Goyle[206]).
Vibrio comma (Balteanu[38]).

(5) Microbic dissociation is observable in the course of "adaptation phenomena" as shown by organisms grown in broth containing antiseptics or other unusual substances:

Gentian violet.

B. coli, B. typhosus (Ainley-Walker and Murray[479]).
B. coli, Bact. aerogenes, B. lactis (Esther Stearn[452]).

Malachite green or brilliant green.

B. coli (Loeffler,[307] Revis[403]).

Sodium acetate, or oxylate.

B. coli, B. typhosus (Penfold,[385] Burnet[87]).

Bile.

Diplococcus pneumoniae (Reimann,[405] Amoss[8]).
B. coli (Adami, Abbott and Nicholson[1]).

Mercuric chloride and cadmium nitrate.

B. prodigiosus and other organisms (Wolf[490]).

Potassium bichromate.

B. prodigiosus and other organisms (Wolf[490]).
B. anthracis (Chamberland and Roux,[94] Surmont and Arnould[455]).

Phenol.

B. anthracis (Chamberland and Roux[94]).
B. coli (Malvoz,[324] Villinger,[475] Lommel[511]).
B. dysenteriae (Arkwright and Goyle,[18] Goyle[206]).
B. enteritidis (Goyle[206]).
B. proteus (Braun and Schaeffer,[66] Braun and Salomon[67]).
B. typhosus (Feiler,[187] Goyle[206]).
Vib. cholerae (Balteanu[38]).

Urea

B. coli, B, typhosus, B. pyocyaneus, B. enteritidis (Wilson[488]).

Sodium chloride—in high concentrations.

(Matzuschita[321]).

Saliva.

B. coli (Adami, Abbott and Nicholson[1]).

(6) Microbic dissociation is concerned in the loss or the gain of capsules by the following organisms and presumably in others:

B. anthracis (Chauveau and Phisalix,[97] Preisz,[393] Hess,[230] Bail and Flaumenhaft[34]).
B. avisepticus (Manniger[317]).
Bact. ozenae (Eisenberg,[153] Hadley[225]).
Bact. pneumoniae (Wilde,[485] Beham,[40] Eisenberg,[153] Toenniessen,[462] Baerthlein,[30] Hadley,[225] Julianelle[305])
Bact. rhinoscleromatis (Eisenberg,[153] Hadley[225])
Diplococcus pneumoniae (Kruse and Pansini,[293] Neufeld,[363] Eyre, Leatham and Washburn,[163] Laura Stryker,[434] Yoshioka,[494] Blake and Trask,[52] Griffith,[215] Reimann,[405] Amoss[8]).
Sarcina tetragena (Eisenberg,[153] Wreschner[492]).

(7) Microbic dissociation is operative in many, and perhaps all, of the slow or rapid changes in the virulence of pathogenic cultures in vitro, and it may be equally in the bodies of immune or naturally refractory animals. The following cases may serve as examples:

B. anthracis (Pasteur,[383] Chauveau and Phisalix,[97] Chamberland and Roux,[94] Hess,[230] Preisz,[391-4-5] Ascoli[20] and many others).
B. avisepticus (Manniger[337]).
B. cholerae-suis (Orcutt,[370] White[487]).
B. diphtheriae (Roux and Yersin,[410] Corbett and Phillips,[110] Hewlett and Knight,[231] di Martini,[125] Zupnic,[498], Lessieur,[302] Haven,[228] Baerthlein,[28, 30] Bernhardt,[49] Schmitz,[434] Heinemann,[229] Goodman,[209] Crowell[114]).
B. dysenteriae (Steinhardt,[453] Arkwright,[16] Fejgin[165]).
B. enteritidis (Topley and Aryton,[464] Goyle[208]).
B. lepisepticum (Bernhardt,[49] de Kruif[119, 120, 121]).
B. paratyphosus B (Baerthlein,[27, 30] Jordan[272-4]).
B. pertussis (Bordet and Sleeswyk[64]).
B. pestis (Gotschlich,[204] Dudtschenko[137]).
Bact. pneumoniae (Eisenberg,[153] Baerthlein,[28] Julianelle[305, 306])
Diplococcus pneumoniae (Charrin and Roger,[96] Kruse and Pansini,[293] Neufeld,[363] Friel,[185] Stryker,[434] Yoshioka,[494] Blake and Trask,[52] Griffith,[215] Reimann,[405] Amoss[8]).
Micrococcus tetragenus (Eisenberg,[153] Wreschner[492]).
Streptococcus viridans, and hemolyticus (Charrin and Roger,[96] Cowan[112, 113]).
Vibrio cholerae (Hamburger,[231] Ransom and Kitashima,[401a] Shousha[515]).

Also in changes in toxicity or toxigenic power of cultures:

B. diphtheriae (Roux and Yersin,[416] Goodman,[200] Bernhardt,[49] Crowell[114]).
B. dysenteriae (Fejgin,[169] Goyle[208]).
B. botulinus—perhaps (Reddish[402] and others).

(8) Microbic dissociation underlies the changes in type of organisms and in their biochemical and serological properties when cultures are grown in, or otherwise submitted to the action of, immune serum (homologous); or sometimes to normal serum if germicidal for the species concerned. This has been indicated in the following cases:

B. anthracis (Metchnikoff,[345] Sacharoff,[428] Sawtschenko,[426] Behring and Nissen[41]).
B. cholerae-suis (Metchnikoff[347]).
B. dysenteriae (Hamburger,[231] Steinhardt[458]).
B. diphtheriae (Bernhardt[49]).
B. subtilis (Soule[450]).
B. tuberculosis (Karwacki[508]).
B. typhosus (Nicolle,[366, 367] Steinhardt,[453] Eisenberg,[147] Rosenthal,[413] Park and Williams,[382] Saquépée,[421] Paul Th. Müller,[355] R. Müller,[356] Feiler,[167] Morishima[352] and others).
Bact. pneumoniae (Julianelle[303]).
Diplococcus pneumoniae (Charrin and Roger,[96] Stryker,[454] Clough,[102] Yoshioka,[494] Griffith,[215] Reimann,[405] Amoss[8]).
Streptococcus viridans and hemolyticus (Yoshioka,[494] Cowan[111]).
Vibrio comma (Ransom and Kitashima,[401a] Hamburger[231]).

Or when organisms are grown in tubercular, pleuritic fluid:

B. tuberculosis (Karwacki[508]).

Or when organisms are grown in urine, as in the case of:

B. coli (Horrocks—cited by Gurney-Dixon,[220] Wilson[488]).
B. typhosus (Hamburger and Czickeli[232]).

Or when they occur in naturally infected urine (as in urethritis):

B. coli (Archard and Renault,[11] Ali-Krogius,[2] Sörensen[449] and many later workers).*

(9) Dissociation is concerned with the spontaneous agglutination of many kinds of bacteria in 0.85% salt solution (and often in even less concentrated solutions) as shown by Nicolle[366] in 1898 for B. typhosus, and later by many others (Savage,[425] von Lingelsheim,[304], and others). In later years the subject has received closer study for members of the

* It is impossible within the scope of this review to consider the voluminous literature relating to the variants of B. coli and B. typhosus occurring in urethritis and related infections. The majority, however, reveal modified forms, morphologically, culturally, biochemically and serologically. Among these the R types can always be detected in abundance, and with them often the intermediate O forms. These are to be found particularly in carriers or in convalescents and have been reported by Fletcher[179] and Gildermeister[198] (for B. paratyphosus), by Lacy[286] (for B. dysenteriae Shiga) and by Krumwiede and collaborators[288] for several paratyphoid strains. In typhoid carriers similar strains of B. typhosus have often been reported.

colon-typhoid-dysentery group (Arkwright[16]); for B. cholerae suis (White[487]); for Bact. lepisepticum (de Kruif[119, 120] Webster[481]); for streptococci (Cowan,[111] Shibley[440]), for certain diphtheroids—together with general considerations of the subject in relation to pleomorphism (Mellon[341]); also for Vibrio comma (Balteanu[38]).

(10) Microbic dissociation is concerned with the serological differences in cultures coming from different colonies in a single pure line as shown in:

B. cholerae-suis (Andrewes,[10] Orcutt,[376] White[487]).
B. dysenteriae (Baerthlein,[30] Arkwright,[16] Arkwright and Goyle,[18] Goyle,[206] Ørskov and Larsen,[313] Benians[46]).
B. enteritidis (Schütze,[436] Topley and Aryton[464]).
B. lepisepticum (de Kruif[119]).
B. paratyphosus B (Sobernheim and Seligmann,[446] Baerthlein,[30] van Loghem[305]).
B. pertussis (Krumwiede, Mishulow and Oldenbusch[290]).
B. proteus (Weil and Felix,[483] Braun and Schaeffer[66] and others).
B. subtilis (Soule[450]).
Diplococcus meningitidis (Griffith[214]).
Diplococcus pneumoniae (Griffith,[215] Reimann,[406] Amoss[5]).
Streptococcus viridans and hemolyticus (Yoshioka[494]).
Vibrio comma (Balteanu[38]).

(11) Dissociation is often, and perhaps always, involved in the changes in biochemical or fermentative reaction occurring in pure line cultures, or in cultures from isolated colonies. It is manifested in changes in gelatin liquefaction in B. proteus (Braun and Schaeffer,[66] Baerthlein[30]), in Vibrio proteus (Vibrio Finkler-Prior) (Firtsch[178]), in Vibrio comma (Balteanu[38]), in B. subtilis (Soule[450]), in B. anthracis (Wagner[476]); also in the coagulation of milk in the case of B. proteus (Baerthlein[30]); also in changes in gas production in B. coli (Revis,[403] Sörensen,[449] Arkwright[16]); also in pyocyanin production in B. pyocyaneus (Hadley[224]); also in toxin production in B. diphtheriae (Roux and Yersin,[416] Corbett and Phillips,[110] Bernhardt,[49] Enderlein,[160] Crowell[114]); also in acid production in B. diphtheriae (Roux and Yersin,[416] Goodman,[200] Bernhardt[49] and others); in toxin production in B. dysenteriae (Fejgin[168]) and perhaps in B. botulinus (McIntosh and Fildes,[323] Reddish,[402] Hall[230] and others); also in fermentation reactions of B. coli mutabile (practically all the workers mentioned under the heading of "daughter colony formation"). Probably also in many other changes in fermentation reaction in which the presence of the dissociative reaction has not been observed in respect to correlated colonial or cultural variations.

(12) Dissociation is probably operative in the splitting of pure line yeast cultures as observed by Hansen, Beijerinck, Saito (from Tanner [456]), Baerthlein and others. Perhaps in the adaptation experiment to sodium fluoride of Fulmer and others. It may occur in the "clumpings" of yeast cells in cultures as reported by Effront (from Tanner [456]), Fulmer [187] and others; also in the loss of sporogenic function and secondary colony formation in Sch. octosporus as reported by Beijerinck.[42] It also may play a part in the "top" and "bottom" yeasts of Hansen.[223] It is unquestionably present in the curious erosive reaction observed by Sonnenschien [448] in a culture of Monilia from the throat. It may be recalled that colonies of yeast often show the secondary colony formation observed among the bacteria (Beijerinck [42]).

(13) Dissociation and its results in the generation of the R type culture is responsible for the rapid or gradual acquisition of mild resistance toward the lytic influence of the bacteriophage, as is indicated in several of the studies of Bordet and of Gratia, as also by many experiments of my own. In this relation it occurs most commonly in cultures which have been carried for long periods in liquid mediums without frequent transfer. It underlies the phenomenon of "serological cosmopolitanism" (Schütze,[436] Goyle [206]) and similar heterologous serological affiliations as suggested by the work of Stryker,[454] Neufeld,[363] Yoshioka [491] and Reimann [405] for the R type pneumococcus and by the study of Torrey and Buckell [466] for their gonococcus strains; also by Julianelle [506] for Friedländer's bacillus; it also probably concerns the "convergence phenomenon" as exemplified by work of Esther Stearn [452] on intestinal bacteria under the influence of gentian violet. It unquestionably furnishes the basis for the "major" and "minor" agglutinins recognized in many groups of bacteria. Evidence supporting these last conclusions, and involving the S, O and R antigens, is presented in later sections of this paper.

(14) Microbic dissociation is probably connected with the generation of filtrable forms of bacteria derived from either normal cultures (or blood); pathologic tissues (or exudates); or cultures under the influence of the lytic principle which, as we shall see later, may act in the capacity of a dissociating agent. The following instances may be mentioned under their respective headings:

Filtrable forms of bacteria not known to be related to the presence of the lytic agent.

B. cereus (Andervont and Simon [9]).
B. (diphtheroid (Mellon [325]).

B. fusiformis (Mellon[337]).
B. of Johne's disease (Morin and Valtis[354]).
B. of Schweineseuche (Lourens[318]).
B. pestis (Et. Burnet[500a]).
B. tuberculosis (Fontés[182] Valtis,[470] Durand,[138] Durand and Vaudremer,[139] Arloing and Dufourt,[19] Veber[473]).
B. typhosus (Almquist,[3] Friedberger and Meissner[184]).
Diplococcus meningitidis (Hort[255]).
Leptospira icteroides (Noguchi[372]).
Spirochaeta recurrentis (Novy and Knapp,[374] Breinl and Kinghorn,[72] Nicolle and Blanc,[369] Wolbach[480]).
Spirochaeta elusa and biflexa (Wolbach[480]).
Streptococcus—from encephalitis (Rosenow,[412] Evans[161, 162]).
Streptothrix—(Mellon[335a]).

Filtrable forms of bacteria related to the presence of the lytic agent.

B. coli (Izar,[264] d'Herelle,[246] Hauduroy,[235] d'Herelle and Hauduroy,[249] Tomaselli[463]).
B. dysenteriae Shiga, Flexner and Hiss (d'Herelle,[246] Hauduroy,[234, 235] d'Herelle and Hauduroy[249]).
B. gallinarum (d'Herelle[245]).
B. pestis (d'Herelle[245]).
B. typhosus (Hauduroy,[234, 235] d'Herelle and Hauduroy,[249] Fejgin[171a]).
Staphylococcus albus and aureus (d'Herelle and Hauduroy[249]).
Organisms from river water (Hauduroy[236]).
"Gonidial forms" of various bacteria (Miehe,[340] quoted from Enderlein[160]).

From this brief review of the extent and multiplicity of bacterial reactions with which it seems probable the phenomenon of dissociation is in some way related, we may conclude it is common not only among diverse types and genera of the lower bacteria, but among some of the higher forms as well. Wherever it occurs it is usually accompanied by changes in cultural growth, colony form, morphology of the cells, biochemical reactions, serologic reactions, antigenic power, virulence and resistance to injurious conditions in general; also to the influence of the bacteriophage.

The Existence of Parallel Trends in the Dissociation Process.—Having now outlined the sort of changes that characterize the phenomenon of active dissociation, it is of interest to observe the extent to which similar types of change may occur in bacteria at large and the degree to which they may run parallel through different, and sometimes distant, bacterial groups (Bernhardt,[49] 1915). It has already been intimated by the literature cited that such cases are numerous; indeed that some form of dissociation seems to be characteristic of nearly all bacteria and perhaps of all single celled organisms.* But it is also important to observe that parallel trends of dissociation are frequent, not only among closely related bacteria (B. coli, B. typhosus, B. dysenteriae), but also among bacteria belonging to diverse groups (B. coli, B. anthracis and Vibrio cholerae). In other words there exists a high degree of correlation in the frequency-distribution of characteristics belonging to the mother types, and a similar high degree of correlation in the distribution of characteristics appertaining to the daughter types, whatever the bacterial species may be. Without attempting to present a complete list of the many characters which run parallel for many organisms, and without meaning to imply that there are not exceptions to the character correlations presented, I may cite the following instances as supporting the point. The evidence for these cases, together with their proper references, will be found in various sections of this work.

CHARACTERISTICS MOST COMMONLY ASSOCIATED WITH "NORMAL" AND "MUTANT" TYPES OF CULTURE

S, NORMAL TYPE	R, MUTANT TYPE
Homogeneous clouding in broth.	Agglutinative growth in broth.
Normal suspension in 0.8% salt sol.	Sedimentary suspension in 0.8% salt sol.
Fair, conservative growth on agar.	Often expansive growth on agar.[1]
Colonies smooth, regular, convex.	Colonies rough, irregular, flat.
May generate secondary colonies.	Seldom generates secondary colonies.
Agar growth soft, opaque.	Agar growth harder, translucent.
Agar growth fluorescent.	Agar growth seldom so fluorescent.
Agar growth pyocyanogenic	Agar growth nonpyocyanogenic
In motile species, active.	In motile species, nonmotile.
Possessing distinct capsules.	Noncapsulated.
Biochemically more active.	Biochemically less active.
Carries double antigen (S and O).	Often pure R; may have some O or S.
Generates the "specific soluble substances."	Lacks the "specific soluble substances."
Flocculogranular precipitate in serum.	Flocculent precipitate only in immune serum.
If a pathogen—virulent (or toxic[2]).	Slightly or nonvirulent (nontoxic[2]).
More common in active disease.	More common in carriers and convalesc.
More common in acute infections.	More common in chronic infections.
Sensitive to aging.	More resistant to aging.
Sensitive to bacteriophage.	Less sensitive to bacteriophage.
Represented by freeliving forms.	Product of adaptations.
Cells of "normal" morphology.	Tendency to short rods and cocci.
Transformed to O or R in S immune serum.	Not transformed in S immune serum.
Not transformed in R immune serum.	Transformed to S in R immune serum.[3]
Resistant to phagocytosis.	Susceptible to phagocytosis.

[1] Apparent exception in case of B. proteus.

[2] Relation to toxicity established only for B. diphtheriae and B. enteritidis; suggested for B. dysenteriae Shiga and for B. botulinus.

[3] Established for B. subtilis (Soule[450]).

In a general way in all bacterial groups, but especially within a single group, such as the colon-typhoid-dysentery (about which our knowledge of dissociation phenomena is most complete) the distinctive characteristics are not, at least in their fullest expression, interchangeable between the mother and the daughter types. As many writers have pointed out,

* Note Middleton[348] on Stolonychia, Jennings[268] on Difflugia, also Hansen,[233] Beijerinck[42] and Saito (see Tanner[456]) on yeasts, and Sonnenschien[448] on Monilia.

however, there often occur intergrading variations so as to produce cultures of "mixed" type; or, as P. B. White[487] has phrased it, variations in the "degree of roughness."

In connection with any attempt to discover these differences in the components of a given culture it also should be pointed out that some of the characteristics of the "mutant" form may not be discoverable at once in the new culture, but only after a period of growth, or sometimes after several selections of the most promising of the daughter colonies which have been continued on the new medium or in the new environment. The process is often gradual, although in other instances the dissociation may complete itself in the production of the R type with astounding suddenness. As a rule, the products of dissociation are found most frequently in old cultures, as shown by Firtsch[178] as early as 1888 for Vibrio proteus.

Although many exceptions to the character correlations mentioned in the foregoing tabulation may occur (and the number may increase with further study), they cannot alter the general truth of the circumstance that each group of closely related organisms is characterized by a set of mutation-like changes which accompany the phenomenon of dissociation in all the species of that group. Other bacterial groups may manifest "mutational" changes, some of which involve quite different features. If there exist any mutational trends that seem to be common property of nearly all the species and groups on which data are available, they are the changes relating to the following characteristics: assumption of a sedimentary or spontaneously agglutinative form of growth; change in size, form, color and consistency of colony; loss or modification in antigenic power and agglutinability; loss or diminution in virulence; increased resistance to unfavorable conditions of environment, including the action of starvation, antiseptics, heat, desiccation and the lytic principle. These are largely constant for all dissociating cultures so far as reported to the present time.

In the present attempt to emphasize the two chief types of culture and the usually parallel trends in the dissociative process as just described, the point must not be obscured, however, that the "mutant form" is not a unit, but comprises a variable group, the components of which may vary from one another while at the same time manifesting the mutant or group characters. Whether the different R types produced by different means from the same sensitive culture possess essential differences is a point on which we possess little knowledge, but one which will be considered later (section 13) and with special reference to similar variations produced by the action of the lytic principle.

Suggestions for Terms of Reference to Primary Dissociates.—If all the variations or so-called mutations among bacteria were such that each species was a law unto itself, there would be little to be gained by making comparisons. But manifestly this is not the situation; for we have begun to see for the first time that, underlying these transformations occurring in very diverse bacterial types, there exist distinctly parallel dissociative trends. It is this mutation-like phenomenon which has already been termed "microbic dissociation," and for which I now wish to propose the term "spontaneous" or "active microbic dissociation," in order to differentiate it from what is apparently another sort of dissociation to be mentioned later.

In dealing with this problem, which as we shall see possesses some complexity in respect to the number and variety of cultures concerned, it is a matter of convenience, if not of necessity, to have available a terminology or mode of reference by which various bacterial stocks or strains can be referred to with some degree of exactness and precision. And the need of this will be seen to be greater when we come to consider the dissociative influence of the lytic principle in producing apparently similar changes of the primary types that have arisen through active dissociation. Such terms of reference as are to be recommended are, I believe, not likely to be of permanent value; but they may serve a useful purpose pending the time when we shall have acquired a deeper knowledge of the nature and meaning of the dissociative process. We may therefore consider a possible terminology sufficient for present needs.

A suggestion of what is demanded of a mode of reference can be obtained in part from a survey of the literature of the past three decades dealing intentionally or inadvertently with "mutations" and with some other apparent incongruities of bacterial behavior. Within this period bacteriologists, consciously or unconsciously, have given much information regarding the nature of the two fundamental culture types, the "mother" and the "daughter," as well as of a third culture type which we shall later come to recognize as a "transitional" form between the first two. It is the last two, but particularly the second, that have commonly been termed "mutants," and which have often received special designations by their discoverers. In the consideration immediately following, however, we shall confine our attention to the normal form and its "mutant" R. The intermediate type, which has been observed less frequently (and for reasons which will subsequently be pointed out), will be considered later.

Although it is impossible to state accurately just when the two chief culture types made their appearance in the early bacteriological literature, it is safe to say that one of the first observations of the dissociative reaction was that of Firtsch [178] (in Gruber's laboratory) on Vibrio proteus (Sp. Finkler-Prior) in 1888. And, remarkably enough, this work still stands as one of the clearest expositions of the dissociative phenomenon so far as the cultural and morphological aspects are concerned. Neither in the study by Firtsch, nor in the later interesting study on B. lactis-erythrogenes by Dyar [142] in 1895, were any special terms of reference employed. Firtsch spoke of his different colony types as I, II and III, while Dyar referred to his normal culture as "smooth" and the "mutant" as "wrinkled." Bordet and Sleeswyk [64] in their study of variation in B. pertussis in 1910 employed the symbol MS (sang), and MG (gelose), in referring to the types of culture grown on a blood medium and on plain agar, respectively. Cecil Revis [404] in his experiments on the adaptation of B. coli to brilliant green and malachite green used the symbols A and B for the variants so produced; and Orcutt [376] used the same symbols for her variants of the hog cholera bacillus, adding AV and BV for later, secondary variants. Weil and Felix and their followers called their proteus X19 forms H and O. Von Lingelsheim [304] mentioned his typhoid variant as the Q form, used also by Gildemeister [198] for a form of B. paratyphosus. Joss [272] referred to his two chief antigenic typhoid types as α and β. Arkwright,[16] in studying members of the colon-typhoid-group, used S (smooth), R (rough), a usage which has been followed by Bordet and by Gratia for B. coli, and by Jordan [273, 274] for B. paratyphosus B. De Kruif,[119] however, referred to his two types of Bact. lepisepticum as D and G; and in this he has been followed by Webster [481] for the same organism as well as for B. typhi-murium. Both Griffith [215] and Reimann [405] in dealing with the dissociation of the pneumococcus employed S and R, while Amoss [8] for the same organism used C and Z. Cowan [111] in her study of streptococci followed the usage of Arkwright. Others have used certain descriptive terms: "regular," "irregular"; "homogeneous," and "agglutinating." More recently Ørskov and Larsen [513] have called their types of paradysentery culture V and B, with the further variants, M and Bu. Eisenberg,[150] and also Kolle [282] have referred to their two chief types of colony in Vibrio comma as "helle" and "dunkel," denoting thus the S and R (or O).

Besides these many other designations have undoubtedly been employed for new or unusual culture types.

In order to effect a common usage in dealing with these two common and manifestly significant bacterial types, I propose the general employment of S and R. If we let these stand for the opposed characters, "smooth" and "rough," or "sensitive" and "resistant," they have the advantage of appropriate reference, in the three most important languages in science, to the most frequent and most closely correlated pair of characters which exists for the respective types.* In addition to these instances X might be used, and could refer to the circumstance that its state and antigenic nature are unknown, as is most often the case when new cultures are encountered. Gratia [211a] has already suggested the use of O for such an original culture; but as we shall see, it seems more desirable to reserve this designation for another form, the intermediate, which will receive consideration presently. Of course an X type culture could be identical with S, or it might be either an intermediate, or an R type, of which (especially the latter) there are many in laboratory collections carried under the designation of normal culture.

One further term may be added to those already suggested. Sometimes a culture of the S type, although having given normal growth for a considerable time, suddenly and spontaneously presents the phenomenon of active dissociation, either on the edge of the fresh growth or at some point in the interior of the bacterial mass. The clearest instances of this phenomenon that I have observed are in B. coli, B. pneumoniae, B. lepisepticum (all marginal dissociations) and in B. pyocyaneus (central dissociation). Instances similar to the pyocyaneus dissociation, at least in certain macroscopic features, have been observed in B. anthracis and in Monilia, as already reported. When such a dissociation once begins, it may be perpetuated indefinitely in subculture so long as the medium remains favorable.

If, however, colony after colony and subculture after subculture are found to perpetuate this tendency to fall into autolysis at a certain point in their development, such a culture may be termed "lytic" or one of the L type. Bordet [60] has already used "lysogenic" for any culture which, though not manifesting lytic phenomena itself, is able to precipitate transmissible lysis in other cultures. In this sense the B. coli

* As mentioned on a previous page, there are many instances which make it appear that the R type is not homogeneous, but may include cultures characterized by varying degree of roughness and correlated characters. If these observations are confirmed, the various R types might be referred to as R^1, R^2, R^3, etc., as will also be proposed for the resistant cultures arising from transmissible autolysis (section 13).

culture of Lisbonne and Carrére [305] was of the lysogenic or Lg type. From this point of view every S type culture is potentially lytic. At present, however, it seems that only the R type cultures are potentially lysogenic, but this point is by no means settled. Naturally a culture may be made up of two fractions and thus be both lytic and lysogenic. The need of further modifying these primary terms of reference will be pointed out in a later section dealing with somewhat similar dissociative changes effected by the bacteriophage or lytic principle.

Conclusion.—From the foregoing considerations it seems probable that the reactions which have been termed dissociative are to be observed at times in every bacterial species. Indeed, Dyar [142] in 1895 expressed the opinion that such colony variation as he observed in B. lactis-erythrogenes (and which we today recognize as being a fundamental manifestation of the dissociative reaction) could be seen in the majority of bacterial species; and, although he apparently had no conception of life stages in bacterial development, he saw nothing remarkable in the facts observed.

It now becomes clear that the dissociative reaction can be manifested by representatives of all the larger bacterial groups—aerobic and anaerobic, pathogenic and saprogenic, sporogenic and asporogenic—and among all morphologic types with the possible exception of the spirochetes, which as a rule have not been sufficiently studied in culture.

We have seen, moreover, that there exists a certain sort of parallelism in the trend of the dissociative phenomena in widely separated species. How close the parallelism is, and to what extent it is dictated by corresponding differences in the cells of the species concerned, remains to be ascertained. At least the parallel trends seem sufficiently well marked to make appropriate the differentiation of the chief culture forms into the S and R types, together with the less frequently observed intermediate, O, to be referred to on a later page. It is not to be assumed, however, that these terms are intended to serve as final designations in the sense that all S or all R organisms are alike morphologically or physiologically. They are merely intended to indicate that the culture concerned seems to become to a certain degree stabilized in a definite stage or type—sometimes highly transient, but at other times markedly permanent in its characteristics. In other words, these terms of reference are acknowledgedly superficial, referring only to certain gross appearances and reactions. With further study of the problem of dissociation they will doubtless give way to a more concise

5. FURTHER CONSIDERATION OF CHANGES PRODUCED BY THE DISSOCIATIVE REACTION: CULTURAL; MORPHOLOGICAL; MOTILITY; ENCAPSULATION; SPORULATION; CHROMOGENESIS

Between the mother (S) and daughter (R) types of culture there exist, as has been roughly intimated, important differences. While, in a general way, it has become a habit to differentiate these types on the basis of smooth or rough forms, a little experience with dissociating cultures indicates that, while these simple criteria are valuable, they may possess less significance in further differentiation than some others which have not as yet been sufficiently studied. I refer here to certain accessory morphological characteristics such as flagella, capsules and endospores; also to certain biochemical features including fermentative, antigenic, serological and immunological reactions. All of these will receive consideration in the course of this paper; but it is the chief object of the present section to describe the fundamental type differences as they relate particularly to the accessory morphological features, and to chromogenesis.

Cultural Characteristics of the Primary Dissociates.—When a culture that has undergone partial dissociation on solid or in liquid mediums, as outlined in the last section, is plated on agar, two (and sometimes three or more) different types of colony usually appear. Of these one is similar in most cases to the normal, mother culture (S), while the other is distinctly different (R). (The third type we shall consider in detail on a later page). The S colonies are usually round, regular, smooth, soft, glistening ("moist appearance") and sometimes highly fluorescent; while the R colonies are irregular, with broken or fimbriated margins, rough, granular or wrinkled, dull in luster, bluish or translucent, usually larger and occasionally tenacious to the medium. Of all these differences, two or more of which may appear in a given case, the most striking and constant relate to size, surface, and density of the colony. In some cases, however, and especially in first isolations, the differences in certain colonies are not clearly marked and there appear gradations between the clearcut rough and smooth forms. Moreover, as has been shown particularly for B. anthracis by Preisz,[393] Wagner,[476] Katzu [276] and others, and by Baerthlein [30] for many species, these colony differences may not be observable during the first 12 hours of growth. In other instances several days or weeks may be required for their differentiation (Baerthlein,[30] Penfold [385]). Toenniessen [462] has shown the same for variations in capsule formation in B. pneumoniae. White [487] particularly has recognized intergrading differences between the R and S types of Salmonella cultures and speaks of the "roughest of the roughs." Such intergradations are not to be wondered at when one recognizes that, although a colony may have been founded by a single S type organism, dissociation in the colony may begin at once, if the conditions are highly favorable; and, depending on the progress atttained, may manifest varying pictures after successive intervals. Further studies of such advanced colonies will invariably show them to be mixtures of R and S, while other rough colonies may yield apparently pure cultures of the rough type. Such cultures compared with the S cultures are likely to be very constant, although they may eventually show S inclusions. Thus, Arkwright [17] has been able to isolate smooth type cultures of Shiga from a number of rough variants, although these "derived smooths" differed in some respects from the original S culture.

The plating from a culture that is undergoing dissociation often gives mixed single colonies. These may be divided diametrically into the two chief growth types, R and S; one-half the colony may be translucent, the other half opaque. Whether such pictures are determined by the circumstance that a colony arises from two cells (one R and one S), or whether dissociation started with the two-cell stage of the colony cannot be ascertained. It is true, however, that cultivation from each half yields its respective culture type. In some cases it should be added, the division is not diametrical but the R type forms a gouge, sector, or wedge-shaped area in the main colony mass, as if dissociation had started in a cell at the margin rather than in the center of the colony. This is often seen in dissociating streptococcus colonies in which part of the margin is ragged while the balance is smooth (Faith Hadley [499]).

When cultures of the type showing active erosive dissociations are plated (B. pyocyaneus for example), at least three types of colony may appear (Hadley [224]). These are: S type (nonlytic and pyocyanogenic); R type (nonlytic and nonpyocyanogenic), and L type (lytic and highly pyocyanogenic. Sometimes colonies such as the L type above have been termed "lysogenic." But Bordet [60] has used this term with special reference to cultures which, without themselves giving evidence of lysis, are able to generate transmissible lysis in other homologous sensitive cultures. It therefore seems preferable to designate as "lytic" a colony which manifests lysis in itself, whatever may be its power of transmitting the condition to fresh culture. It can of course happen that a colony or culture can be both lytic and lysogenic, as Canzik [89] and myself [224] have found to be the case in certain strains of B. pyocyaneus.

This was apparently not true, however, with the lytic anthrax colonies studied by Pesch,[387] Katzu [276] and others; nor apparently in the lytic culture of Monilia reported by Sonnenschien,[448] nor with the B. cereus dissociation of Andervont and Simon. The case made out for transmissible autolysis by means of lytic filtrates of B. pyocyaneus has been doubted by d'Herelle,[248] Sonneschien and others. Further reference to this matter is made on another page of this paper. Suffice it to say for the present that there is need of more detailed study of the action of filtrates of dissociating cultures, and particularly those of the acutely lytic type.

When transfers are made from smooth types of colony of intestinal organisms into broth, as Arkwright [16] and others have pointed out, the growth is uniformly turbid, as is also the case with suspensions in 0.85% salt solution. The rough colony transferred to broth, however, often forms a pellicle and yields a sedimenting or agglutinative growth at the bottom or sides of the tube. According to Arkwright, Cowan, Shibley [440] and others such cultures can be kept in suspension, for purposes of serological work, by reducing the salt content to 0.4 or even 0.2 %. Stable culture suspensions have been obtained by some workers by heating the antigen at 100 C. In broth the growth of the R type is usually slower than that of the S. When lytic (or lysogenic) colonies are inoculated into broth, sometimes no growth occurs. If it does occur it is likely to be of the agglutinative form and when examined later only organisms of the R type are present. Bordet [61] has called special attention to the rapid disappearance of cultures of the lysogenic type when inoculated into broth and left even for short times. He has also pointed out the varying proportions of S and lysogenic type colonies that may arise on plates under varying conditions of growth of S organisms in a lytic filtrate (bacteriophage).

In some instances, however, when colonies which manifest the cultural appearances of S are seeded into broth a condition of growth appears which might be termed "intermediate"; some agglutination occurs in a tube giving at the same time homogeneous clouding. This is due to a mixture of S and R, or of S and O, organisms, one or the other of which in repeated transfers usually gains the ascendancy. Although the colony appearance is usually a good criterion of the type of the organisms contained, it is not infallible. There is apparently greater chance that an old colony giving the superficial characters of the S type will contain some R than that a colony giving the appearance of R will contain S organisms.

In addition to the common R and S colonies, Arkwright [16] has mentioned another sort, "Rv," a variant from R. These colonies were small, coherent and sticky. They arose from an R culture of B. dysenteriae which, when grown in horse serum and then picked to broth, almost invariably failed to grow. When transferred to agar the growth was thin and endured for only two or three days. But after several short interval subcultures on agar a culture was obtained that was still viable after seven days. Further variants from R have also been reported by Orcutt [376] for the hog cholera bacillus and for B. coli by Gratia.[211a] These data show, as Arkwright has already pointed out and as was shown by Firtsch for Vibrio proteus, that the R type itself may be capable of further variation. To what extent there may occur variation among different colonies of the S type is not known.

Regarding the nature of the R types of a single species, one important question arises. Are they all the same regardless of the manner of their production? The nature of the problem can be best shown by reference to the work of Braun and Schaeffer [66] on the O variants from Proteus X19 and X2. As stated earlier in this review, these investigators produced the OX19 in two ways: one by growth of HX19 on phenol agar, the other by growth of the same on agar impoverished in the broth content. The O types so obtained differed markedly from the H form from which they arose; but morphologically and culturally they also differed from each other. We shall see later that these modified types were in all probability not real R forms of culture, although the one produced by starvation seems to have been nearer the true R than was the strain modified by phenol. Gratia [211 b] has shown the range of difference in R forms produced by the action of the lytic principle in different concentrations. That different serologic reactions may exist in the case of different R forms from the same culture has been shown by both White [487] and Arkwright. These observations concern a problem of considerable interest about which at present little may be said. It is sufficient for present purposes, however, to indicate not only that the S forms differ from the R forms, but also that the R forms differ among themselves. We shall find that this is also true of the intermediates.

Dissociation in Relation to Cell Morphology and Motility.—In our considerations thus far the influence of the dissociative process on the individual cell has escaped notice. The effect in this respect, however, is no less significant than that on colony form or culture. We must accept the truth of Baerthlein's [30] view that colony variations are closely correlated with variations in cell morphology as well as with more essential physiological characteristics of the species. Omitting many references which are not found sufficiently related to known cases of dissociation to be of present value, we can still observe in the older literature some cases that merit citation.

Perhaps the most striking feature observed in the R individuals in different species is that some may be distinctly shorter and some longer than the cells of the S type. The S form of B. anthracis, for example, derived not from the "typical" anthrax colonies, but from the white, compact colonies, is composed almost exclusively of single rods, united at most into chains of two or three elements as pointed out by Preisz [393] in 1904 and confirmed by others. These organisms possessed something of the morphology of the cholera spirillum, that is, with rounded or slightly pointed ends and often a slightly curved body, quite different, as I myself have also observed, from the typical anthrax culture. The colonies of the R type, which were larger, more translucent, diffuse and of a dull luster (apparently conforming to the requirements of the "typical" medusa anthrax colony) gave quite a different picture. The cells were long, united into chains of considerable length or extended into filaments lying parallel for the most part and giving the common appearance of the "Medusa head" in the curled-edge colony. The ends of the cells were blunt or square cut, and never pointed. The cells and filaments were usually poor in spores. Sometimes, as Eisenberg [149] has shown, they may be quite absent. Whether the whiteness of the S colony as contrasted with the R is attributable to the richness of spores in the former, as Preisz believed, is uncertain, but on the whole doubtful, since the same point of differentiation occurs in the S and R colonies of nonsporeforming species.

In dealing with the relation of dissociation to cell morphology the work of Barber [37] in 1907 is of interest. Using the single cell isolation method that bears his name, Barber started about 140 pure line cultures of B. coli. The original organisms were selected for greater than average length. In all but one instance the population of each culture returned to the normal length of the strain as one would expect in the case of fluctuating variations. In one instance, however, he obtained a new race characterized by individuals of greater length. They also suffered some loss of motility and the cultures showed some changes from the normal type. This strain bred true for 32 months, undergoing frequent transplants to fresh medium. By similar methods of selection from other coli cultures two other "long races" were established. From the descriptions given it seems probable that Barber was dealing with dissociative variants belonging to the O or early R type. The extreme R type of B. coli is most commonly short, or even coccoid.

The dissociation of several cultures of B. subtilis has quite recently been accomplished by Soule.[450] For this organism he has obtained and studied in considerable detail the two chief colony types. He has also isolated a third type, presumably an intermediate, which has as yet not been studied sufficiently to permit of conclusions regarding its exact nature or behavior. Of the chief forms, one is smaller, whiter, more compact, and possesses a distinctly glistening luster. The other is slightly larger, more translucent (at least in young colonies), diffuse in manner of growth and possessing a dull surface. The first type is "domed" and about its edge the organisms are so arranged as to give the typical appearance that Fraenkel has termed the "bayonet front" (quite comparable with the "Krausköpf" anthrax colony of Preisz [393]). The second sort of subtilis colony, on the other hand, is thin and flat, with a strong tendency for the outer fringe of colony growth to curl under, exactly as in the "Medusa head" colony of B. anthracis. It is the culture yielding the first colony type that Soule has found dissociates readily into the latter. The reverse transformation, however, though less common, was obtained in a number of ways (section 11, on Incitants). Soule concluded that the small, white colonies showing motile bacteria represent the S type, while the larger, translucent colonies, showing nonmotile bacteria, represent the R form. The S form occurred mostly in singles, doubles or short chains and were actively motile, while those of the R type occurred in long chains and filaments. Motility could be observed only in rare instances and then was sluggish. When inoculated into broth the S type gave from the start a homogeneous clouding but with subsequent pellicle formation and partial clearing. The R form gave from the beginning an agglutinative growth. Thus it appears that the S and R forms of B. subtilis may be clearly differentiated by colony form, morphology and grouping of the individual cells; also, as we shall see later, serologically. Regarding the change in motility it may be added that the complete transformation did not occur in the first rough colony that was picked, but only after several other colony selections had been made from the R stock. The first two or three selections yielded a much diminished rate of movement which practically disappeared in the next few transfers. In its present state the culture is nonmotile, and highly stable. The behavior of the S and R forms of Soule's culture in the presence of normal and immune serums is presented in sections 9 and 11.

Evidence is presented later in this review demonstrating the identity of the H form of Weil and Felix's proteus X19 with the S type, and establishing the O form as an independent type not identical with the R. Although the O form of culture was first obtained by natural means, Braun and Schaeffer [66] produced it by growth on phenol agar, also by starvation (growth on agar with diminished nutrient substances). The morphology of the modified cells in this case thus seemed to depend on their manner of production. In the case of phenol agar the new cells were long, swollen and filamentous, while on starvation agar they took the form of minute coccobacilli, "not unlike Pfeiffer's bacillus." Neither type resembled the original HX19. Braun and Schaeffer add an interesting comment on what appears to them as an adaptive reaction in the respective changes: the reaction type to phenol (poisoning) is one which affords the greatest volume to the least surface, while the reaction type to impoverished agar (starvation) is one which affords the greatest surface to the least volume.

Although in the R cultures of B. coli and other members of the intestinal group there may often be recognized long or filamentous forms of the organism, especially in the earlier passages of R culture, the usual R type is a coccobacillus often approaching a coccus, as reported by numerous authors for B. coli, B. dysenteriae and B. typhosus (Arkwright [16]). This variation holds true whether the dissociation is produced in vitro by artificial means or occurs "naturally" in the urine in cases of cystitis, pyelitis, pyelonephritis, etc. (Mellon,[328] Zdansky,[495] Hamburger and Czickeli,[232] and others). It may be added that it was a considerable time before these atypical forms were recognized by serologic tests as related to B. coli; and as will be shown in section 9, even in such tests, the reactions are far different from those of normal B. coli. These variations in morphology remain constant over long periods. Similar types, it may be added, result from the action of the bacteriophage on a sensitive culture of members of this group (d'Herelle, Gratia). This subject will be considered in section 13.

One further point regarding the morphological aspects of the cell transformation from S to R is of special importance—namely, that it is not a sharp and direct change, but one that passes through an intermediate stage in which the variability in the morphology of the cells is much increased over that of the S type. This has been pointed out particularly by Feiler [167] in his report on the dissociation of B. typhosus, but that the phenomenon occurs rather commonly, and perhaps in all cases, appears from much other evidence. Feiler, among other dissociation provoking methods, made use of phenol agar. The population of the original culture was fairly homogeneous in morphology, but on the first phenol agar slant he reports that many peculiar and bizarre forms, lacking motility, developed. These, however, disappeared after two or three passages on the same medium and thicker and plumper rods gradually came into prominence. This and other similar observations indicate, I believe, as will be mentioned later in greater detail, the existence of an intermediate culture type through which the culture passes in the course of transformation from the S to the R form, which we may doubtless regard as the end product of dissociation. It seems probable that many of the intermediate colony types that have been described are not necessarily mixtures of S and R but mixtures of R with the transitional form of culture, O, to be described later.

Turning now more fully to the relation of dissociation to motility we find as one of the clearest instances the change occurring in B. proteus in its passage from the H to the O form, as also reported by Braun and Schaeffer [66] in 1919. The passage of the H form of X19 of Weil and Felix on phenol agar, or on starvation agar, gave cultures which varied from the normal not only in cell morphology, as already mentioned, but also in motility and flagellar equipment. After one or two passages the loss of flagella was noticeable, but after many passages no flagella could be found on any organism. The loss was the same, whether the O form arose from phenol agar or from the "starvation agar."

Regarding the R type of various members of the colon-typhoid-dysentery group, Arkwright [16] has stated that they possessed little or no motility. Baerthlein [30] showed that the variant colonies of Vibrio cholerae contained only nonmotile organisms. Krumwiede and collaborators [289] showed the same for R, and mucoid, paratyphoid variants. Gratia,[211] on the other hand, stated that the S type of B. coli studied by him was nonmotile while the R form was motile. It is also to be noted that Gratia reported the R type to be more virulent for guinea-pigs; also less phagocytable. These findings are quite remarkable and require confirmation. I have found the R type of several B. coli strains all nonmotile.

Feiler [167] in 1920 gave a clear picture of the effect of dissociation on the motility of B. typhosus. The reaction was produced by the same methods employed by Braun and Schaeffer for the dissociation of

B. proteus. Cultures of the actively motile S type were grown on starvation agar, phenol agar and in homologous immune serum. Feiler observed that the starvation R strains differed the least from the normal type and the immune serum R strain differed the most. In all cases, however, the difference was increased by continuous passage on the respective media. The starvation R, though losing its motility, failed to lose all of its flagella, although the flagellar system was much damaged. The phenol R lost all motility and all trace of flagella. In both of these cases a return of the cultures to normal culture medium was followed by a return to the normal culture type and with it the normal flagellar apparatus. Continuous growth in immune serum, however, not only determined the loss of flagella but also produced an absolute R form which, when it was returned to normal medium, failed to revert to the normal flagellated S type. Feiler thus regarded it as a permanent modification. He developed particularly the conception of the double antigen hypothesis for his S type culture contrasted with the single antigen of the R. The explanation of the double antigen was developed by Feiler (as it was developed independently by Wege Weil, and by Felix and Mitzenmacher,[173] apparently without the recognition of still older observations on this point) on the basis of endoplasmic and ectoplasmic agglutinogens, the latter comprising mainly the flagellar protein.

These instances could be multiplied but a sufficient number has been presented to indicate that the dissociative reaction is ordinarily accompanied by deepseated changes in both morphology and motility. The R type is ordinarily nonmotile and nonflagellated. Whether there are other types besides O that are nonmotile but at the same time not R forms, cannot be stated. That the loss of flagella is a gradual process, however, is indicated by a number of reports, but most clearly by Braun and Schaeffer. Regarding the changes in cell morphology, apparently their appearance is not easily predicted, but depends upon the nature of the organism undergoing dissociation (for instance, B. coli compared with B. subtilis) and upon the nature of the stimulus to dissociation (for instance, phenol agar compared with starvation agar). In most nonsporebearing bacteria the direction of the change is certainly toward the foreshortening of the rod, with the consequent production of coccoid and even coccus forms. The types of B. coli found in cystitis, pyelitis and gallbladder infections serve as good illustrations of this point. The situation seems to be different, however, in the sporogenic forms such as B. subtilis and B. anthracis.

Dissociation and Encapsulation.—The same dissociative mechanism operative with respect to cultural characteristics, cell morphology and motility can be detected in the modification of other characteristics of bacteria. We may take, for example, capsule formation exemplified by Friedländer's pneumobacillus as it has been studied by Wilde,[485] Eisenberg,[153] Baerthlein [30] and Toenniessen,[462] and more recently by Julienelle [505] and myself.[225] Baerthlein as others just mentioned started by plating out old cultures and detecting various colony types, among which we can now clearly recognize the S and R forms, the former encapsulated, the latter not. In my own case I used mass cultures on agar plates only a few days old and selected material from the translucent wedge-shaped marginal invaginations which make their appearance sooner or later along the free edges of the culture mass, and which are made up almost exclusively of organisms of the R type. Selecting from normal areas yields normal capsulated bacteria, characterized (when plated) by a form of colony quite different from that of the R type. One may however obtain the dissociation from plain broth cultures. If we start with a culture fresh from the infected animal, or one grown in a blood medium, we observe that the growth is highly viscous and microscopical examination of stained films reveals heavy capsules on all bacteria. If now we make passages through an impoverished broth medium, such as extract broth or peptone water, we observe that the viscosity diminishes with every transfer. If, however, at intervals in this series, we examine microscopically a stained preparation, we do not find that each organism has lost "a little of its capsule," but that some organisms have lost none and others have lost all. If the transfers through the impoverished medium are continued we reach a stage in which all observable bacteria are noncapsulated. These represent the secondary, resistant type; they have also lost their virulence. The same result can be determined by streaking agar plates and selecting cultures from the translucent marginal areas of completed dissociation in dissociating colonies. Here again, therefore, we are apparently confronted with no slow change overcoming equally all the organisms in the culture, but with the sudden appearance of a "new" pneumobacillus type. It is not necessary to state that these changes are reproducible in cultures coming from single colonies, after several repeated platings; or that, starting with a culture predominantly of the noncapsulated type, it is possible, by return to a favorable medium, or by animal inoculation, to regenerate the capsulated form in abundance. If the capsulated type has been once lost, however, as in an apparently "pure" R type colony I find no evidence denoting the possibility of its return, although I have not studied the matter extensively.

Although the loss of capsulation accompanying dissociation in Bact. pneumoniae seemed to be accomplished quickly and decisively in the cases that I have studied, Toenniessen, on the contrary, has presented a different picture, demonstrating the loss of capsule by gradual steps. In order to enforce his changes he made use of chemical antiseptics and by this means was able to "fix" the degree of capsule formation in different grades—complete, intermediate, or slight—each grade remaining constant in further cultivation on a favorable medium. These results, while differing somewhat from my own findings, correspond well with the manner of loss of motility in B. proteus and in B. subtilis during dissociation. In addition they conform with the views of White, Arkwright and others regarding the existence of degrees in the expression of the "rough" characteristics in R cultures coming from different colonies.

In 1921, in an article dealing with the biological significance of capsules in Micrococcus tetragenus, Wreschner [492] made an important contribution to the phenomenon of dissociation in the sarcina group. The culture concerned came from the mouth and grew at first in the form of gray-white, slimy, convex colonies. In the course of further cultivation on ordinary agar the colonies lost their sliminess by degrees and came to resemble the colonies of a staphylococcus. Along with these colony modifications went changes in the morphology of the single cells. By the fourth transfer the capsules, previously abundant, had begun to disappear and by the twelfth transfer they had vanished. If platings were made of the culture in this stage the colonies were pure white, flat and, when young, somewhat smaller than the capsulated type. In the center was usually a depression. While the original colonies could be removed en masse by reason of their viscosity, the new form was butyrous. In broth the noncapsulated form showed much greater growth energy and, when planted in equal volumes with the capsulated, in 24 hours outgrew the latter "by eight to nine times." In plating such mixtures, sometimes hardly a capsulated colony could be discovered. Propagated in ordinary broth the new form was "absolutely constant," but reverted to the capsulated when blood was added to the medium. The capsulated type was highly virulent, while the noncapsulated form lacked virulence. (For further consideration of this point see section 8.)

Although Wreschner has given the clearest picture of dissociation in M. tetragenus, the case is also apparent from the earlier study of Eisenberg [153] in 1914. As in the majority of Eisenberg's studies on bacterial variation, he plated an old (40 day) agar slant culture. Two sorts of colonies appeared: One was large (2 to 6 mm.), round, even convex, slimy and half-transparent. The other was smaller at the start (0.5 to 2.5 mm.), even, round, porcelain white and opaque. The first colony type gave well capsulated packets and tetrads. The second type gave only noncapsulated forms, or organisms with "rudimentary" capsules. The capsulated form of culture was much more virulent for mice. After one month in broth or alkaline peptone water the capsulated type had become completely transformed to the noncapsulated, while under the same conditions of cultivation the noncapsulated organisms underwent no change.

In reviewing these observations of Eisenberg and Wreschner there can be no doubt that both were dealing with dissociation phenomena and that the capsulated and noncapsulated organisms represent respectively our S and R dissociates. Intermediate types (O) have not been observed. The subjects of reversion and virulence in capsulated bacteria are treated in greater detail in section 8 of this work.

In the case of the pneumococcus the capsulated nature of the S type and the noncapsulated nature of the R have been clearly pointed out by Griffith, Reimann, and by Amoss. I have noted the same phenomenon in both B. ozenae and B. rhinoscleromatis. The latter observation confirms the earlier report of Eisenberg for the same organism.

The relation of capsules to the chief dissociates of the anthrax bacillus as we at present understand them is not so clear as in the case of Friedländer and M. tetragenus. If it were permissible to reason from analogies, the situation existing among other capsulated bacteria would lead us to predict that the S type of anthrax would be capsulated and also virulent. We shall see that, although the virulent type of the organism is usually regarded as the capsulated form, according to our present interpretation the virulent form of anthrax is not the S but the R. But we may turn to review the actual evidence.

Although it might be possible to read into the older work of Chauveau and Phisalix [97] a delineation of dissociation, the picture is not clear so far as capsules are concerned. Suggestions of a correlation between capsules, virulence and nonphagocytability appear in the later studies by Löhlein,[309] Deutsch,[126] Donati,[132] Kodama [281] and Preisz.[394, 395] The observations of Preisz [394] (1911) are of special interest. For a number of years previous to this date he had control of the Pasteur vaccines employed in anthrax immunization in Budapest. During this time he checked them for purity by the plating method. Occurring among the normal anthrax colonies on these plates he observed more or less constantly two abnormal colony forms which he at first took for contaminations, and the appearance of which made him hesitate to use the vaccines, either I or II. Further study of these new colony types, however, showed that they were in reality modified anthrax types. Of these the preponderant form was moist, glistening, transparent, "structureless" and slimy in varying degrees. When touched with the needle it would often spin into threads of 30 to 40 cm. Since he saw many normal colonies apparently becoming transformed to the slimy type, but since this form could also give rise to the normal, he termed it the "Uebergangsform." Although large animals could not be infected with this culture, small animals died, and from these he could recover normal culture. The degree of sliminess showed considerable variation; while some colonies were dense others were almost liquid in consistency. When in this form they did not give well isolated colonies on the plates but ran together into irregular groups and masses. These often appeared like "Schleimtropfen." The organisms in these slimy colonies were commonly capsulated but the degree of capsulation varied with the density of the sliminess of the culture. Cultures of greatest density showed bacteria surrounded with firmly adherent capsules, while those of more watery constitution showed organisms less definitely capsulated, as if the capsule substance had run off from the bacteria to produce an interbacterial slime. Of these cultures, as well as of the individual organisms, Preiz presents good photographs. They were found in both vaccine I and II, and perhaps represent the anthrax O types.

The second type of variant from the normal medusa form was represented by white colonies that were perfectly round and with even or slightly fringed edges. These colonies which one would be inclined to call the S type, not only by reason of their form but also from analogy with Soule's [450] results on the dissociation of B. subtilis, were not slimy and were composed of bacteria that were noncapsulated and entirely lacking in virulence. Even small laboratory animals failed to manifest infection, and when killed later no organisms of any kind could be found in their blood. This was true for the white colonies obtained from both vaccine I and II. It is apparent that this is the same type of colony described later by several investigators, particularly Bail,[32] Eisenberg,[140] Markoff,[319] Wagner [476] and Gratia,[212] but without clear reference to capsule formation and with somewhat contradictory reports on the subject of virulence. Preisz was able later to duplicate these results by heating normal virulent anthrax cultures and expressed the view that the reason why earlier workers of the French school had not observed these unusual culture forms was because they worked largely with broth cultures, rather than with B. anthracis on solid culture mediums. It may be added here, however, that the slimy cultures and the white cultures of Preisz do not necessarily owe their existence to the heating of normal virulent cultures. Mr. Nungester [499] in our laboratories has obtained, in the course of colony selection in normal dissociation, cultures which are in every respect identical with those described by Preisz. Of the fact that they represent transitional forms between S and R there can be no doubt.

In the work of Bail and Flaumenhaft [34] in 1917 we also find data bearing upon the relation between capsules and type of organism in B. anthracis. They found that, if a virulent culture was exposed to a temperature of 42 C. for a period that just fell short of depriving it of ability to form capsules, certain strains were produced which on plating yielded a mixture of colonies. These we can easily recognize as the S (Bail's α) and R (Bail's β), the latter being more virulent for guinea-pigs. Neither form, however, showed a tendency to return to the capsulated state. Hess [250] in 1921 described two forms of anthrax colony and indicated the presence of capsules in both. The number of capsulated bacteria was however less in the culture (Stamm III) which we may regard as the S type. The rough colonies gave cultures in which all of the organisms were more or less heavily capsulated. Continuous cultivation in horse serum broth caused a loss of capsules in all cases. Hess concluded that, for the development of capsules, a special stimulus is required. For normal bacteria a weak stimulus may be effective, but for modified bacteria "the stronger stimulus of the living body is required." In other words, for modified bacteria, serum in vitro is not effective. From these somewhat confusing results relating to capsules in B. anthracis, we can perhaps make out that they stand in some relation to virulence, which itself, peculiarly enough, seems related to what we have considered as the R type of culture. The whole problem of dissociation in B. anthrax is in an unsatisfactory state and much in need of further study.

From the foregoing considerations it is apparent that, although capsule formation is often correlated with the S form of culture, it also appears in the R type of anthrax. Our conclusions on this point, however, cannot be final. In addition, capsules often appear in the intermediate culture type lying between the extremes, S and R. As we shall see a little later, other bacterial species also manifest a transitional form showing the same slimy characteristics. The Q form of typhosus of v. Lingelsheim [304] may have been one of this sort since he emphasizes its slimy qualities; also the Q form of B. paratyphosus described by Gildemeister,[198] which was clearly in a lytic state. Other slimy cultures occurring in species which do not ordinarily produce definite capsules were seen by Ledingham [300] who suggested that the mucus-like substance might be related to the peculiar agglutination features also observed. Arkwright [16] reported sliminess for the assumed R type of certain Shiga cultures and was inclined to connect it with the presence of "swollen-looking, large, broad, irregularly-shaped bacteria, sometimes showing bud-like and branch-like processes. . . ." The sliminess, he attributed to changes in internal composition of the bacilli and not to capsule formation. These forms were probably mixtures of O and R, rather than pure R types. Such forms we shall see later often exist among the intermediates. It is safe to say that the "extreme R" is never capsulated. It may be added here that many cultures, arising as secondaries to the action of the bacteriophage, possess mucoid characteristics, but it has not been clearly shown that they are capsulated. Indeed, Kimura [278] has apparently shown that the B. coli from mucoid colonies produced by the bacteriophage, has no capsules but that the slime is a "secretion product." He believed this slime had the power of protecting the bacteria nonspecifically against the lytic principle. I [225] have shown, however, that, in Friedländer's bacillus, heavy capsules are no obstacle to bacteriophage action.

Dissociation and Spore Formation.—That nonsporogenic strains of bacteria can be isolated from sporogenic cultures has been recognized for many years, and the same is true for yeasts. Moreover, it is a matter of common knowledge that such strains, for a time at least, tend to breed true. In view of our present knowledge of dissociation, it might be suspected that this reaction is in some way connected with the phenomenon of sporogenic versus nonsporogenic types. This possibility indeed seems to have some support in the observations of Mellon and Anderson in 1919 [344] that spore substance and vegetative cell substance possess marked antigenic differences in the case of B. subtilis. The older literature on variations in sporeforming bacteria, particularly B. anthracis, although in no single instance revealing the entire truth of the situation, affords elements of evidence which are mutually complementary and to a large measure confirmatory. These may therefore be reported in such form and order as to give a picture of the situation as it exists at the present moment.

We have noted in many instances that the ability of a colony to generate papillae or daughter colonies seems invariably to be correlated with reactions of dissociative significance (Neisser,[362] Massini,[320] Müller,[356] Thaysen,[460] Penfold,[386] Baerthlein,[30] Eisenberg[148] and others). That such a phenomenon occurs in anthrax colonies was shown certainly among the first, by Preisz[393] in 1904. He points out typical and atypical strains, and presents the colony differences in detail. He also states that in the case of weak spore formation the colonies are blue (transmitted light), while in rich spore formation they are whitish. In the larger, blue colonies, which seem to represent the more typical form, there arose after a few days to a week small papillae (Knötchen), "the size of a sand grain," and these increased in size to form half-moon shaped colonies, yellowish-white in color and appearing like contaminations. If numerous, they imparted an uneven appearance to the colony. These "Knötchen" were found to be made up of spore-forming bacilli, while spores were absent in the surrounding culture; and Preisz concluded that they arose from the germination and subsequent multiplication of the spores of bacilli of a new type and different from the old culture (colony). They were shorter and occurred largely as single cells or in short chains, whereas the nonsporeformers were long, thin and united into chains of many elements. The ability of cultures to give the secondary (or even tertiary) colonies was held for a long time in cultures characterized by slow and rare spore formation; and such cultures were always blue or translucent as compared with the whiter sporeforming cultures. Preisz found no sporeless races but many in which spore formation was "almost lost."

In 1912 Eisenberg[149] reported studies on spore forming and sporeless races of B. anthracis, dealing chiefly with the possibility of transformation, one to the other type. He pointed out that in many anthrax colonies, after 12 to 16 hours of growth on agar, a differentiation occurred in which some became whitish with a glistening surface while others were dull and eventually became bluish and transparent "as if self-digested." Microscopically the whitish colonies showed chiefly sporulating rods and free spores; and to this he attributed the white appearance of the culture, as Preisz had done earlier. The second culture type was composed mainly of organisms without spores or with only a few. Such asporogenic cells, however, contained many "fat globules" while the sporogenic cells lacked them; he therefore assumed that these granules were normally used up in spore formation. Signs of an autolytic process were often evident. With reference to the mutual transformation of the two types, it was found that neither rapid passages on agar, nor successive passages in guinea-pigs (apparently the organisms possessed some virulence), served to change a sporeless race into a sporogenic. Indeed the only successful method reported was to heat the asporogenic culture at 70 to 90 C. This procedure yielded a strain rich in spore production. He found it possible, however, to transform a spore race into a sporeless by five to 20 passages on glycerol agar, or by the addition of grape sugar to the medium; this last method, however, gave less constant results. Some sporeless races produced by the glycerol agar method are said to have remained constant. Eisenberg did not employ single cell isolation methods but he did take the precautions to make eight successive colony isolations preceding his study; and this circumstance lends special significance to his results.

The significant study of Wagner [476] in 1920 has been reviewed in another section and needs reference here only to the extent of recalling that he, as well as Markoff,[319] Baerthlein [28] and Gratia,[212] pointed out the two chief colony types of B. anthracis and also suggested the difference in sporeforming ability although this point did not receive special consideration.

Most recently Pesch [387] and also Katzu [276] have reported for B. anthracis a peculiar disintegration phenomenon simulating the action of the bacteriophage, but which they concluded was of different nature, although perhaps belonging to "the same group of phenomena." The appearance of the anthrax cultures, it may be noted, was similar to that pictured by myself [224] and by Sonnenschien [448] for B. pyocyaneus; by Sonnenschien for Monilia; also by Andervont and Simon [9] for B. cereus; and possibly by Lawrence and Ford [298] for B. megatherium. The phenomenon concerned macroscopic erosive action in colonies and in mass cultures. Pesch, working with B. anthracis, obtained the usual two types of colony from two-year old agar slant cultures. He stated, however, that there existed no difference in the morphology of the organisms nor in the extent of spore formation. Neither form was virulent for mice. He was able, however, to obtain the same dissociative reaction in several virulent cultures and regarded the changes as "related to variation phenomena."

The results of Katzu's study were more or less similar. He obtained two anthrax cultures, one a heated Pasteur "vaccine strain," the other fully virulent. The former culture, after standing for about two months on sheep agar, showed round, bare areas ("Löcher") which penetrated through the culture mass to the agar. Over the floor of the pockets was a thin, transparent film of growth. Daughter colonies (secondaries) were observed along the border of some of the larger colonies. Whether subcultures were made from the pockets themselves, or from areas of culture between the pockets, the resulting cultures yielded similar erosions. I have demonstrated exactly the same circumstance for lytic cultures of B. pyocyaneus. Katzu showed, as Preisz had done much earlier, that the pockets could be traced back to small granulations ("Kolonienknöpfen," Knötchen) in which a different type of organism possessing marked capacity for autolysis was being generated, thus confirming the older and highly significant observation of Preisz on the same point. Katzu also described again the "Krausköpfähnliche" (S type) colonies and pointed out that these did not contain erosions. Although the fully virulent anthrax culture studied by Katzu did not at first present these dissociative phenomena, growth of this culture at 42 C induced similar erosive manifestations. Attempts on the part of both Pesch and Katzu to obtain a filtrable agent capable of producing similar changes in normal culture were without success. Katzu concluded that the reaction studied merely involved the splitting off of a transparent variant from the mother culture.

Other early contributions of interest bearing upon one aspect or another of this problem are those of Altmann and Rauth[7] (1910), Burri,[88] Phisalix[389] (1893); Sobernheim and Seligmann (in Kolle and Wassermann) and Wolff.[490] A good review of studies on the variability of B. anthracis is given by Prigsheim.[397]

From this brief review of observations on dissociation in B. anthracis with special reference to spore formation it seems probable that the sporogenic and asporogenic cultures correspond respectively with the two chief colony types: the first being small, whitish, round, smooth and regular; the latter larger, bluish or translucent, somewhat rough ("matt") and irregular or spreading. The latter form manifestly constitutes the "typical" anthrax culture, giving the Medusa head colonies, but the direction of the dissociative trend is clearly from the atypical to the typical form, seldom the reverse. These observations suggest that the former (atypical) is the S type while the latter (typical) corresponds with the R. This view is further supported by the following facts: that the nature of commonly observed laboratory transformations (over long periods of time) are from the sporogenic to the asporogenic type of culture; that it has been commonly reported an easy matter to transform a sporogenic into an asporogenic race; and that it is more exceptional and difficult to produce a spore race from a culture which, after repeated, careful microscopic examination, is seen to contain no sporebearing cells. It may be added in passing that the same circumstances hold for many species of yeasts, in which it seems probable that similar dissociative reactions occur.

In conclusion it should be said that the observation that anthrax cultures can be split into these two types, one sporeforming, the other sporeless, lends weight to some recent views (Mellon) that endospore formation may possess a significance somewhat distinct from that of a reaction to unfavorable environment as commonly assumed. Whether this phenomenon, like arthrospore formation among cocci, may be regarded as a "cover" for nuclear reorganizations, that demand physical protection, as suggested by Mellon, must be left a question. If this should prove true, however, we should come to regard asporogenic strains of sporeforming bacteria as forms which have lost some power of sexual reproduction, and which in this respect might be analogous to the asporogenic yeasts.

Dissociation and Chromogenesis.—In closing this section, one other interesting point relating to the R type cultures may be noted. This refers to the marked tendency for the appearance of a yellow or brownish chromogenesis among these forms. The fact was observed perhaps first in the case of Vibrio proteus (R variants) by Firtsch [178] in 1888. Eisenberg [150] and others have described the same phenomenon in the "dunkel" forms (probably both O and R types) of the cholera vibrio. A yellow colony form was reported by Balteanu [38] in 1926. Yellow or yellowish R types of B. diphtheriae have been described by Roux and Yersin,[416] Corbett and Phillips,[110] Bernhardt,[40] Heinemann [239] and many others. I have observed these chromogenic forms many times in the R types of the Park 8 strain of B. diphtheriae. Baerthlein,[28] many years ago, described the brownish white variant of Micrococcus citreus, presumably an R, and E. M. Brill, a student in my laboratory, recently obtained what is apparently the same form through the influence of a lytic agent which he isolated from sewage contaminated water. In this connection, I may add a curious feature relating to this culture. When growing in pure culture on agar, it forms flat, brownish white colonies; but, when growing in a mass of citreus S, it forms small, spider-shaped colonies which make distinct depressions in the mother culture. Almquist [8] has described a yellowish form of B. typhosus. In addition to the above cases, Atkin [21] described as yellow or yellowish his type II and IV meningococcus colony variants. Novy and Soule [409] have observed an orange yellow form of B. malleus which undoubtedly represented an R type and which later gave rise to the normal, nonchromogenic culture through the formation of a regeneration fringe. Seligmann [437] in 1919 observed, in connection with a study of the cause of slimy bread in Berlin, a culture of B. mesentericus which gave off "mutations" of a distinctly yellow type and was believed to be identical with B. berolinensis.

In the case of B. proteus, yellowish colonies possessing unusual characteristics and perhaps representing R type culture, were observed by Braun and Schaeffer.[66] Fejgin [169] also has found five types of B. proteus X19 secondaries to transmissible autolysis which in colony form on agar plates were characterized as being, respectively, white, round, opaque, irregular and chromogenic. Two of the latter type were bright yellow and one was canary yellow, the pigment of all three being soluble in alcohol. In serological tests the antigenic character seems to have remained close to that of the original X19, but in two of the yellow types the agglutinability was much diminished. All were gram-positive and the bright yellow strains were found to contain long filaments along with the rod forms. The growth of the canary yellow strain only was spreading, the organisms small and nonmotile. D'Herelle [247] believed that all of these cultures were "mutations" produced by the bacteriophage. It is clear, however, that the presence of the bacteriophage is not necessary for Mr. Weaver, working in our laboratories, has produced from B. proteus apparently similar yellow strains without the use or presence of the proteus bacteriophage. Whatever the cause of this curious phenomenon may be, there is no doubt that a careful survey of the literature would yield other instances in other bacterial species.

Conclusion.—In concluding this section it may be said that the commonly observed characteristics of culture form, cell morphology, motility, possession of capsules, possession of endospores, and of yellow chromogenesis (within limits), are distinctly correlated with the bacterial type coming into prominence in the dissociative reaction. These observations should impress upon us the great danger involved in making use of such characteristics, without reservation, as a means of species differentiation. When we say that a certain organism is "motile" or is "capsulated" or is "a short rod," or "forms endospores," or "is a yellow chromogen," all that we can truthfully denote by these descriptive terms is that the organism is motile or capsulated, or has the form of a short rod, etc., at a certain stage in its life history. It thus begins to appear that for purposes of systematic bacteriology the registration of the patent characteristics of an organism or species is by no means so simple a matter as we have been accustomed to believe. The old descriptive terms may still possess a value if we know the definite state or stage of the culture to which we refer. If we do not, these terms are meaningless.

6. FURTHER CONSIDERATION OF THE NATURE OF THE DISSOCIATIVE REACTION AND OF THE "THIRD COLONY INTERMEDIATE"

While, as has been noted earlier, it is the S and R colony types that come into greatest prominence in the majority of dissociation reactions so far observed, in other cases the situation is complicated by the appearance of at least one, and sometimes several other variants, also manifesting themselves as do both S and R by their characteristic colony form and cell morphology. Baerthlein [30] for instance in dealing with colony variation in certain bacterial species presents so many colony types that one can easily become confused and unable to recognize the actually significant forms, much less their sequence of appearance or disappearance in the original culture. The same is true of the serologic variants of B. proteus described by Felix; [172] also of the colony variants of Vibrio cholerae pictured by Eisenberg.[150] The occurrence of such a degree of variability observed in many of the early studies was unquestionably sufficient almost entirely to obscure the view that through it all existed a definite trend of cultural modification. But we are now coming to see (as indeed was strongly suggested by Gruber and his pupils [178] as early as 1888) that, in the kind and degree of variation, there exists a certain constancy and orderliness, involving not only what we recognize today as the S and R forms, but the socalled intermediate forms as well. Indeed it becomes apparent from such studies as those of Firtsch[178] (Vibrio proteus), Eisenberg[150] (Vibrio cholerae), Bernhardt[49] (B. typhosus, B. diphtheriae), Preisz (B. anthracis) and many others, that there exists a distinct parallelism in different bacterial species with reference to the appearance and behavior of the intermediate culture types and particularly with reference to the "third colony intermediate." When one has reviewed carefully the detailed descriptions of the culture changes in many bacterial species, it becomes entirely clear that the chief difficulty of interpretation in the past has been the circumstance that in some cases the S $\rightarrow$ R transformation occurs directly and sharply without intermediate steps, while in other instances there enters into the dissociative reaction a transitional type. This may be represented by a single colony form, as the third colony intermediate, which possesses characteristics quite different from either S or R, or by a fairly closely connected series of such intermediates, or "Zwischenformen," as they have been commonly termed. Some of these Zwischenformen may more closely resemble S while others are more like the R type culture. The chief of these, however, and the one which is constant, even though the others are lacking, is the third colony intermediate. The evidence that this form is actually intermediate in nature will be presented shortly.

As to the probable reason why in some cases only the two chief types of colony, R and S, can be recognized, and why in other instances we observe either a single or a variety of intermediates, with S at one end of the series and R (extreme variant) at the other—I believe it lies in a differential death rate of the various cells. It was the greatest merit of Bernhardt's study to present the first evidence for this conception which, as we shall see, contributes to a better understanding of many peculiar phenomena in the field of microbic dissociation. To this subject we shall return.

Sudden or Gradual Appearance of the R Type.—Among those lytic and variational phenomena which we now include under the head of dissociation, one of the important aspects has been the question of their sudden or gradual manifestation. And here the reports have often been contradictory. While Baerthlein has emphasized the sudden nature of the S → R transformation, amounting practically to a mutation in the usual sense, Bernhardt, working with the organisms of typhoid and diphtheria, has equally emphasized (and, one may add, more conclusively demonstrated) its gradual and progressive occurrence. I may use as an illustration of this, however, the dissociation of the capsulated bacteria, such as Friedländer's bacillus or M. tetragenus. In Baerthlein's and my own experience with the former culture the change from a capsulated to a noncapsulated form has been quick and sharp. The former passed directly into the latter apparently without the appearance of transitional or intermediate forms; and this was also true of the colony picture. Toenniessen,[462] on the other hand, emphasized the occurrence of intermediate degrees of encapsulation which he related to a slow process of transformation. He reported even that he was able to "fix" the grades of encapsulation more or less permanently. De Kruif [119] pointed out that the dissociation of Bact. lepisepticum is very abrupt and he mentioned no intermediates; while P. B. White,[487] on the other hand, represented the transformation of the hog cholera bacillus as gradual, characterized by "degrees of roughness." The gradual nature of the change is also brought out forcibly by the study of variation in B. diphtheriae by Roux and Yersin [416] in 1890. Also by Braun and Schaeffer and by Friel for proteus dissociation, so far as its morphological and cultural features are concerned; and both Weil and Felix [172] in case of Proteus X19 have indicated in great detail the wide range of serologic types which lie between the H and the O forms. The same has been shown by Weil and Felix,[484] Furth,[188] Breinl and Fischer [71] for B. paratyphosus, by Breinl [69] for dysentery organisms, by Feiler and by Furth [190] for the Aertrycke bacillus and by Grushka [217] for B. enteritidis, confirming the important work of Sobernheim and Seligmann [446, 447] on B. enteritidis transformation. Felix [172] particularly has emphasized the importance of the transition forms ("Uebergängsformen") which in B. proteus may show all degrees of serological relationship between the "Ausgangsform" and the "Endform." In this respect his work is in agreement with the earlier work of Eisenberg on intermediate forms of colony variation in the cholera vibrio, as well as with the "Zwischenformen" detected (especially in colonial variation) by Bernhardt in the case of B. typhosus, B. paratyphosus, B. diphtheriae and other species.

White [487] has noted that the change from S to R in certain Salmonella cultures is sometimes abrupt and sometimes gradual. He stated (p. 72): "In some cases roughening of a culture seems to be a gradual, insidious process; in others it seems abrupt, mutative. Sometimes rough colonies picked from a plate culture show every grade of variation in their agglutinative behavior; at other times the culture is sharply divided into perfectly smooth translucent colonies and rough colonies which never flocculate." It can scarcely be doubted that some of these colonies come under the head of what we have termed the intermediate type. Several recent writers on serological subjects connected with bacterial variation have gone far astray in their conclusions as a result of their failure to comprehend these fundamental aspects of the dissociative reaction.

Postponing for the moment the earlier and more striking work of Firtsch in this field, which possesses points of special interest, we may consider the later study of Eisenberg [150] on variation in the cholera vibrio. Among the four colony variants described by Eisenberg (three of which had previously been described by Baerthlein) one ("helle") appears as the S, another ("geschwulste dunkel") is analogous to the R and the third ("Ringform") has an uncertain place. The fourth type ("Uebergängsform"), however, is of special interest in our present consideration. This form of colony had a size equal to, or larger than, that of the "helle" but was opaque, highly convex and had precipitous edges. Eventually the center became depressed and an outgrowth of the "helle" form appeared from the margin, thus showing reversion. In aging cultures this "Uebergängsform" disappeared and was replaced by the "dunkel." Regarding the frequency of appearance of the transitional type it is shown by Eisenberg's tabulation (p. 6) [150] that it may be quite absent on plates poured from cultures left to age. The chief point for our present interest is, however, that Eisenberg gave clearly the details of the transitional trend of the S type culture through an intermediate form to the "extreme variant" R. He thus confirmed the earlier report of Firtsch for Vibrio proteus, to be mentioned shortly. The transitional form of B. anthracis as described by Preisz,[394] involving a slimy type of culture made up of organisms capsulated in varying degrees, has already been referred to in section 5 and does not require further mention here except to state that it is manifestly analogous to the other transitional forms being considered.

For the typhoid, paratyphoid and diphtheria species Bernhardt [49] described "Zwischenformen" occurring between the normal type and the "extreme variant," that is, between S and R. The following concerns mainly the intermediates of B. typhosus and paratyphosus. The colonies of the intermediates, which possessed characteristics quite different from those of the S or R forms, did not arise from young, normal cultures nor from old cultures, but appeared with the apparent beginning of dissociation, increased for a certain time and then disappeared again. Thus, at the beginning of dissociation or prior to its appearance, there were only normal (S) colonies while at the end of the reaction (quite old broth cultures) there were only the "Endformen" (R). At a certain intermediate stage, however, one could detect all three colony types. As will be seen presently, this circumstance of the sudden appearance and immediate disappearance of a transitional culture type is a duplication of the observations made by Firtsch at an earlier date on Vibrio proteus.

But the most curious and perhaps most significant aspect of Bernhardt's work with these transitional colony forms and correlated cell forms was the peculiar growth characteristics of certain members of the intermediate group. This aspect of his study merits the presentation of further details, and for reasons that will later become apparent.

In B. typhosus, as in other species, Bernhardt observed not only the normal colony type (S) but also its extreme variant (R). Along with these, however, were several intermediate forms. Although most of these intermediate culture types grew well on gelatin as also on other mediums, there was one colony type in particular, representing one of the "Zwischenformen," which on further transfer failed to grow on either gelatin or agar at 22 C. in two days. Neufeld according to Bernhardt had already pointed out a similar phenomenon. This form of culture was apparently extremely "sensitive." There was one way, however, in which such colonies could be made to perpetuate themselves. If the dilutions from the original colony were made in broth, plating yielded no growth. But, if the dilutions were made in salt solution, colonies often appeared on the plates, gelatin or agar. This culture therefore possessed the remarkable characteristic that it was destroyed in broth but lived in salt solution. To explain this curious circumstance we must assume either that the broth was germicidal for the culture (or otherwise unfitted for promoting growth), or that it assisted the organisms to destroy themselves. In view of the lytic tendencies of the original colonies on agar the latter view seems the more probable. If this is true we have here certainly a lytic and possibly a lysogenic culture type. And at this point we may remind ourselves of the interesting work of Bordet [61] dealing (in this case with bacteriophage action) with those culture forms of the lysogenic type of B. coli which disappeared in broth although they might yield a lysogenic growth on agar if transplanted from agar.

It may be noted, furthermore, that these peculiar colony forms of Bernhardt which were incapable of further cultivation did not occur in every culture, nor were they present in young, "normal" or old cultures. Their time of appearance was rather narrowly limited. Bernhardt was of the opinion, however, that this type might be present more commonly than appeared, but that it was destroyed in the process of culturing. Bordet has employed exactly the same argument to account for the disappearance of certain cell types in the case of B. coli in relation to the lytic agent. A consideration of the possible significance of these observations for the problem of transmissible autolysis is reserved for a later page.

To the foregoing I should like to add that in certain cultures of B. diphtheriae and staphylococcus I have observed bare plaque-like areas which I have termed "invisible colonies." They occur on a background of apparently normal culture and resemble lytic areas produced by the bacteriophage. On close examination, however, there can often be detected over the surface of these spots an extremely thin film of growth which at first does not increase. Eventually a sprinkling of secondary colonies may arise on such sites and it may be noted that the picture so produced bears resemblance to lytic areas with their secondary colonies as produced by the bacteriophage. Similar "invisible colonies," barely detectable, develop occasionally in isolated form and not surrounded by other culture material. These also may present, after a time, the phenomenon of secondary colony formation. Microscopically such colonies are composed of extremely small forms and usually give no immediate growth in broth. Transferred to slants they may yield a delicate growth that seldom survives for more than a few transplants. These colonies may have something in common with the nonviable forms of Bernhardt; also with similar faint growths of slight vitality described by Arkwright for Shiga. They may also stand in relation to the secondary colonies (lytic stage) of Preisz, Pesch and Katzu (B. anthrax) and of Andervont and Simon (B. cereus); also of Sonnenschien for Monilia, as mentioned earlier in these pages.

We may now return to the question regarding the probable reason why the intermediate colony forms appear in some cases of dissociation but not in others. That one or more of them occur quite commonly in natural dissociations is indicated by the frequency with which the third colony type has been reported in the literature. In the first place it is strongly suggested by the work of Bernhardt [49] on typhosus and paratyphosus that the possibility of revealing this intermediate or transitional form will depend in part on the particular stage in the dissociative reaction at which the colony sampling is made. If too early the majority of the organisms will still be in the S form; if too late the majority will have passed over to the R. There is a point, however, when the transitional type may be obtained, sometimes mixed with S, at other times mixed with R. It is indicated by many observations, however, that once the transition has started the tendency for its completion is strongly manifested by the culture or colony. What lies behind this tendency is a question of much interest.

Although the time of sampling a dissociating culture is thus indicated to be one controlling factor in the revelation of the "third colony intermediate" there are doubtless other factors; and the nature of one of these is also suggested by the valuable contribution of Bernhardt. It seems to depend, as has been earlier intimated, upon a differential death rate or transition rate, among organisms of the S and transitional type. There exists, as Bernhardt has shown, a stage of transition as indicated by a certain colony form in which the organisms are not viable when transplanted to other mediums; and this point has been confirmed by observations by Bordet for B. coli, and by Mellon [325] for B. fusiformis. These observations may possess the significance of suggesting to us that the dissociating culture is not merely a culture undergoing transformation, but a culture, certain cell components of which, are experiencing complete annihilation, at least when surrounded by a fresh growth medium. We can perhaps see added evidence of this in the acute dissociations shown by B. pyocyaneus, in the acute lysis of B. anthracis and Monilia, as already described, as well as in the socalled suicide cultures. But the point which I desire especially to make in this connection is that, in different bacterial species, there may easily exist differences in the degree of stability of this highly unstable "third colony intermediate." And this circumstance can assist in explaining many of the variable phenomena observed.

The Intermediate Forms and the Sequence of Types.—Considerable reference to the existence of the intermediate or transitional colony types lying between S and R has now been made, but without the presentation of sufficient evidence supporting the fact that they are actually transitional in character and behavior. This important point may now be considered.

Although Bernhardt presented valuable data bearing on this point with special reference to B. typhosus, B. paratyphosus and B. diphtheriae, and although similar observations have been made by others and most recently by Mellon and Enderlein, it is in the old work of Firtsch [178] in 1888 in Gruber's laboratory that we find, for those early years, the deepest recognition of the actual problem involved and thorough-going experimental attempts to throw light, not only on the cultural features of the variants themselves, but also on the trend of the transitional process. Firtsch studied the variants of Vibrio proteus (Sp. Finkler-Prior) occurring, as we should now say, under the conditions of natural dissociation in old gelatin stab cultures. In such cultures he observed the production of several variants from the normal vibrio and these he termed variants I, II and III. All could be differentiated both from the normal culture and from each other by growth in gelatin and on agar or potato, by colony formation and by the cell forms. The rapidity and sequence of their appearance depended on the time of aging of the cultures concerned. A brief review of the most interesting of his results follow.

Although gelatin previously seeded with normal culture, and sampled by plating methods, up to eight days gave only "normal" colonies and cultures (N), beginning with the 14th day, variant I appeared. In colonies it was brownish, glistening and thrown into projections and folds. On agar it reverted to N growth; in gelatin stabs it showed slower liquefaction than N. When plated on gelatin it gave brownish colonies which kept the same form if transfers were made at rapid intervals. If, however, the plates stood four, six or eight days between transfers some of the succeeding colonies developed a delicate, extending halo of growth, and transferring from these fringes gave normal culture again. Firtsch stated that variant I was always present in old cultures of a certain age. It showed no motility—to which fact Firtsch attributed its different colony form.

If the gelatin stab cultures made from N, instead of standing 14 days stood 48, 73 or 94 days, Firtsch found present in such cultures not only N and variant I but also a second type, variant II. The number of organisms of this type increased with further aging of the culture: at 14 days, 0; 48 days, 8% of total organisms; 73 days, 23%; 94 days, 73%. He ascertained, moreover, that variant II could arise, not only from N, but also from variant I if such a culture stood 56 or 67 days or more. Variant II gave colony characters (yellow, flat, granular) quite different from both N and variant I. It also showed a different form of growth in gelatin.

If the standing culture was observed over still longer periods, N and variant II survived but variant I disappeared. Thus, after 70 days in gelatin stab:

Top layer gave—		Bottom layer gave—	
Normal colonies	208	Normal colonies	10
Variant I colonies	0	Variant I colonies	0
Variant II colonies	18	Variant II colonies	7

From these and many other similar experiments we may conclude with Firtsch that variant I was highly unstable and was able to give further variation either into N (slow passage on gelatin or on agar) or into variant II (old gelatin stabs). Variant II, on the other hand, was stable. Neither gelatin stabs, nor slow gelatin plating, nor beef broth, nor agar, nor potato, nor either slow or rapid transfers on any medium, nor growth at 37 or 21 C., nor transfers in series of many separate colonies, accomplished any sign of reversion either to variant I or to the normal form (N).

But, notwithstanding this apparent stability against reversion, variant II ultimately did undergo further transformation into a form which we may safely regard as the "extreme variant," the variant III of Firtsch. A gelatin stab had been inoculated with N culture. After 375 days the colony count gave: some N, also some II, and a few III, but no variant I. After 476 days there were found, among the then greatly reduced population in the same tube, still some N forms together with some variant III, but no I or II forms. The growth of variant III on gelatin plates was very slow but ultimately gave—first, small blue-white, but later brownish, colonies with a "crumbly" surface and finely dentated margin. It was commonly nonmotile, did not grow on potato and reversion to any of the earlier types was never observed. Firtsch therefore regarded it as a permanent modification. From the foregoing exposition we can see clearly the relation of the vibrio types observed by Firtsch to our chief colony forms. His normal type (N) is the normal S; his variant I is the intermediate (O), which may generate either S or R, depending on age and circumstance; his variant II is the first R type, incompletely stabilized; his variant III is the "extreme variant," R—or, as we shall later propose, the R^n.

The splendid work of Firtsch at the time of its appearance was, as we should expect, relegated to the field of useless observations along with other similar work of the day. At the present time, however, I believe it stands as one of the most noteworthy contributions to our knowledge of the dissociation phenomenon; and, fortunately, more than a quarter of a century after its publication, when the bacteriological world had become a safer (though not too safe) place for the variationists, was made the basis of Eisenberg's comprehensive, though less critical, study of variation in the cholera vibrio.

Firtsch concluded his contribution with the following pertinent words: "In all dem heissen Kampfe, der über Monomorphismus und Pleomorphismus der Bakterien geführt wurde, bleib bisher Eines vollig unberührt, wurde Eines von allen Seiten als der 'ruhende Pol in der Erscheinungen Flucht' angesehen: die Form der Kolonien auf festen Nährboden."

Returning to the general consideration of the intermediate types and the reasons for their presence or absence in dissociating cultures, the influence of the medium may be mentioned. Since, as we have seen, slight differences in the reaction have an effect on the dissociative and lytic reactions, and an alkaline condition favors the rapid and complete transformation of type, the appearance or nonappearance of colonies of the transitional form may depend to some extent upon this circumstance. But, whatever the explanation may be, I believe that there now exists ample evidence to assure us that, standing between the S and R types in the course of transformation, there exists a transitional form of culture which manifests itself in the "third colony intermediate" or in a closely related colony of the transitional form. This culture type, for which I have proposed the symbol, "O," is apparently both the "progeny" of the S and the "progenitor" of the R. It is such an unstable culture form (which we can picture to ourselves as ready to lean in any direction) that Mellon[326, 327] may have observed when he has spoken of the marked influence of the environment in determining the nature of the succeeding growth of certain cultures. In any case, this group of transitionals embraces certain members characterized by extreme instability and often difficult, or sometimes impossible, of further cultivation, especially in broth medium, although they are more stable in physiologic salt solution or on agar. It is probable that those cultures which afford the remarkable "fading-away" appearances, characteristic of the "suicides" and amounting in some cases to an almost complete annihilation of growth, belong to this group of transitionals. If the lytic tendency is acute, such cultures may usually be depended upon to leave amidst their lytic residues their natural transformation product, namely, the R type culture. In all this discussion, however, it should be borne in mind that these peculiar transitional forms represent a series of apparently closely related types, some of which are more stable and susceptible of perpetuation, for a time at least, on ordinary culture medium. Of this varied group, the "third colony intermediate," which has been most frequently described, is only one form.

In concluding this subject of the intermediates, a word may be added regarding the relation of these culture forms to various incitants to dissociation, a fuller consideration of which is presented in a later section. It has been demonstrated in numerous instances, including B. anthracis, B. diphtheriae, B. proteus, and members of the colon-typhoid-dysentery group, when observed in the dissociative reaction, that the first observable sign of the influence of the unfavorable condition or environment is a transition from the normal S type organism to elongated, filamentous or fungoid forms of growth, accompanied by beaded or swollen rods, large and small coccus forms and various "involution" structures often including the so-called "giant coccus" which, from the recent descriptions of numerous observers, is manifestly identical with the "Ferran'sche Körperchen" observed and pictured (Taf. VI, fig. 12) in the work of Firtsch (1888) on Vibrio proteus. It is moreover this varied assortment of bizarre cell forms that constitute the major population of the colonies or cultures of the transitional type—a type which has for its chief characteristics a heavy and voluptuous growth accompanied by a mucous or gelatinous consistency. Such colonies are probably related to, if not identical with, the colon colonies which Mellon [331] has regarded as the mother form of B. coli mutabile. They have been described by investigators too numerous to mention. Some of the clearest references to them, however, may be found in the works of Hort, Mellon, Eisenberg, Feiler, Bernhardt, Baerthlein, Fletcher and Enderlein, not to omit the earlier study of Massini.

Conclusions.—In summarizing this section it may be pointed out that, although this review deals largely with the two chief types (S and R) involved in the dissociative reaction, microbic dissociation is not so simple a phenomenon as the "splitting" of the R type away from the normal S, as we might expect in an ordinary mutation. As we watch this reaction going on before us on plates or in culture tubes, superficially considered it seems a simple process; but we may remind ourselves that there must lie behind it all a mechanism. What this mechanism may involve, we do not yet know because we have not been able to detect or to isolate all the factors. But one thing is apparent: The R type is not a direct, but an indirect product of S; its production comes about through the functioning of an intermediate stage of culture development. Sometimes we are able to detect this transitional form by the use of appropriate cultural methods, while at other times it quite evades detection and one might thereby be led to conclude that it is absent or lacking in the transformation process. That, on the contrary, it is present in all cases is strongly suggested by several important observations which show that the transitional form (particularly in the guise of the "third colony intermediate") is a highly unstable and sensitive phase in the life of the culture. Why this is so, we do not know. But it is certain that its vitality on ordinary culture mediums, particularly in broth, is often slight. It disappears quickly from our view and leaves us—if anything —a modified culture of the R type which it seems to be at least one of its functions to produce. More than this cannot be said at present regarding the transitional type, although some of its antigenic characteristics are considered on a later page. It may eventually appear, however, that while other aspects of dissociation are given more prominence in this paper, it is the "third colony intermediate" and its nearby colony relatives among the transitional forms, that carry the greatest significance in microbic dissociation.

7. DISSOCIATION AND BIOCHEMICAL REACTIONS

Variations in the biochemical behavior of microorganisms, especially with reference to fermentative reactions, were naturally the first to be studied after morphological and cultural variations; and biochemical disparities were clearly recognized long before the study of serological and antigenic eccentricities had gained much headway. In later years, however, the increasing conviction that serologic and antigenic criteria of the constitution of bacterial species and varieties give information of a more fundamental and dependable character, has resulted in a correspondingly diminished activity in seeking out and studying fermentative or other biochemical variants. Unfortunately, morever, in most cases where clearcut serologic variants have been discovered, there has been a noticeable lack of effort to correlate these with biochemical changes in the same cultures. And the reverse also is true: that, when outstanding fermentative variants have been obtained, there has been slight attempt to correlate these with serological changes. Many of the variations in fermentation have been brought together by Gurney-Dixon.[220]

It is quite true that fermentative divergencies from the normal reaction have been numerous; but the slight degree of the departures, the apparent failure to work with pure line strains, and the lack of data bearing upon correlated differences, cultural, morphological or serological, usually leave us in doubt regarding the relation of the observed changes to the dissociative reaction as we have come to know it. This circumstance renders the majority of such contributions of slight value for the purposes of the present review. For this reason the majority of them have been omitted, and there consequently appears a lack of balance between the biochemical and serological aspects of the dissociative phenomenon. Such a discrepancy might seem to imply a certain lack of significance of the purely biochemical reactions in dissociation. I believe, however, this would be an unwarranted view of the matter. From the few cases which merit presentation by reason of their clearly observed relation to the phenomenon, it appears that the biochemical changes involved, though perhaps often less significant than the serologic, are in reality of much importance; and, when further studied, may add new and valuable facts to our knowledge of the dissociative reaction.

*Dissociation and the Production of Pyocyanin.**—We can often observe that a pyocyaneus culture showing erosive action on a solid cul-

* This illustration of a certain biochemical aspect of dissociation also enters the field of transmissible autolysis, but is none the less applicable to the case in point. The details of the reaction have been reported in full in another paper by the author.[224]

ture medium gradually loses its power to produce the blue pyocyanin and, little by little, comes to the stage of a fluorescogenic, but non-pyocyanogenic organism. What is happening to the population of such a culture? If, at varying intervals in the progress of the change, we plate out a suspension in broth, it is first ascertained that perhaps 99 out of every 100 organisms yield a rich blue-green colony. But gradually these bacteria are replaced by organisms (colonies) which produce no pyocyanin whatever. We may thus have the changing ratio: 99:1; 75:25; 50:50; 25:75; 1:99; and, in all probability, even 0:100. The change in this biochemical character therefore does not involve "a little loss" of pyocyanogenic power on the part of all the organisms present, but the sudden appearance of a new type of organism (the nonpyocyanogenic) and an increase in this form to the greater or less (and in some cases complete) exclusion of the "normal" blue type. I possess several such cultures which have been dissociated for over three years and none of them shows any tendency as yet to produce even a trace of pyocyanin as demonstrated by the chloroform test. It seems safe to say that whereever pyocyanin production can be awakened in old cultures which may appear to have lost the ability to produce it, it is usually due to the regeneration of a relatively few organisms of the sensitive type which have managed to survive amidst the overwhelming population of non-pyocyanogenic forms. But, when the culture is founded on such a "mutant," or has through aging or an unfavorable medium lost the last sensitive organism, then the bluing character has, so far as I have been able to ascertain, been lost beyond power of recovery. This circumstance is opposed, however, to other studies clearly indicating the possibility or reversion of the R to the original form in other bacterial species.

Dissociation and Proteolytic Power.—Another biochemical characteristic in which the R dissociate is likely to vary from the mother type is proteolytic ability, and this has been observed in two markedly proteolytic species, B. proteus and B. pyocyaneus; also Firtsch [178] in 1888 pointed out the diminished gelatin-liquefying power of his variants (O and R forms) of Vibrio proteus. Braun and Schaffer [66] in 1919 reported the diminished power of gelatin liquefaction of their O forms of Proteus X19, whether produced by the influence of phenol or of starvation. Baerthlein's "second variant" of B. proteus (the O type?) is also reported to have given very slow liquefaction compared with the normal type. The same was true of Balteanu's [38] opaque variant of Vib. comma. Its coagulating action on milk was correspondingly slow.

I have already reported the same phenomenon in B. pyocyaneus, in which the R type is inactive biochemically as compared with the S form. Soule [450] has shown a retarded rate of gelatin liquefaction for his R type strain of B. subtilis.

Dissociation and Fermentative Ability.—To be certain of the relation between products of dissociation and fermenting power of organisms in many of the cases reported is a difficult matter. This is occasioned by the well recognized differences not only between different strains of the same organism, but also in the same strain at different times in its history. To what extent these differences may be due to dissociative reactions that are too slight to reveal themselves in physical modifications of the culture we do not know, although it cannot be doubted that unrecognized dissociations are occurring in perhaps all cultures. It is only when the dissociates have been revealed, isolated, and studied as separate cultures that we are justified in entertaining any conclusion regarding the effects of dissociation on fermentation power. And naturally the same conditions hold for the study of virulence, serologic activity, and indeed of all biochemical reactions. Up to the present time, as has been said, there have appeared few reports which can justifiably be employed for the purpose of this review, although citations of the most diverse variability in fermentative reaction are numerous in the literature.

Interest in modified fermentation power of variants dates largely from 1906 and 1907 when Neisser and Massini made their first observations on the "mutations" of B. coli mutabile. Neisser's [362] observations showed that an organism of coliform type, previously lactose-negative, suddenly threw off colonies demonstrating lactose-fermenting power, and retaining this ability with considerable tenacity. Massini,[320] one year later, studied the same organisms more thoroughly and confirmed the observations of Neisser. On lactose-fuchsin agar the culture produced at first only white colonies but, after the second or third day, many of the better isolated colonies showed papillae ("Knöpfe") which were white at first but later red. One-tenth per cent of lactose was sufficient to produce this result, but it was not produced by dextrose, mannitol or other carbohydrates. When white colonies were subcultured within 24 hours they gave only white colonies, which produced red papillae later. When the red papillae of the white colonies were subcultured to similar mediums, they give both red and white colonies. When the red colonies on these plates were further plated they gave red colonies only. These were fairly permanent in their new character after they had been carried for some transplants on lactose medium. The red colonies themselves never gave papillae.

These observations were carried forward by Burk,[84] by Kowalenko,[286] who succeeded in confirmation tests by Burri's single cell method; and by Reiner Müller.[357, 358] The latter obtained similar papillae-forming colonies from feces. He found that nearly all of 120 typhoid cultures gave this colony reaction on rhamnose agar while it was obtained in only a few of 200 coli cultures. The newly acquired character seemed permanent and he pointed out that the delicacy of the test compared favorably with the agglutination test as a means of diagnosing the bacterial type; also that this reaction was the most typical cultural feature of the typhoid bacillus. All Müller's work was controlled by single colony isolations performed by Burri's method. In addition Müller showed that, although cultures of B. paratyphosus B gave no daughter colonies on rhamnose or lactose agar, they were produced on raffinose agar. The daughter colonies were thus able to ferment a sugar which the mother type was unable to utilize. These new raffinose fermenters were never seen to revert to the original form.

In 1910 Jacobsen[265] isolated from a typhoid case a typhoid-like bacillus which gave a typical growth on Conradi-Drigalski medium. Secondary colonies appeared, however, which represented the typical typhoid form. Jacobsen therefore called his original culture, B. typhosus mutabile. These results were confirmed by Reiner Müller in 1911. Similar observations were made by Schröter and Gutjahr[435] on B. dysenteriae Shiga and Kruse which acquired the ability to ferment maltose and sucrose after undergoing prolonged growth in these mediums. The changes seemed to be permanent. Other contributions bearing upon this subject are those of Sobernheim and Seligmann[447] and of Thaysen[460] for a similar type. In 1910 Penfold[385] also confirmed the observations of Müller and furthermore studied the "mutations" of B. typhosus on isodulcite agar, of B. coli on lactose, of B. paratyphosus B and B. suipestifer on raffinose; also B. coli on chlor-acetic agar. He found that the acetic agar caused many intestinal bacteria to throw off "mutants" characterized by greater resistance to still greater concentrations of the same antiseptic. With reference to raffinose he showed that, while paratyphosus B formed papillae, the Aertrycke bacillus did not; and that this was the only cultural distinction between the two species. The papillae were formed only after seven to nine days and gave no acid colonies.

Burri [88] in 1910 made an unusually interesting study of a paratyphoid-like organism obtained from fermenting grass and termed Bact. imperfectum. This culture was at first unable to ferment sucrose, but when grown continuously on sucrose mediums some of the colonies gained this power. This sucrose-fermenting type Burri named Bact. perfectum. Both of these cultures Dobell [130] regarded as belonging to the paratyphoid group. The change undergone by the original culture manifestly resembles the change in B. coli mutabile in respect to lactose. As a result of plating different numbers of organisms of the original type (imperfectum) Burri concluded that all of the cells of the imperfectum type were able to mutate directly into B. perfectum. When the individuals were sufficiently separated on plates all behaved in the same manner; they became transformed. As Dobell pointed out, Dreisch has expressed this in the sense that "every individual had the same prospective potency." It may be noted in passing that Burri also observed the existence of transitional stages between imperfectum and perfectum. The ability to ferment sucrose was acquired only by gradual steps. Burri regarded it probable that the ability to ferment sucrose was latent in all the cells of the imperfectum type, probably in the form of a proferment.

Baerthlein [29] in 1912 obtained results in the study of 13 different strains of coli-like organisms from the normal intestinal tract which are not in exact conformity with the results mentioned above. All the strains showed the same reactions as observed in Massini's culture which showed colorless mother colonies containing red secondaries on Endo's agar. Baerthlein found, however, that after long cultivation on nonlactose agar there occurred a reversion to the nonlactose-fermenting type. Even after six to seven days there was a partial reversion on Conradi-Drigalski medium; that is, some white colonies appeared among the red. But Baerthlein introduced additional complicating features. He showed that both B. coli mutabile and typical coli strains on plain agar were transformed into two different types, characterized by colonial, biochemical and serological differences, also accompanied by different cell-forms. All of these types might revert to the original.

In reviewing these preceding cases, one further point is of interest. In the study of Massini [320] and others working with B. coli mutable it is clear that a form of culture unlike typical B. coli "mutated" into the typical form, capable of fermenting lactose. For this reason Massini believed that the mutating strain was a typhoid-like organism capable of throwing off coli races more or less constantly. In the case of Müller, however, working with B. typhosus (typical) we see that this culture threw off colonies differing from the original and typical typhoid strain. The same was true of the typical typhoid of Penfold which threw off isodulcite-positive "mutants." And we observe a similar case in the daughter colonies of typical paratyphoid B produced on raffinose as described by Müller. Thus we seem to have the possibility, not only of a typical form of culture becoming transformed into the atypical, but also the reverse. It is only when these studies have been repeated and carefully checked with the cultural features and serological reactions, that we shall be able fully to understand the reactions observed. In the meantime, however, we can conclude that they deal with the problem of dissociation and involve biochemical changes of considerable significance.

One of the earlier references dealing with changes in fermentative power accompanying dissociation is found in Goodman's [200] study of variation in B. diphtheriae in 1908, mentioned subsequently in connection with virulence. Goodman made a double selection series for most acid reaction, and least acid reaction. After 36 selections he obtained from the original one strain which gave high acidity in dextrose and a second which failed to ferment dextrose at all—which in fact made it more alkaline. In both strains the ability to ferment maltose was diminished, while the ability to ferment sucrose was increased. The acid culture was virulent, the nonacid culture nonvirulent, a point confirmed by Bernhardt [49] in 1915. Here the assumed change to the R type (nonvirulent) was accompanied on the whole by a reduced fermentative power. Essentially the same point was demonstrated by Roux and Yersin [416] in 1890.

Revis [404] showed fermentation differences in his B (R) type of B. coli following dissociation through the use of brilliant green. He had shown earlier that new types produced by malachite green had lost the power of gas production. Esther Stearn [452] in 1923 reported gradual fermentation changes produced in water-borne bacteria under the influence of gentian violet and probably concerned with dissociation, although the accompanying cultural changes are not mentioned. Cultures taken from the gentian violet broth after 48 hours, 120 hours and 5 months showed transitions in communis A and B, communior A and B, aerogenes and B. acidi-lacti. At the end of 5 months all had assumed the fermentation type of B. communior A. The results of the dissociative process thus seem to have been cumulative.

In the recently reported changes in a paradysentery culture Ørskov and Larsen [513] showed that their variants, V and M gave fermentation reactions different from those of B and Bu. The latter pair fermented lactose, maltose, sucrose and xylose while the former pair did not. If we may regard the V form as equivalent to S, as shown on another page, the new forms in this case gained rather than lost in fermentation power.

A loss in the ability of B. lactici to ferment milk when the culture was grown in the presence of phenol has been shown by Schierbeck.[431] We know that in general phenol mediums are able to produce, or at least to cause the initiation of, the dissociative reaction; but, whether such occurred in the present case, is impossible to state since the cultural features of the modified phenol strain are not reported.

Certain references to the relation of dissociation to the production of pigment by bacteria are presented in section 11 under the heading of "Action of Antiseptics"; also in section 5.

In relation to the streptococci Eugenia Valentine [469] quite recently observed that the green (alpha) forms may be divided into two distinct types, X and Y, on the basis of peroxide production and methemoglobin formation in 18 to 24 hour broth cultures. The X type gave a positive peroxide and methemoglobin reaction, while the Y type, in cultures of the same age failed to yield peroxide and formed methemoglobin only after the addition of sterile broth or serum. These differences were not observed in four hour cultures. The growth activity of the two forms was also different, the X strain reaching its maximum growth point after 12 to 14 hours, while the growth of the Y strain began to fail after six hours. The X forms appeared to approach the pneumococcus in their peroxide producing and methemoglobin forming properties. Although these differences were not correlated with different colony types or with different cultural growths, serologic and fermentative differences were observed. It therefore seems probable that the dissociative reaction was involved in these cases, which also may have a bearing on the studies of Webster [481] on dissociation in Bact. lepisepticum.

Conclusion.—From the foregoing references to the biochemical changes accompanying the dissociative reaction, and from other references which are to be presented in section 11, it appears that with dissociation are concerned, not only variations in the physical aspects of the cells, but also physiological modifications of considerable significance.

Fermentation reactions in particular may undergo profound alterations accompanying the transformation of the S culture type to the R. It can scarcely be doubted that in many instances in which changes in fermentation capacity have been observed in cultures maintained in the laboratory over long periods, these are due to dissociative reactions. Of the cultural characteristics, however, which have their basis in the biochemical aspects of cell behavior, perhaps the most important is that of virulence—due, as we now have cause to believe, to specific substances such as the aggressins of Bail, the antiphagines of Tschistovitch or the virulins of Rosenow. To this subject we may now turn our attention.

8. DISSOCIATION AND VIRULENCE

One important aspect of the dissociative reaction relates to oscillations in the character of virulence in pathogenic bacterial species, present not only in the animal body but also in vitro. Although it may be unwise to picture to ourselves too far-reaching a significance in the small number of observations that merit being reported under this heading, it must now be accepted as a fact that they yield for the first time clearcut instances demonstrating the correlation of virulence with a definite form of culture—or as we shall see later, with a definite cyclostage in the genetic history of the species.

To the bacteriologist, as to the epidemiologist, the character of virulence has always been an elusive thing. Within limits, it is true, he has been able to control it; but he has seldom been able definitely to correlate it with other cultural, biochemical or serological features. Moreover, erroneous views have grown up regarding the distribution of virulence among the different cells of the culture. The most common conceptions relating to virulence are: that a highly virulent culture is one in which all the organisms equally possess virulence; that a moderately virulent culture is one in which all the organisms equally possess an intermediate grade of virulence; and that a nonvirulent culture is one in which all of the organisms are equally nonvirulent. It is quite remarkable that this view has persisted for so long a time, despite much evidence to the contrary. Although a different situation had been suggested by the early study of Chauveau and Phisalix [97] on B. anthracis, and still more clearly by the work of Bordet and Sleeswyk [64] on B. pertussis in 1910, as well as by many observations on the diphtheria bacillus and typhoid organism, de Kruif was the first to demonstrate clearly the relation of virulence to what we now recognize as microbic dissociation.

Bact. lepisepticum and Other Pasteurella Forms.—De Kruif [119] studied this subject in Bact. lepisepticum (B. cuniculicida), a highly virulent species and one of the best known members of the Pasteurella group; an organism long noted for its strongly aggressive characteristics. In this case, as de Kruif was able to point out, nothing like what we have postulated above regarding the nature of virulence really obtains. What does happen, at least in Bact. lepisepticum, is a transformation in cell type in which the number of fully virulent bacteria (de Kruif's D type) gradually decreases, while organisms of a new type of nonvirulent organisms (G) make their appearance and gradually increase in the bacterial population. In other words, lessened virulence in this case does not mean a similar fractional loss by each and every organism in the culture but a dilution of the number of virulent bacteria (S) by an increasing number of bacteria possessing no virulence whatever, or at least a markedly diminished virulence (R). The new type, appearing under the conditions which usually produce attenuation, is the nonvirulent "mutant." In the actual comparative virulence tests, as first reported, although massive doses of his G type (R) killed rabbits, doses of one cc. or less usually produced no lethal effect. In the rabbits that succumbed it is important that the same G type was obtained from necropsy. On the other hand, type D cultures were often fatal in doses of 10^{-7} cc. and usually in doses of 10^{-5} cc. for 1500 gm. rabbits.

Although not successful in earlier tests, de Kruif [121] reported later (1922) that relatively large amounts of type R culture were fatal to young rabbits of 600 to 700 grams. He also demonstrated that the virulence of R could be raised by passage through young rabbits to such an extent that 10^{-4} cc. might be fatal for a 750 gram animal. In some of these cases a few type S organisms were found at necropsy in the peritoneal fluid or heart blood. It is of special interest that, notwithstanding the increase in virulence, the other characteristics of the culture remained of the R type. De Kruif therefore concluded that the granular form of growth and lack of virulence are not invariably concomitant. The R cultures possessing the raised virulence were not, however, tested on adult rabbits.

To the above exposition of de Kruif's important work it should be added that the dissociative reaction in Bact. lepisepticum had been recognized, although not fully studied, by Bernhardt [49] in 1915. Bernhardt reported the production of two forms of colony, mutant from the normal type of culture. One was his socalled "gerippt" form; the other his "durchschleimig" type. The latter was more virulent for rabbits.

While the description of these two types is very incomplete it seems probable that the "durchschleimig" type was not the virulent D (S) form of de Kruif, but the intermediate or "Zwischenform" of many writers which is referred to in greater detail in section 6 of this paper. Although de Kruif did not report this intermediate there can be little doubt but that it is present at certain stages of the dissociative process of Bact. lepisepticum. It may be added that also in the closely related form of Pasteurella, the B. avisepticus of fowl cholera, Bernhardt recognized several colony types among which we can detect the R and S. These, however, were not studied by him in detail.

B. avisepticus (Pasteurella avium): Interesting data on the dissociation of B. avisepticus are, however, available from the work of Manniger,[317] published in 1919. It is well known that this organism seems to survive in broth cultures better than on agar slants. Manniger undertook to study this peculiarity and at intervals for one and one-half years prepared agar plates from an old broth culture. Among the colonies that appeared more or less normal he observed some that were dark brown in color and showed other irregularities. These he believed at first were contaminations, but, on further study, they were found to be mutants from the original stock. The fact of special interest, however, is that they were nonvirulent, while cultures in broth from normal colonies were highly virulent. The nonvirulent organisms were larger and plumper and possessed no capsule, while those of the virulent type were minute and well capsulated. Of the nonvirulent culture, rabbits and hens easily tolerated one-half of an agar slant culture. A whole agar slant sometimes killed pigeons. When the culture was recovered from them and then injected into mice, 0.000,01 oese was usually fatal in two days. Thus the loss of virulence was not permanent when large amounts were injected. Obviously, such large amounts of the mutant culture still contained some of the organisms of the virulent S type. These propagated in the body of the injected animal while the nonvirulent mutant organisms were destroyed. A study of the relative phagocytability of the two types showed that, while the virulent organisms were not phagocytable, the mutant organisms were quickly taken up by the white cells. Manniger presents other interesting data on the immunizing value of the mutant type; also on the serologic relationship between the virulent and the nonvirulent types of culture. These subjects are presented in detail in sections 10 and 9, respectively, of the present work. It can scarcely be doubted that the two Pasteurella forms with which Manniger and de Kruif worked were closely related; and that the former investigator was concerned with the S (virulent) and the R (nonvirulent) forms of the culture which de Kruif was able to recognize more clearly. It results, however, that each of these studies clearly confirms the other with respect to the unequal distribution of virulence among the various colonies of the respective original cultures.

Bacillus pestis: Gotschlich[204, 205] has reported dissociation phenomena in connection with his studies on plague in Alexandria in 1899-1900, thus involving another member of the Pasteurella or hemorrhagic septicemia group, B. pestis. From human buboes and also from cat pest he obtained, in addition to organisms of the normal type, a variety which, as he states, would not be regarded as B. pestis if one did not know the source. The growth on agar was characterized by heavy, shining colonies which did not grow on gelatin at 15 C. These cultures, as opposed to the normal, possessed no virulence for rats, guinea-pigs or rabbits, the latter easily tolerating 10 agar slant cultures at a single inoculation. The variant cultures showed diminished agglutinability and yielded an immune serum which agglutinated the normal culture only slightly. After some weeks in the icebox these atypical cultures reverted to the normal form and resumed virulence. It seems probable that Gotschlich dealt with the normal S type of B. pestis and, not with the R variant, but with the intermediate, O.

Probably the forms of the pest bacillus reported by Dudtschenko[137] in 1915 represented similar types. It may be added that Klein[279] in 1909 had, according to Gotschlich,[205] also called attention to two regularly occurring forms of the plague organism, one from the active human infection, characterized by possessing cylindrical rods and high virulence; the other form more common in rat pest and characterized by short, ovoid rods of greatly diminished virulence. Shibayama[439] in 1904 had also noted in plague cultures certain forms showing different cultural features and significant discrepancies in agglutination reactions, making diagnosis difficult. From these records it is thus apparent that dissociation is not limited to the rabbit and avian type of Pasteurella, but is also present in the human form (B. pestis), although here the matter has received little attention. In this disease interesting results might follow attempted immunization with the R type cultures as de Kruif has studied the matter in Bact. lepisepticum, and as I have observed the immunological results in the related organism, B. avisepticus.[223]

B. anthracis.—Although, as has been said, the studies of de Kruif have given us one of the clearest pictures of the close association between virulence and distinctly recognized type S cultures, the conception of virulent and avirulent organisms existing side by side in the same culture was not entirely new. In 1895 Chauveau and Phisalix [97] obtained from lymph glands, but not from heart blood, of cattle that had died of anthrax infection an atypical and nonvirulent organism which, according to our present view, must have been the S type.* This new culture apparently bred true. In 1904 Preisz,[393] and later Baerthlein [28] and Eisenberg,[149] recognized two distinct colony types. The first was large, transparent, somewhat indistinct and seemed to represent the normal medusa form. The latter was smaller, succulent and cloudy white, with a more glistening luster. The latter was undoubtedly transitional, as this form was made clear later by Preisz. In 1912 Markoff [319] obtained five colony variants all of which bred true for 10 to 30 generations. Among these we can recognize, not only the S and R forms, but also the more or less succulent and often slimy transitional or intermediate type (O).

In 1911 Preisz [394] showed more conclusively than in his earlier publications on anthrax the relation of colony form to capsule formation and virulence. Since this study has been mentioned in some detail under the heading of capsules in section 7, a brief statement here will suffice. When Preisz plated on agar Pasteur's first and second vaccines, he noted the appearance of three sorts of colony. The first was much like the normal medusa type and was virulent for mice. It possessed the ability to generate capsules in the body of the animals. The second colony form was a succulent and slimy type in which many of the organisms were poorly capsulated. When injected into mice death usually resulted and normal anthrax organisms (and colonies) were obtained from the necropsy material. The third type of colony differed markedly from the other two. It was round, white and fairly compact. The single organisms showed no capsules and the culture was nonvirulent upon inoculation. When the animals were killed later no bacteria of any sort could be found in the blood or tissues. This noncapsulated and nonvirulent form bred true. In the light of other dissociation phenomena we should be inclined to call the first type mentioned above the R form; the second the transitional and the last the S. Such an interpretation corresponds with the results obtained by Soule [450]

* We shall see in the following exposition that there are certain discrepancies in the apparent relation between virulence and type in anthrax. The situation cannot be made fully clear until much further study has been conducted on this problem. The statements which follow represent merely tentative conclusions. Serological tests must eventually decide the matter.

on the dissociation of B. subtilis. But, as we shall see, there are various complicating factors in the dissociative reaction as related to the anthrax bacillus.

Regarding the correlation of type and virulence Bail and Flaumenhaft [31] in 1917 described what I have termed the R and S colony forms, pointing out that the one which we regard as S was the variant from the normal (medusa) type. In a limited number of virulence tests the R form of culture showed the greater virulence for guinea-pigs. In cultures both types bred true, but in the animal body a transformation occurred from S to R. In this work it is of interest to note that in the necropsy of the pigs that had received the S culture ā type of Bail the culture form varied with the location in the tissue. From the edema came only R (β type), from the heart blood mainly S, while in the spleen and liver the numbers of the two forms were about equal.

Again in 1920 Gerhard Wagner [476] obtained from human sources two colony types of B. anthracis which he termed the typical A and the atypical B. The atypical differed from the normal in being smaller (although denser and heavier), "cloudier" and more of a grayish white. The individual organisms were shorter and plumper. In gelatin the culture grew in short, thick processes, not in fine threads (inverted fir tree growth) as the typical form, confirming the earlier observation by Bail. The virulence of both for pigs seemed to be the same, but for white mice the atypical form was the more virulent. The reversion of the atypical to the typical occurred after several years. After seven years the two forms could no longer be differentiated. Excellent photographs of these types accompany Wagner's paper. They show clearly the usual characteristics of the R and S types and strongly resemble the R and S cultures recently obtained by Soule in his dissociation of B. subtilis.

The latest contribution to the study of the relation of dissociation to virulence in anthrax, although having another purpose in view, is the brief report of Gratia [212] in 1924. Gratia also noted the two chief types of colony (R and S) so often described before. One (A) was convex with regular margins and gave a homogeneous clouding in broth. Microscopically it showed short, well isolated organisms forming only short chains. The other type of colony (B) was flat, more translucent and filamentous. Microscopically it contained long chains and filaments made up of elongated bacilli. In broth it gave an agglutinative form of growth. When the A type was inoculated into nine rabbits intravenously four survived while the others died after a prolonged period. When the B type was injected similarly into twelve rabbits eleven of them died in two to four days. When freshly isolated and then left to themselves these two culture types were reversible; but after longer cultivation each became more stable. These findings compare favorably with the results of Priesz and of Bail and Flaumenhaft, mentioned earlier.

Mr. Nungester in our laboratory has ascertained, from a long series of inoculations into guinea-pigs of pure R and S cultures, that the average time until death of the R inoculated animals was 49.5 hours while the time for the S inoculated animals was 78.3 hours. This difference was not observed in inoculated mice. The assumed transitional type, characterized by slimy growth on agar, we might conclude from the tests of Preisz and others, holds an intermediate degree of virulence. As a matter of fact Mr. Nungester found that a short series of guinea-pigs inoculated with his slimy culture type gave an average time until death of 58 hours, intermediate between the times for R and S. Of the other four culture types obtained by Mr. Nungester, all of which remained constant for 20 or more passages on slanted agar, the virulence has not yet been satisfactorily tested.

These results, suggesting a greater virulence for the R type of B. anthracis than for the S form, are opposed to the results of Wagner. With the exception of the last observer there is general agreement among the workers cited that, what we are inclined to call the R form of anthrax culture, is the more virulent, while the noncapsulated S form, giving the compact, white colonies, is nonvirulent. The transitional type, characterized by slimy growth on agar manifestly holds an intermediate position so far as virulence is concerned. So far as the R and S types are concerned, this interpretation is opposed to the findings of de Kruif for Bact. lepisepticum; and, as we shall see as we proceed with this review, it is opposed to the majority of reports dealing with other pathogenic bacterial species, where it is the S form that carries the virulence. According to Enderlein [160] the virulent stage of B. anthracis is the "Phytascit." It is also of interest to note that, if our present conception of these types should hold true, it would be the R type of anthrax, rather than the S, which has become established through many years as the "typical" anthrax culture. The final interpretation of the results already presented can be accomplished only when the whole subject is submitted to a more thorough-going study than it has yet received; and particularly we must await an answer to the question as to which form of the culture carries the double (S-O) antigenic load; for, so far as our present knowledge is concerned, this is the criterion of last resort.

B. diphtheriae.—In the case of B. diphtheriae the question of relationship between the Klebs-Loeffler bacillus (clinical type) and the Hofmann types has been a subject of perennial interest and one on which a vast amount of inexact investigation has been conducted. While Loeffler himself adhered to the view of two distinct species and converted many to his view, other early workers, particularly Roux and Yersin [416] (1890), maintained that all the variable forms belonged to a single species and that transmutation between virulent and nonvirulent types was common. Although acknowledging the similarities in many respects, Roux and Yersin were among the first to point out distinct differences between the types, the more significant perhaps since their experiments were in accord with a belief in a species unity. These differences lay in the virulence and morphology of the organisms, rapidity of growth in broth, density of growth and acid production. They thus recognized the avirulent type as shorter, plumper and less evenly stained, giving quicker and stronger clouding, and alkaline reaction rather than acid in sugar broth. Starting with a pure line culture of the virulent type they showed clearly the origin of nonvirulent forms; also the influence of growing cultures at 39 to 40 C on this transformation. They did not, however, associate virulence with form of colony, cultural features on agar or, naturally, with serologic reactions.

More convincing experiments were reported in 1897 by a small group of workers, Hewlett and Knight, Zupnik, De Martini and particularly Corbett and Phillips in England. Corbett and Phillips [110] showed the splitting of a pure line culture to yield two sorts of colony —one thin, gray, "inconspicuous," acid-producing and highly virulent; the other whitish or yellowish white, heavier, more opaque, nonacid-forming and nonvirulent. These characteristics may easily be recognized as representing respectively the cultural features of the virulent Klebs-Loeffler bacillus and the pseudodiphtheria bacillus. I have studied the strain, "Park 8," in this connection and have found that it dissociates readily, especially on agar and under the influence of ascitic fluid, into the forms described by Corbett and Phillips and into a distinctly coccoid type.

In 1897 the unity of the diphtheria bacillus type was questioned by Zupnik.[408] He found cultures from collections, as well as from the throat, which showed on plating two different growth forms. One gave on agar a relatively heavy, flat, dull colony with irregular borders; the organisms were gram-positive and virulent for guinea-pigs. In beef tea the growth was at the surface and with no clouding. The other type gave smaller colonies which were quite round, convex and glistening. The organisms showed Babes-Ernst corpuscles but did not stain well with Gram. Injected into guinea-pigs subcutaneously they produced infiltration and necrosis but never death. The beef tea cultures showed at first diffuse clouding but later pellicle formation and clearing.

Hewlett and Knight [251] (1897) observed that pseudodiphtheria organisms often made their appearance in normal virulent cultures and reported the change of virulent and typical Klebs-Loeffler forms into pseudo types by 17 hours exposure at 45 C. The resulting organisms were short and plump, without granules, and nonvirulent for guinea-pigs. They also transformed typical pseudo forms into virulent by guinea-pig passage and by passage through a serum medium. This has been often reported since. In the same year de Martini [125] stated that the pseudo form would grow in immune serum while the true Klebs-Loeffler type would not. This is an important observation, but so far as I am aware it has never been confirmed. Additional data favoring the view of transformation in culture have been reported by myself [221] (1907).

The most comprehensive early picture of variability in the diphtheria bacillus, however, appears in the study of Slawyk and Manicatide [443] in 1898. These authors studied in great detail the behavior of 30 strains. Anyone with patience to review the mass of data embodied in the 42 pages of tables accompanying the article can obtain a clear picture of Klebs-Loeffler variability. The same two colony types appeared: one, transparent, flat, dull; the other, opaque, convex, glistening. Between these were many intermediates. On the whole the former were more virulent for guinea-pigs, but this subject receives slight consideration.

Gorham,[202] on the basis of cultural examinations, came to the conclusion that there occurred in the throats of convalescents a transformation from granular and virulent diphtheria bacilli to solid-staining nonvirulent forms. He interpreted this change as due to the influence of body fluids of the immune individuals. This view, although unsupported by direct evidence at the time, is now amply justified in the light of more recent studies of the effect of immune serum and body fluids on the morphologic type and pathogenic characteristics of the diphtheria bacillus, as well as many other microorganisms.

Lesieur [302] in 1903 found that passage through three rabbits changed Hoffman-Wellenhof types to the morphology of the true Klebs-Loeffler, while cultivation in diffuse daylight for eight months changed typical clinical forms to strains indistinguishable from pseudodiphtheriae. Denny [124] in the same year showed that virulent throat strains changed to pseudo forms. Others have maintained equally the invariability of the pseudo type, and Clark [100] has alleged the separate nature of the Hofmann forms.

A further interesting and convincing contribution to the dissociation of the diphtheria bacillus with respect to virulence is that of Goodman [200] in 1908. Goodman made the attempt to split a pure line culture of a virulent strain by means of repeated selections for high and low acidity produced in sugar mediums. Thirty-six selections were made in series and thus two bacterial types were obtained, one of which gave a strong fermentation of glucose while the second not only failed to ferment but gave an alkaline reaction. The acid forming culture was the more virulent for guinea-pigs. We shall note presently that this correlation between acid production and virulence and the correlation between alkali production and nonvirulence was further confirmed by Bernhardt [49] in 1915, as also by many other observers. Unfortunately Goodman did not present data dealing with the correlated morphological features of his two culture types; but with the work of Corbett and Phillips and others in mind there can be no doubt that Goodman succeeded in effecting the dissociation of his original form, at least to the stage of an intermediate, through resort to selections for high and low acidity. It is of interest to note in this connection, however, that according to Crowell [114] Meader in 1919 was unable to observe correlation between acid production and virulence. Of course it must be taken into consideration in these studies that, as Bernhardt has pointed out, there may exist pseudodiphtheria forms which are not true R types (diphtheroids) but which may produce acid without at the same time being virulent. In this circumstance we have an analogy with many other bacterial species.

In a highly comprehensive study of variability in pathogenic organisms George Bernhardt [49] in 1915 presented a mass of valuable data including transformations in B. diphtheriae as well as in some other species. One essential aim of the work was to demonstrate that the transformations mentioned were not, as believed by Baerthlein, sudden changes (saltations), but that they were produced gradually and progressively through a considerable range of transitional colony types lying between the original culture and the extreme variant (R). Here he established the basis (as Firtsch had done for Sp. Finkler-Prior and as Eisenberg did for the cholera vibrio) for the "Zwischenformen" referred to later by Weil, Felix and others in discussing the transitional types of B. proteus and other bacterial species. Bernhardt plated out on agar a culture of virulent B. diphtheriae coming from an old broth tube. After a week or more he obtained two colony variants: type I, large, thick, whitish colonies containing ordinary diphtheria rods accompanied by involution forms; type II, small, transparent and delicate colonies with indented margins containing small thick bacilli, often with pointed ends, and among these some club shaped or gourd shaped "degeneration forms." These differences disappeared, however, when the cultures were returned to Loeffler's blood serum. The type I cultures were highly virulent while type II was only slightly virulent in young broth cultures. In seven to ten day old broth cultures, however, there was no difference in toxicity. Bernhardt next attempted to obtain by selection an "extreme form" of type II which should have lost all toxicity and virulence. He selected type II colonies which showed no secondaries (as most of them did) and after plating further secured two types of transparent colony (type II, a and b). Type II b yielded organisms which were nontoxic and nonvirulent. They were Neisser-negative and showed polar staining. They possessed the characteristic morphology and staining properties of the Hofmann-Wellenhof pseudodiphtheria type and were constant in further propagation. Bernhardt was also able so to dissociate a highly virulent "American strain." He could, moreover, observe the secondary type of culture in nasal diphtheria and in urine. He succeeded, furthermore, in inducing the type I (S) → type II b (R) transformation under the influence of immune serum. By animal passage he changed the S form to an intermediate type ("Zwischenform") possessing reduced virulence, even if not entirely to the nonvirulent (extreme) R form. In view of these results Bernhardt concluded that the true diphtheroids (Hofmann types) are the mutation product of the clinical diphtheria type of bacillus, but that there are also "pseudodiphtheria" bacilli which are not necessarily related to the causative agent of diphtheria.

In 1917 Heinemann [239] made an interesting observation on the relation of toxin production to morphologic type and certain cultural features of the Park 8 strain of B. diphtheriae. His results as we can clearly observe are interpretable on the data supplied earlier by Bernhardt, as just reported. The incentive to Heinemann's brief study was the circumstance that in his laboratory variations appeared in toxin production, intermittently and apparently without cause. Film formation always occurred and no contaminating organisms could be detected. Heinemann observed, however, that the nature of the film varied. Sometimes it was thick, rough and yellowish, occasionally brownish in patches; and under these conditions toxin formation was low. Again it was more delicate, friable and light gray in color; and in this case the toxin was potent. It is well known that it is possible for those experienced to predict with considerable accuracy the amount of toxin forming by observing the nature of the film. In films of these two types Heinemann observed marked differences in the morphology of the cells. In the nontoxic cultures coccus and coccoid forms had largely taken the place of the rods. These cocci "did not appear as well rounded forms resembling typical staphylococci, but might easily be mistaken for streptococci in diplococcus form or in short chains." Intermediate forms were also observed, which "still retained the outline of a bacillus that seemed to be swollen, then broken up into cocci of various shapes and sizes." It was proved that these coccus forms were really variations of the original bacillary form. On Loeffler's (horse) blood serum typical bacilli appeared, but club and granular forms were always among them. When such a growth was transferred to veal-glucose-agar the coccus forms appeared. Transferring back to Loeffler's medium caused a return of the bacillary forms in 24 hours. Heinemann recognized the similarity between these results and those reported by Mellon [326, 327] for diphtheroid bacilli. I have seen these same changes occurring in the Park 8 strain and there can be no doubt that they involve the dissociation phenomenon. Successful plating on favorable mediums reveals the two colony types together with "intermediates," as first shown by Corbett and Phillips [110] in England thirty years ago, and by many others since that time.

The most recent study involving dissociative aspects in B. diphtheriae is that of Crowell [114] in 1925. From his original "single cell" culture Crowell was able to obtain both toxic and nontoxic strains. The toxic cultures, on further plating, gave rise to both toxic and nontoxic forms; while the nontoxic cultures produced only nontoxic. These, Crowell believed, were permanent in their lost heritage of toxicity. Unfortunately we find in this work slight record of the differential cultural, colonial or morphological features of the toxic and nontoxic strains. It can scarcely be doubted, however, in the light of earlier studies, that Crowell was dealing with the S and R dissociates.

In concluding this consideration of the dissociative reaction of the diphtheria bacillus, it seems to me that there can be little doubt that the Hofmann type represents the R form of the clinical diphtheria type; or, we might better say, represents one of the R forms. The actual position of the coccoid type which was referred to by Bernhardt as the "extreme variant," which has been described by Mellon and by Heinemann, and which I have seen in several dissociations of the Park 8 strain of the diphtheria bacillus, is still a question. At present I am inclined to agreé with Bernhardt that it represents the "extreme variant" R. In any case, the remarkable persistence with which the diphtheroid forms hold to their type helps to explain the long conflict of opinion regarding the transmutation of one form to the other. At the same time, I believe that we should hold ourselves open to the view that there may exist in nature culturally and morphologically typical diphtherial organisms that either never possessed virulence or have become nonvirulent without at the same time having passed over to the R form. For such organisms it seems desirable that "pseudodiphtheria bacilli" should be used rather than "diphtheroids." It may be noted, in addition, that these pseudo cultures also may have their correlated R types, which might well be termed "pseudo-diphtheroids." According to these views, while the pseudodiphtheria culture may be differentiated from the clinical type on the grounds of virulence, the pseudodiphtheroid may be differentiated from the true diphtheroid on the ground that the latter only has the power of reversion into the virulent, clinical type. To put the matter in another light—the clinical type possesses actual virulence, while the diphtheroid carries potential virulence. On the other hand, neither the pseudodiphtheria type nor the pseudodiphtheroid possesses either actual or potential virulence. From this point of view the great mass of socalled Hofmann types comprise two distinct forms: the true diphtheroids which are potentially virulent, and the pseudodiphtheroids which are neither actually nor potentially virulent. At the present stage of our inquiry I know of no means of differentiating between these two types except by attempting to enforce a reversion. In such attempts it may be that the observation by Soule [450] in the case of B. subtilis (that a serum immune to the R form of culture has the power to effect a reversion of R to S) may play an important part. Whether the S form of the nonvirulent, pseudodiphtheroid can, under any conditions, ever acquire virulence, or whether it ever possessed any, must still remain a question; but both seem doubtful.

One further point, and one of considerable interest, attaching to these studies on the dissociation of the diphtheria bacillus is the circumstance that, not only the factors for virulence but also the factors for toxin production, seem definitely to be correlated with the S form of culture. How close this correlation actually is, remains for further study to demonstrate. But we shall observe presently that the same fact holds true for the dysentery bacillus, for B. enteritidis, and it may be in the case of B. botulinus and other toxic anaerobes.*

B. botulinus.—In further reference to the possible relation between dissociation and toxin production, as just illustrated by B. diphtheriae, the situation in B. botulinus is of some interest, although the actual facts still remain to be ascertained. The significant points to which I wish to call attention have been introduced by McIntosh and Fildes [322] in 1917, and by Shippen [441] in 1919; but particularly by Reddish [402] in Rettger's laboratory in 1921. McIntosh and Fildes pointed out that the deep colonies obtained from plating B. botulinus were always likely to yield impure cultures and emphasized the point that it is only by the appearance of surface colonies that the purity of anaerobes can be tested. They obtained from Kral's laboratory a culture of B. botulinus which they alleged was contaminated with B. sporogenes. Shippen found that some of his botulinus cultures were atypical and nontoxic. Reddish subsequently examined 19 supposedly pure cultures of B. botulinus from various laboratories in the United States and reported that 18 of these were "contaminated" with an organism giving the cultural and fermentative characteristics of B. sporogenes. He believed the other culture also would have shown the same impurity if he had made a second examination. While the "normal" botulinus colonies were smooth and regular, and gave rise to toxic cultures, the contaminant colonies were like typical B. sporogenes; that is, more diffuse, translucent and with irregular or fimbriate margins, as described in the textbooks. These last were nontoxic. The toxic cultures fermented lactose and sucrose, in addition to glucose, maltose and levulose; while the nontoxic cultures fermented the last three only. Both types of cultures gave putrefactive odors and decomposed egg-meat medium, but the nontoxic more extensively and rapidly than the toxic. Here it is perhaps of importance to note that neither van Ermengen himself, nor his followers among the

* Our consideration of the dissociative aspects of B. diphtheriae would not be complete without reference to the work of Enderlein.[100] This author has seen in the cycle of development of the Klebs-Lœffler bacillus many cyclostages ("Cyclostadien") through which the culture passes in its life history. Of these it is the highest stage in the cycle (the "Kulminante"), marked by the socalled "Cystascit" phase, which he believes carries the supreme virulence. Further consideration of Enderlein's work is presented on page 280.

European observers, have ever described the putrefactive odor that characterizes the botulinus strains of American origin, as isolated and described by American workers. Reddish also reported that "the surface colonies which answered the description of typical colonies of B. botulinus contained, besides the botulinus organism, at least a small number of B. sporogenes." He remarks further, "On the other hand, some of the typical B. sporogenes colonies were proved to enclose B. botulinus or spores." He thus apparently demonstrated that, in the majority of cases at least, his toxic strains, coming from isolated colonies, contained both botulinus and sporogenes, while he states elsewhere that his nontoxic strains usually contained sporogenes only.

Reddish concluded that, in view of these results, no toxic strains of B. botulinus should be considered as pure cultures. His explanation of this curious phenomenon (which seems to discredit common methods of pure culture isolation by plating and colony selection), was that, "when B. botulinus and B. sporogenes are present in the same material, they are so closely associated that it is impossible to separate them." He concluded, morever, that the reason why American workers thus deal exclusively with "contaminated" botulinus cultures is because they are forced to sample, in the main, "spoiled products," while van Ermengen and his followers in Europe usually obtained their cultures from fairly clean ham.

It seems to me more probable, however, that the experimental data presented by McIntosh and Fildes, Shippen, Dickson [127] and others offer a quite different and fairly clear explanation of this somewhat remarkable phenomenon: B. botulinus of the normal S type, like the clinical type of the diphtheria bacillus, is toxic; but it readily dissociates into the R form which is nontoxic and gives a colonial, cultural and biochemical picture which (to express the point most conservatively) approximates that of B. sporogenes. The reason why the R type of B. botulinus was the invariable contaminant of the cultures studied by Reddish and other American workers, while it is apparently seldom observed in European cultures, may depend on the circumstance that the exclusively American method of isolating B. botulinus involves heating the culture before dilution and plating are carried out, a course which, so far as I am aware, is seldom followed in European laboratories. We shall note in a later section the rôle of heat in enforcing the dissociative reaction. The proof or disproof of the hypothesis presented above may afford an interesting problem for those interested in botulinus research, and may have a bearing on the biology of B. tetanus and other anaerobes. The early work of Schattenfroh and Grassberger [427] in 1900, dealing with another anaerobe, the butyric acid bacillus, has established the lead into this field, but as yet I believe they have had no followers, although the work of Ida Bennington [47] in .1922 on toxic and nontoxic strains of an anaerobic organism isolated from larvae of the green fly may bear closely upon this problem.

To the above instance it may be added that a circumstance which may possess similar significance has recently been noted by Hall [230] in connection with the examination of a considerable number of cultures of various species of spore-forming anaerobes. Hall observed that the majority of these cultures were "contaminated," and chiefly with B. sporogenes. In some cultures this organism only was present, and the only pure culture was one labelled "B. sporogenes." Of course we know that contaminations often creep into cultures of spore forming anaerobes, or may have been there from the beginning. At the same time we know that spore forming organisms (both aerobic and anaerobic) possess the ability of dissociating into a form so unlike the original, in colony structure, cultural growth and biochemical and serologic reaction, that it may readily be mistaken for a contamination by an inexperienced observer who has not made a special study of the R types of the cultures concerned. In the case of Hall, as in the case of Reddish mentioned earlier, it might be interesting to ascertain to what extent, if at all, Hall's "contaminations" are merely the dissociated R forms of the original S cultures, perhaps, as is so often the case, manifesting the phenomenon of bacterial convergence, which in this instance may have brought them together into a culture type bearing the characteristics of B. sporogenes. In this case reference to the results of Esther Stearn,[452] involving apparent convergence of certain members of the coli group might be in order; also to the work of Grassberger as mentioned above.

In concluding the relation of dissociation to toxicity of various forms of culture, reference may be made to the results obtained by Bronislawa Fejgin [168] in 1923 on the Shiga dysentery bacillus. From normal Shiga cultures she obtained three strains that were resistant to the action of the lytic principle and possessed other divergent physiological characteristics. While normal cultures on solid media died after a few weeks, these forms lived several months. While the normal culture was toxic for rabbits, the modified type, inoculated in amounts of four to five cc., gave no characteristic lesions. These new forms grew in broth as a sediment, gave different fermentative reactions and manifested altered agglutinative reactions. They undoubtedly represented R types.

Colon-typhoid-dysentery Group.—Although Babes [25] in 1890 had pointed out the variability of B. typhosus in colony form and certain biochemical features, Steinhardt [453] in 1904 showed much more conclusively the existence of the two chief colony types of the typhoid bacillus and correlated them with their respective forms of growth in broth. She also reported the slight virulence for animals of the spontaneously agglutinating cultures (R) compared with those cultures which clouded homogeneously (S). In 1921 Arkwright [16] first recognized clearly the S and R forms of B. dysenteriae and some other organisms, and showed a somewhat greater virulence for the S cultures. His results were confirmatory of the limited evidence pointing to the existence of two cultural forms and the greater virulence of one of them as presented by Steinhardt in 1904. In 1918 Baerthlein [30] laid the basis for the recognition of dissociation in the paratyphoids and B. enteritidis, but the virulence of the S and the nonvirulence of the R form were not made clear until the work of Topley and Ayrton [464] appeared in 1924. Jordan [273] in 1926 reported virulence for the S form of B. paratyphosus B and its disappearance in the R culture. When the reversion of the R type to S was obtained, after many generations, the virulence reappeared. Orcutt [376] in 1923 indicated the same relation between type and virulence in B. cholerae-suis; and this was confirmed in a complete and thorough-going study by White [487] in 1925 on the same species. Similar results may have been obtained by Baerthlein [27] in 1912. Gratia's [211] results on B. coli, indicating less virulence for the S form and greater for the R, are opposed to the usual findings for members of this group, but seem to be in harmony with findings in B. anthracis. We observe also in Gratia's study that it was the R form that was most actively motile, while the S form was nonmotile. In both instances these results differ from those observed by most other investigators and are in need of confirmation. In B. enteritidis Gärtner the relation of virulence to definitely recognized S and R forms has been revealed most clearly by Goyle.[206] When the respective cultures were tested on mice he found that the S form was highly virulent while the R form was notably attenuated. Of 17 mice inoculated with the S type culture all died, while from a group of 14 inoculated with the R type culture only six died; and in these the period was longer than in any of the S inoculated mice. Goyle also studied the subject of the relation of toxicity to the two culture types and was able to show that, as in the case of B. diphtheriae and B. dysenteriae, the filtrates of the type S cultures possessed greater toxic power; and this appeared whether the filtrates were or were not heated at 100 C.

B. pertussis.—There should also be mentioned here the early work of Bordet and Sleeswyk [64] who presented data on B. pertussis, alluded to on a future page with reference to serological reactions. It can scarcely be doubted that their MS culture grown on a blood medium was the S type while their MG culture grown on agar without blood represented the R. The cultural details are not complete but they report the MS culture as more "toxic" than MG for rabbits that were undergoing immunization.

Capsulated Bacteria.—The correlation between virulence and the type of culture is of special interest in connection with the capsulated bacteria. The older literature contains many references to this subject, as in the case of Wilde [485] (1896) and Beham [40] (1912); but Baerthlein [30] in 1918 was the first to correlate the older data and to bring them into relation with the more or less uniform colonial and cultural transformation occurring with great constancy in many bacterial species. He showed that, while the capsulated form of the Friedländer pneumobacillus was highly virulent, the noncapsulated form was entirely lacking in this quality; and these types we have already observed represent the S and R, respectively, as I also ascertained in 1925. Julianelle [505, 506] has most recently pointed out the virulence of the S form and the nonvirulence of the recognizable R form (noncapsulated) of the pneumobacillus. He observed that the S type culture possessed the specific soluble substances which were lacking in the type R culture. The situation thus resembles that found in the pneumococcus by Griffith in 1923, and by others since that time. Although Eisenberg [153] in 1914 had demonstrated the more virulent character of the capsulated S form of Micrococcus (Sarcina) tetragenus, Wreschner [492] in 1921 again studied the matter in this organism and, though not relating it in any way to microbic dissociation, gave unmistakable evidence of the existence of the S and R types, the former capsulated, the latter not; and also of the correlation of virulence with the former type. This organism, it may be noted in passing, is the only pathogenic sarcina on which we have clear data on the dissociative forms. Wreschner demonstrated that while 0.002 oese of the S culture killed mice in 24 hours and 0.000,01 oese killed in four to six days, the R (noncapsulated) form was entirely lacking in virulence when in the "absolute" R state; one-half of a slant culture killed mice in four to five days but by toxic action rather than by reason of virulence. The noncapsulated organisms were rapidly phagocytosed in the body. When, as occurred in some cases, apparently pure R culture gave fatal results it was because of its ability to undergo transformation to the S type in the body of the animal; and in such cases the S form was isolated at necropsy. When, however, the R form was no longer reversible, as indicated by tests in vitro, all virulence was lost. This form of culture he regarded as the "absolute" variant. Wreschner held, as have most others, that the capsules were the "cause" of virulence.

Although the Pasteurella form, B. avisepticus, does not properly come within the capsulated group, this organism has often been reported to form distinct capsules; and in broth cultures usually gives a viscous precipitate. The observations of Manniger [317] on dissociation of B. avisepticus have been mentioned earlier in this section, but attention may again be called to the fact that it was only the virulent or S type of the organism which possessed capsules—the nonvirulent R type lacking them entirely. The correlation of capsules and virulence is thus not limited to the members of the socalled capsulated group.

In addition to what has been said in the present section and in section 5 one concluding point may be mentioned here regarding the association of capsule formation and virulence. It has been quite commonly recognized by bacteriologists for many years that capsule formation by bacteria seems to give them a protection against the body defenses, both humoral and cellular; and that, ipso facto, capsulated bacteria are better qualified to remain in the tissues and to create infections. And this would seem to be true in infections caused by the pneumobacillus, pneumococcus, M. tetragenus and probably the anthrax bacillus. It must be pointed out, however, that a fallacy can easily be involved in this manner of interpretation. It may not be the resistance that the capsules offer to the body defense that renders more dangerous the organism possessing them. In the light of our present knowledge regarding the interrelation of dissociation, virulence and capsule formation, it appears more probable that capsules are merely correlated with virulence; and that the factors for virulence do not lie in the capsules. In other words, in the body of the animal, the pathogenic organism is not virulent because it forms capsules; but rather, it forms capsules because it is of the S type; and it happens to be the S type that is virulent. In capsulated organisms it has not yet been demonstrated that the specific soluble substances underlying virulence and the capsular material are identical, unless we accept the recent work of Julianelle [505, 506] as evidence of this fact. Although this view is opposed to the mass of textbook opinions on the subject it finds support in the circumstance that Bail and Flammenhaft [34] could actually produce anthrax infections in guinea-pigs by means of cultures which never attained a high degree of encapsulation in the body of the animal. They conclude the matter with the following statement: "Die Ausbildung der Kapsel ist vielmehr nur eine Erscheinungsform des Infektiousität des Milzbrandes, nicht aber ihre wesentliche Ursache. Beides geht auf eine gemeinsame, aber tiefer liegende bisher unbekannte, Eigenschaft des Bazillus züruck."

Streptococcus.—In both hemolytic and greening streptococci Cowan [111, 112] (1922) in England established the association of greater virulence with the S type. In addition she pointed out the cultural, colonial and morphological differences between the R and S forms. The S type gave bluish, translucent colonies having an even outline and showing a finely granular texture. These cultures also produced an even turbidity in broth. The organisms themselves grew in shorter chains than the R form, were smaller, and showed slight variation in size. The R type, however, formed white, opaque, coarsely granular colonies with irregular outlines. The individual R organisms were distinctly larger and united into longer chains. In broth the growth occurred in the form of a precipitate. In animal inoculation tests involving these types a distinct difference in virulence was observed, although the distinction was not so clear as de Kruif had found it to be in Bact. lepisepticum, and as Eisenberg and also Wreschner had found it to be in M. tetragenus; or as others have subsequently observed it for the pneumococcus. When large doses (0.25 cc. of broth culture) of S and R streptococci were administered intravenously the comparative death rates were not far different—65 per cent for R, and 85 per cent for S. When, however, the dosage was reduced to 0.05 cc. the rate was 71 per cent for R, and 100 per cent for S. When the dosage was further reduced to 0.04 cc. the rate fell to 35 per cent for R, and 71 per cent for S. Each of these tests involved 14 mice. There was a difference, however, in the medium used in the first and second tests: in test 1 serum-broth was employed, in test 2 serum-agar slants. We know the greater tendency to dissociate in liquid mediums, and this may explain Cowan's discrepancies. In tests 2 and 3 the same medium was employed (serum-agar) and here we observe the expected conformity of results. In any case we can observe that the R type was apparently not destitute of virulence. When the injections were not fatal, abscess formation usually occurred, and these seemed to play a part in the production of immunity. Unfortunately in Cowan's reported results we are not informed regarding the exact proportions of S and R (determined by plating) in the inoculum used. It is possible that her R strains contained mixtures of S as is often the case. De Kruif's reports are more conclusive because he presented more adequate control data on the actual S-R composition of the inoculum.

Pneumococcus.—Regarding the dissociation of the pneumococcus and its relation to virulence, although the actual facts were not fully established until the work of Griffith [215] appeared in 1923, certain earlier observations pointed clearly in this direction. Even in 1892 Kruse and Pansini [293] showed the existence of two forms of this organism: One formed chains of spherical elements with little tendency to produce capsules and possessing slight virulence; the other appeared as a lanceolate diplococcus, heavily capsulated and highly virulent. They pointed out that, while the former type was favored by growth on artificial medium, the latter was increased by animal passage. In 1891 Roger [410] pointed out that growing virulent pneumococci in immune serum rendered them nonvirulent and produced certain modifications in the type. But Issaeff,[263] who in 1892 repeated these experiments, was led to believe that such results as Roger and others had reported were due to the protective action of the pneumococcus immune serum (which accompanied the inoculum). When he took the trouble to free from the immune serum the organisms that had been grown in it, and then injected them, he found them still virulent. He thus concluded, as did many others including Metchnikoff and Sanarelli, that immune serum possessed no power to cause the attenuation of the virulent pneumococcus. Although Issaeff stated that he employed concentrated serum, we are not informed regarding the length of time the organisms were in contact with it; but one may infer that it was only to the extent of a single passage. This was apparently sufficient, however, to bring about a change in the cultural features of the pneumococcus. The growth became self-agglutinative and the organisms grew in the form of chains of many elements. Issaeff was not able to observe the curious types of cells reported by Arkharoff, although he noted that the organisms lost their capsules as a result of the growth in immune serum, a point often since confirmed.

Neufeld [363] in 1902, however, showed more conclusively that virulent pneumococci grown in immune serum lost their virulence; and, in addition, their specific agglutinability and their power to form capsules. These properties were regained by repeated passage through animals. Kindborg [509] in 1905 pointed out again the two chief types and demonstrated that the less virulent form was the one best adapted to artificial culture mediums. In 1906 Eyre, Leatham and Washburn [163] again drew attention to the two distinct types of pneumococcus as distinguished by manner of growth and by the type of lesion produced in rabbits upon subcutaneous injection—the one giving fibrinous exudates, the other cellular. The organism possessing the lower vitality on artificial mediums was the more virulent, while the form possessing the greater vitality was the less virulent. They could, however, transform the former into the latter by cultivation on artificial mediums, and then restore the original characters by passage through animals.

In 1915 Friel [185] reported that pneumococci grown in homologous immune serum became less virulent for animals; moreover, that they became at the same time more phagocytable, even in normal rabbit serum. Laura Stryker [454] in 1916 demonstrated the same influence of immune serum in producing avirulent forms of pneumococcus which thereby lost their specific agglutinating power as also their capsule-forming ability. These characters were regained by animal passage. These avirulent cultures were produced by growth in 10% homologous, pneumococcus immune serum-broth, transfers being made every two to seven days. In virulence tests on rabbits Stryker found that they often tolerated an amount of modified culture roughly a million times greater than the amount of unmodified (normal serum-broth) culture required to produce death. Fifteen cc. of modified culture was harmless on inoculation, while 0.000.005 cc. of the unmodified culture killed in 24 hours. After 55 immune serum-broth passages the culture was ordinarily so changed that it remained nonvirulent after 27 passages in plain serum-broth. The MLD of such cultures for mice was sometimes as great as 0.5 cc. while the normal virulent strain, passed similarly through normal serum-broth (control), killed mice in 0.000,01 cc. amounts. In another instance the modified nonvirulent culture was given 61 transfers in plain serum-broth and at the end of this time was still nonvirulent for mice while the unmodified culture killed in amounts of 0.000,001 cc. In still another case the modified strain was nonvirulent after 75 passages in plain broth. Although the character of reduced virulence was maintained so long as the cultures were maintained on plain culture mediums, Stryker ascertained that passage of the modified strains several times through mice resulted, sooner or later, in the recovery of virulence. This point is considered further in the section dealing with the permanence of the R types (section 12).

Yoshioka [494] in 1923 demonstrated that a similar change in the form of pneumococcus could be produced by growth on unfavorable mediums, by growing at 39 C., and by drying. The changed culture type manifested itself by colony alteration; also by modified serological reactions as described in section 9. The modified cultures were particularly characterized, however, by loss of virulence. These changes did not appear simultaneously in all the bacteria of the culture but only in those of certain colonies. Other colonies remained constant to the normal and virulent pneumococcus type. Blake and Trask [52] in 1923 showed that growth of pneumococcus in immune serum produced comparatively rapid and complete changes in certain individuals but not a gradual change in all bacteria of the culture. These changes were expressed in three colony types with which were correlated variations in virulence and agglutinability, again confirming Baerthlein's fundamental postulate.

By 1922, Cowan's clearcut results on the dissociation of the streptococci, and the definite recognition of the two chief types, R and S, in many bacterial species by other workers, served to direct attention to the pneumococci from a different point of view; and in 1923 Griffith [215] in England first succeeded in demonstrating the relation of the two chief pneumococcus types with the R and S forms as earlier reported by Arkwright, de Kruif, Cowan and others for other bacterial species. Griffith regarded the S as the "unchanged culture" and R as the "variant" produced by growth in immune serum. He pointed out the colony differences (smooth and rough), and stated that, while S produced the pneumococcus "specific soluble substances," these were lacking in the R strains. While the S form, moreover, possessed high virulence (10^{-8} cc. killing mice in 2 days), the R form lacked it, even in 0.25 cc. amounts. He showed that a partial degree of attenuation existed in "partially rough" colonies; but, when these killed mice, only S cultures were obtained at autopsy. The same general results were obtained with the pneumococcus standard types, 1, 2 and 3. The subject of reversion of type is considered in a subsequent section.

In 1925 the results of Griffith were largely confirmed and extended by Reimann [405] who covered the field of pneumococcus dissociation very thoroughly; also, about the same time, by Amoss,[8] who studied the same types under the designations, C and Z. As in other cases already reported both Reimann and Amoss obtained their dissociations primarily by growth of Pneumococcus 1 in homologous immune serum; also by growth in medium containing bile. Reimann reported the S type colonies as flat, thin, greenish, translucent, smooth, shining and sometimes "sticky;" while the R colonies were raised, opaque, less greenish, dull and sometimes dry, friable and coherent, so that they could be removed from the medium in toto, as also happens with some of the type R streptococcus colonies (Cowan). The S culture possessed the specific soluble substances and capsules, both of which were lacking in the type R cultures. The R type culture was also less easily soluble in bile. The difference in virulence between R and S was very striking. While the S type culture killed mice in doses of 10^{-6} cc., even 2.0 cc. of the R culture was without lethal effect. The virulence of "whole cultures" (S and R mixed) was less than that of "pure" S cultures. Overgrowth of the S type was induced by cultivation in normal serum or by passage through mice. The R type was stated to be irreversible (section 12). The results of virulence tests with type R as reported by Amoss agree with those of Reimann. The injection of 3 cc. killed 14 to 18 gram mice, but mice receiving 2 cc. or less survived, while 10^{-5} cc. of type S culture killed in 24 hours. The serologic and immunologic characteristics of types S and R pneumococcus are presented in sections 9 and 10 of this work.

Falk, Gussin and Jacobsen [165] in 1925 showed that there existed a correlation between virulence and electrophoretic potential in variants of pneumococcus cultures. Quite recently Jacobsen and Falk [266] carried this matter further to study the potential of S and R strains obtained from various cultures through the use of immune serum added to the broth medium. The rough strains obtained as a result of 12 serial passages in immune serum-broth showed the same electrophoretic potential and virulence as the original cultures. These authors conclude that "strains of pneumococci which differ significantly in virulence are not necessarily correspondingly separable into S and R categories." From the data presented in the brief communication in question, it seems that the R characteristics were not well stabilized in the rough cultures employed. The ease with which reversion occurred might indicate that the R cultures made use of were still far from the "extreme" variant. Repetition of these studies with such stable R cultures as those produced by Reimann, Amoss or Griffith might yield different results.

Meningococcus.—Since microbic dissociation has been found to play so prominent a part in the variations in virulence of streptococci and pneumococcus, the question of dissociation in the meningococcus naturally presents itself. That this organism is highly variable has been amply shown from the older experiments of Lepierre [510] in 1904 and others. Lepierre noted that, when fresh from the spinal fluid, the typical Weichselbaum form appeared as a gram-negative diplococcus with flattened cells; but that, after cultivation in ascites bouillon, chains appeared increasingly in successive generations, and greater growth energy developed. Accompanying this change was a loss of virulence for rabbits. The gram-reaction sometimes changed to positive. The final result was the production of a coccus resembling the Jaeger-Heubner type. Sorgente [516] in 1905 was also able to demonstrate an interchange of characters between these two types but stated that there was no change in the serological reactions. In 1909 Dopter [183] described from the nasopharynx strains of meningococcus-like organisms which were not agglutinated by antimeningococcus serum. These he called parameningococci and similar cultures were reported by him in 1911 from sporadic (but not epidemic) meningitis. Griffith [214] in 1918 reported a study of 40 strains which he investigated with reference to agglutination and biochemical reaction. On the basis of agglutination he could recognize two main groups, each of which gave its corresponding fermentative reactions. One fermented glucose more than maltose, while the other fermented maltose more than glucose. As a result of their serological studies Gordon and Murray [201] in 1915 proposed four types of meningococcus. Since that time, however, the weight of opinion seems to have been in favor of regarding a close relation existing between 1 and 3 and between 2 and 4. European workers have always laid stress on two main serologic types, following more or less the division into meningococcus and parameningococcus as proposed by Dopter. In 1920 Arkwright [15] pointed out that two main types of the organism, corresponding with Dopter's groups, had been recorded from all epidemic centers in England. Thus it appears that numerous observations concur in demonstrating that among the meningococci there exist at least two forms of culture which can be recognized by serologic and sometimes by fermentative reactions. In all these studies, however, at least up to 1915, no observations had succeeded in demonstrating the existence of any morphologic or cultural distinctions among this group; and much less a correlation of any such characteristics with such serological features as had already been pointed out. Indeed in 1918 Nicolle, Debains and Jouan,[372] as pointed out by Atkin,[21] dismissed the possibility of correlated cultural variation by the following statement:

"Considérés au point de vue de leurs charactères généraux (morphologie, aspect des cultures, propriétés biologiques), les méningocoques offrent des traits communs, que chacun connait aujourd'hui et des différences individuelles. Ces différences, légères et sans relation entre elles, ne permettent pas de créer des groupes distincts."

But already three years before this date Bernhardt [40] had made an isolated observation which, if its significance had been appreciated, would have made a fair start toward the solution of the problem of the meningococcus groups, and their actual association with cultural changes of considerable importance. Bernhardt's contribution was merely a random observation occupying but the fraction of a page in the midst of other interesting deliniations of dissociative phenomena in B. typhosus and B. diphtheriae. When three to seven day old cultures of the meningococcus were plated on ascitic-grape-sugar-agar, there arose large, flat colonies having a thinner margin in which were imbedded a wreath of daughter colonies of quite a different type. When these "Knötchen" were cultured independently, they showed a different type of growth containing large, well stained diplococci in contrast to the much smaller and poorly stained cocci in the mother culture. In the course of time the mother culture mass was destroyed, but the mutant form remained alive on the same medium up to 14 days. Eventually the mutant might revert to the mother type of culture. This observation of Bernhardt was thrown in merely to demonstrate that the peculiar type of culture reaction being described was not limited to typhosus and diphtheriae, and the study was not carried further. And so this first really significant lead to the conception of meningococcus dissociation died at its birth.

Fortunately, however, the conception of a possible correlation between serological type and cultural form of the meningococcus remained alive in the mind of the English bacteriologist, Atkin,[21] who recognized the importance of the time element in such experiments as he was about to perform. Though apparently unacquainted with the earlier work of Bernhardt, Atkin commenced a study of colony variants of various strains of Gordon's standard type strains 1, 2, 3 and 4, being at first concerned with the question of viability. Without entering into the details of this work it may be pointed out that the examination of the type 1 and 3 strains showed large, irregular, rough colonies dotted with papillae (daughter colonies), while the type 2 and 4 strains gave smaller, circular colonies having a smooth surface and no papillae. The actual colony differences are shown in Atkin's tabulation (table 1).

The details of these two cultural types, comprising respectively serologic types 1 and 3, and 2 and 4, were not further reported by Atkin in his 1923 publication; but in 1925, in his paper on the gonococcus group, he [22] presented further data. Here he calls attention to the form of growth which was characteristic of cultures coming directly from the cerebrospinal fluid of meningococcus cases. This was a thin, clear, transparent sheetlike growth, which disappeared quickly on continued artificial cultivation. On the surface of such colonies there arose, after a time, whitish papillae, sometimes discrete and sometimes clustered in masses. When these papillae were cultured separately they gave the typical whitish or yellowish and more or less opaque colony type. It was clear that the Gordon type strains which he had used for his earlier study had changed considerably to a new type of meningococcus. Here we

TABLE 1

SHOWING SOME OF THE COLONIAL CHARACTERISTICS OF ATKIN'S MENINGOCOCCUS TYPE STRAINS ON TRYPAGAR DEEP PLATES.

Type	Shape	Size	Color	Surface	Papillae	Halo*
1	Irregular	Larger	Whitish or yellowish	Rough	+	+
2	Round	Smaller	Yellowish	Smooth, glistening	0	0
3	Irregular	Larger	Whitish, pinkish	Lumpy	+ or 0	0
4	Round	Smaller	Yellow	Smooth, glistening	0	0

* A reaction product in the medium, believed to be due to precipitation of salts of legumin.

undoubtedly have the S → R (or S → O) transformation, other important aspects of which were not further reported by the author. We shall see later, however, that even the data in hand present several points of independent interest in connection with the respective colony types.

Although I have intimated above that the dissociated form of the meningococcus as reported by Bernhardt and by Atkin suggests the S → R transformation, I believe it is necessary to qualify this tentative conclusion. The assumedly R type as described by Atkin differs considerably from the sort of culture that we should expect if we reason from analogy with other bacterial species, and particularly if we have in mind the R type colonies of streptococcus and pneumococcus. Although Atkin's R form certainly marks a wide departure from the normal S type, as indicated by both appearance and serological tests, there is still the possibility that the "extreme" R has not yet been found and that the variant represented is merely one of the intermediates in the transition process. One of the difficulties in its recognition lies in the circumstance that, on the strength of any data as yet available, we do not yet know the appearance of the transitional forms of the pneumococcus or streptococcus, since these have not yet been reported. The clearing up of this difficulty will come only when we possess the results of experiments in which the S type meningococcus is grown in successive passages through its homologous immune serum, since experience with other pathogenic forms has indicated that it is by this method that the extreme R variant can most quickly be obtained.

Gonococcus.—Regarding our knowledge of the constitution of the gonococcus group it is becoming increasingly clear that attempts to classify the various components in terms of standard serologic types have met only with failure. To be convinced of this one needs only to examine critically the many results published during the past fifteen years. Neither agglutination nor absorption nor complement fixation methods have revealed anything in the way of permanent types, and the results obtained by no two observers have been in agreement. In the older body of literature, involving so much wellmeaning but hopeless endeavor, there are, however, two investigations which have contributed something definite to our knowledge of the group, although it is only certain most recent studies that enable us to grasp their full significance. The first of the two is the work of Louise Pearce [384] in 1915; the second that of Torrey and Buckell [466] in 1922.

Pearce showed, briefly, that when a number of strains of the gonococcus coming from infantile cases (vulvovaginitis) and adult cases (urethritis) were examined by agglutination and complement fixation methods, they divided themselves clearly into two groups correlated with source. Torrey and Buckell demonstrated that when several strains of the gonococcus had been maintained on artificial culture medium for fourteen years and were then tested by agglutination and absorption methods, there appeared to have occurred a convergence toward a common serological type. They stated, "We believe that the whole tendency under conditions of artificial culture is for reversion to our regular group, and that a strain having attained that disposition of its antigenic components remains in a comparatively stable condition." These results of Torrey and Buckell, as also those of Pearce, become further significant in the light of the most recent and highly important study by Atkin (1925).

As already pointed out, Atkin [21] had previously shown the possibility of papilla formation (secondary colony formation) in various strains of the meningococcus, and these were subsequently observed [22] also in gonococcus. Fresh strains from various points of infection and also old laboratory strains (Gordon's original types) were grown on deep layer plates of "trypsin-broth-pea-agar" having a reaction of P_H 7.8. Under these conditions the colonies grew to a large size and after about three weeks two colony types were observed, I and II, which the reader must not confuse with the so-called standard gonococcus types 1, 2, etc. Indeed, to avoid such possible confusion, I shall refer to them as A and B, although this was not done by Atkin. Type A colonies were most prominent in material from urethritis and were large, spreading and transparent, with irregular borders. In this growth papillae finally appeared, due to secondary colony formation, sometimes few and again so numerous as to crowd the colony. In this case they coalesced into masses. The second or "extreme" colony type (which of course represented the pure growth of the papilla culture) was usually smaller, round, raised, opaque and whitish or slightly yellow; the surface was smooth (although Atkin's photographs show radial markings) and such colonies showed no papillae. This colony type was more abundant in material from old stock strains, which were sometimes composed exclusively of this form of growth. Between these two chief types, however, there were many intermediate forms which Atkin regarded as mixtures of the chief elements. Moreover there were also observed colonies showing opaque centers, but also transparent edges, in which papillae might subsequently develop.

Regarding the conditions favoring this diversity of culture form Atkin observed that alkaline (P_H 7.8) agar was necessary and that in such a medium papillae might arise in five to eight days. If the reaction of the medium was P_H 7.4 or 7.5 standard type 1 meningococcus cultures died out in three to five days leaving no papillae. When the papilla masses were subcultured it is noteworthy that the first culture obtained was not far different from the original. With further consecutive culturing from papillae, however, the papillated type of culture became increasingly stabilized and then resembled the B form. In serological tests Atkin observed that aging cultures showed a tendency to lose their agglutinability in type A immune serum; while at the same time they showed no tendency to increase in agglutinability in type B immune serum.

Atkin, whose work was purely objective, did not carry his study sufficiently far to give answers to many interesting questions; nor did he attempt to establish analogies between his results and the earlier results of Cowan on streptococcus or Griffith on pneumococcus. We can easily see, however, that the same factors for dissociation were at work in Atkin's cultures, and that he had isolated what we have come to regard as the S and perhaps R culture types, together with certain intermediate forms. He naturally concluded that his type A (I) represented the active clinical type, which Pearce had shown was the chief agent in the vulvovaginitis of children, while his type B (II) was a form more commonly isolated from adult urethritis cases, presumably in a more chronic state of infection. The convergence phenomenon observed by Torrey and Buckell manifestly finds its explanation in the circumstance that artificial cultivation, together with aging, had reduced all their gonococcus cultures to the R form in which, as we can see from numerous investigations, the antigenic similarities are always intensified. These results therefore increase the already rich evidence that the antigenic matrix which underlies type specificity is mainly, and perhaps exclusively, the property of the S form of culture.

Bacillus Proteus.—In the X19 strains of Weil and Felix we have already noted the occurrence of the two chief types of culture which Weil and Felix termed the H and O forms; and we shall see in their behavior, serologically at least, some of the characteristic features of dissociation. In addition, the existence of many "Zwischenformen" or intermediates has been noted by Felix.[172] The circumstance that it is the O type that gradually appears in old cultures, and in cultures grown at high temperatures, while it is the H type that resembles the common, spreading form, that easily gives rise to the O and carries the "double antigen," might well suggest that the H is equivalent to the S, while the O is equivalent to the R. This view was at first considered by Arkwright and Goyle,[18] but discarded in favor of the view that the H was equivalent to R and the O to S. Even this conception, however, was opposed by White,[487] who was able to present a clearer view of the situation. We shall, on a subsequent page, consider the respective merits of these presentations (section 9), and content ourselves for the moment by pointing out that it is the O form which Weil and Felix and others have shown to be the more pathogenic for rabbits. Added evidence for this might appear in the circumstance that it is the O agglutinins that appear chiefly in the serum of typhus fever patients.

Other Bacterial Species.—That there may be definite stages in the life history of still other bacterial species associated with virulence has been indicated by various writers; but the correlation of this characteristic with other features—cultural, biochemical and serological—has not made the matter sufficiently clear for us to recognize the types concerned. In the case of the cholera vibrio the relation of dissociation to virulence was perhaps indicated in the work of Lumbroso and Gerini [312] in 1911, and by Shousha [515] in 1924. Enderlein [160] also made the statement that the virulence of this species is fixed in what he termed the "anaphitit" stage.* In addition Enderlein pointed out a stage in cyclogeny marked by greater virulence in the case of M. aureus, B. influenzae, the meningococcus, the gonococcus, B. pestis, Treponema pallidum, Treponema dentium, B. tuberculosis and L. buccalis, which Enderlein places among his pathogenic species. Other instances mentioned by Enderlein have been referred to in this section—namely, the pneumococcus, B. typhosus, B. coli, Bact. pneumoniae, B. anthracis and B. diphtheriae. These statements, however, are not supported by direct evidence. The probable relation of the culture changes involved in microbic dissociation to the cyclogenic aspects of bacterial growth, founded by Enderlein on a purely morphological basis, is considered in greater detail in a later section dealing with the biological significance of microbic dissociation.†

Conclusion.—In reviewing the data presented in the present section on the relation of microbic dissociation to virulence, we perceive that many observations uphold the view that among pathogenic organisms at large virulence is not distributed evenly through all the stages of growth through which a culture passes in its normal development, but is commonly restricted to one stage of culture growth; and this, at least in many of the cases that we have examined, is what has been termed the S form. There is also evidence that the toxigenic function is similarly restricted to the same culture state. The form of culture designated the R type, on the other hand, we have commonly found to be nonvirulent, or at least less virulent than the S, although exceptions have been noted, and there are undoubtedly others. This appears not only in many references in the older literature, but also in more recent reports in which we are better assured that the chief dissociates, R and S or R and

* For further reference to Enderlein's nomenclature in his complex exposition of comparative bacterial cytology see page 280.

† It may be noted here that the correlation between nonvirulence and the type R culture is duplicated in the lack of virulence of the SR cultures arising from bacteriophagic action. Instances of this are presented in section 14 of this work.

O, were actually concerned in the type differences. One of these exceptions is B. anthracis in which, so far as present observations are concerned, it is the R form that carries the greater virulence. Another is B. proteus in which it will appear (see section dealing with serologic reactions) that the S form is not the more virulent. In the streptococci, moreover, some degree of virulence still attaches to Cowan's R forms, although to a much slighter degree than to type S. In this connection, moreover, de Kruif showed a slight residual virulence still attaching to his type R of Bact. lepisepticum for young rabbits. In both these and similar cases it seems probable that the experiments were not carried on with an "extreme" R form, but with one lying fairly close to S, or perhaps containing some S elements. It must also be pointed out that no comprehensive study has yet been made of variations in virulence among different type S or type R strains coming from different colonies. Moreover, we do not know that S type cultures may not lose their virulent character without at the same time suffering transformation into R like cultures, or into some other possible form of culture different from either S or R; nor that some R cultures may not gain in virulence without manifesting obvious transformation to the S form, although the latter seems more improbable. In short, the view which the mass of present evidence seems to support is, that both type S and type R are to be found in nonpathogenic, as well as in pathogenic cultures; but that, when found in pathogenic cultures, it is the S type that often carries the greater virulence.

Regarding the underlying reason for the frequent loss of virulence in the R type cultures, little can be said at present. It can only be pointed out that, as shown by the work of de Kruif on lepisepticum, of Manniger on B. avisepticus, of Friel, Griffith, Reimann and Amoss on the pneumococcus, the type S organisms do not undergo appreciable phagocytosis while the type R microbes are quickly consumed. In other words S seems to possess an antagonistic action which R lacks. It seems probable that the substance upon which this antiphagocytic action depends is analogous to the substances designated "antiphagines" by Tschistovitch, "aggressins" by Bail and "virulins" by Rosenow, and that it is this substance that conditions virulence. It seems certain, moreover, that it is related to the socalled "specific soluble substances" referred to by several recent workers. The fact that these findings regarding the relation of culture type to virulence and to phagocytability relate particularly to a member of the Pasteurella group and to the pneumococcus is of special interest because it was mainly in connection with the Pasteurellas that Bail [31] and Edmund Weil most clearly established their views on aggressins and antiagressive immunity; and, as I [223] suggested some years ago, the situation in pneumococcus-pneumonia is not far different.

These matters also may have a bearing upon the problem of the nature of serum resistance and the production of the socalled "serum fast" strains reported by numerous writers, not only for bacteria but also for protozoa, and particularly for the trypanosomes. The reported observations of growth of microbes in their homologous immune serum, and of socalled adaptations thereby resulting, have been numerous; but aside from the cases already mentioned there are available few records from which we can ascertain the extent to which "serum fastness" was correlated with distinct modifications in type, or with increase or loss of virulence. From the data in hand only certain probabilities are established, and the answer as a whole must await further study. I believe, however, there are involved questions of considerable importance, not only in relation to our theories of immunity, but also in relation to our methods of practical immunization. These aspects of the matter are discussed further in section 11.

In concluding this phase of the subject it may be remarked that the relation of dissociation to the phenomenon of virulence in its broader aspects must be left an open question. For certain bacterial species, however, and among them some of the most important, I believe that it is fully established that virulence is dependent on certain definite phases or stages of culture growth; and that fully virulent and absolutely non-virulent organisms may, and commonly do, exist side by side in the same culture. In many pathogenic bacterial species at least, it now appears that exaltation of virulence is dependent upon the increase in relative frequency of S type organisms over R; and, conversely, that attenuation of virulence involves the increase in relative frequency of R type organisms over S. Whether attenuation may occur without cultural transformation to R (or O) types is still unknown, but we have no evidence of it unless it be in B. anthracis. Similarly, we have no evidence that virulence may be regained unless one of two things happens: regeneration of S culture from S elements that remain; or, transformation from R to S in cultures from which the S has quite disappeared. The second is certainly the least common, and seems never to occur in certain cultures. I believe, however, that we should hold ourselves open to the view that it may occur in all cases, provided that the proper cultural conditions are present.

In all this consideration of the relation of bacterial type to virulence, and the designation of a culture state no doubt related to that called by Enderlein [160] the "Virostadium," I believe one other reservation should be made, and this is based to a large extent on Enderlein's data. We may well hold open the possibility that all of the cultures that we have designated as belonging to the S type, and which are commonly virulent, or at least more virulent than the R forms, may not comprise a homogeneous group so far as representing a single stage in the specific cyclode is concerned. The smooth colony form of different bacterial species, we may well believe, holds within itself possibilities for variation so far as its morphological, serological and biochemical characteristics are concerned. Caution permits us to conclude only that the smooth culture type often contains, but does not necessarily comprise, the virulent forms. And I believe that the same reservation may well be made for the rough forms with respect to nonvirulence. Indeed, as already noted there are certain instances, as in the case of B. anthracis for example, in which the most virulent type of culture seems to lie in a stage of cyclogeny quite apart from the smooth type; and the same may be true of B. proteus and of the streptococci—although in the latter we know that the S form is often sufficiently virulent. In reality we have made thus far in our studies on the relation of culture type to virulence only the superficial beginning of an analysis of bacterial types which, we can scarcely doubt, will be richly supplemented and deepened by future studies. There presumably exist many pathogenic bacterial species for which the cyclogenic stage or stages possessing the more or less exclusive powers of virulence have not yet been recognized. For some this most virulent form may be a filtrable one, for our new conceptions of bacterial cyclogeny must, as indicated on a future page, be made sufficiently broad to accommodate invisible forms of bacteria as well as visible.

In final reference to the relation between culture type and virulence, it may be observed that we now seem to possess for the first time a partial explanation for common observations regarding the influence of certain cultural procedures long recognized for their effectiveness in maintaining the elusive character of microbic virulence. And here I refer especially to the use of blood, serum, or fresh tissues in culture mediums; or to the value of animal passages at least at intervals in the life of the pathogenic species. In many cases at least these methods support, or cause to develop, that particular cyclogenic stage which happens to be correlated with virulence. To what extent this conception of the relation of type to virulence may be carried over to the toxigenic characteristics of microorganisms, as already established for B. diphtheriae, B. enteritidis and suggested for B. dysenteriae Shiga, as well as for B. botulinus, must be left for future study to demonstrate.

The significance of the restriction of virulence to the "Virostadium" of Enderlein or, as we may now term it, the virostage, in the cyclogeny of the culture, is of fundamental importance in medicine, as Enderlein has well pointed out. Indeed, I know of no discovery in bacteriology within recent years that is likely to have a more intimate bearing on many problems in immunity, susceptibility and specific therapy. We now know that, whether an organism can become dangerous to us, depends upon its ability to enter the virostage in the course of its normal progress through the cyclode. Whether it can be made to cease from its dangerous, invasive activities depends, either upon our ability to destroy it outright (which result is highly doubtful), or to force it into another cyclostage in which it is naturally harmless. The control of the cyclostage in the patient thus becomes a matter of fundamental importance, for upon it is certain to depend the issue of the infection. In another section of this work I have pointed out, as Enderlein has also suggested, that important questions will arise as to adequate methods of dealing with the virostage in specific infections; and that much future therapeutic endeavor will be concerned, not with attempts to destroy the parasite in the body by means of germicides, but with attempts to force it into another stage of development in which it is nonvirulent—or at least less virulent.* This transformation, I believe, is fundamental in the mechanics of immunity and involves one of the primary features of microbic dissociation, as will be considered further in a later section of this work.

9. DISSOCIATION AND SEROLOGIC CHARACTERISTICS, INCLUDING "SPONTANEOUS AGGLUTINATION"

Although a number of investigators as Biggs and Park, Block,[55] Delepine,[123] Fison,[177] Steinhardt,[453] and others, during the latter part of the last century and earlier years of the present, had pointed out the

* That the further course of such a "forced" dissociation would be the same for all pathogenic organisms is to be doubted. Enderlein has already suggested that in some instances in which the virostage is identical with the socalled "Kulminante" (the peak, or point of highest development of the cyclode), the natural course would be to force the organism into a "lower" stage, marking the beginning of the return to the basic form. If, however, the virostage were represented by a form still anticipating the attainment of the "Kulminante" (which in this case we may assume to be nonvirulent) then the natural course would be to force the organism, either beyond the virostage, or into the "lower" gonidial stage which, as a resting form, according to Enderlein possesses no virulence; and with which the protective cellular factors of the tissues would react as toward a foreign body—that is, with resulting phagocytosis. It is of interest in this connection, however, that Mellon[325] has pointed out in the case of B. fusiformis that the gonidial stage in a filtrable form is able to produce infection while the common form is unable to do so.

existence of self-flocculating cultures of B. typhosus and related organisms, it was probably Nicolle [366] who first in 1898 (and later, in 1902 with Trenl [370]) presented the problem more clearly and demonstrated some of the antigenic differences in various subcultures from the same typhoid strain. Nicolle, moreover, showed the possibility of producing the self agglutinating type by submitting normal cultures to the action of typhoid immune serum, an experiment later confirmed by Steinhardt and many others, not only for B. typhosus but for the cholera vibrio and some other organisms as well. In 1901 the same problem was studied by Savage [425] who concluded that "clumping can be set up by a number of feeble chemical substances," as well as by immune serums. At this time Widal and Durham had stated that there was no difference in the bacteria concerned in these reactions, although this view was opposed by Loraine Smith and by van der Velde. Somewhat later Savage reached the conclusion that there might be some difference in the bacteria but that it was slight and ephemeral. The correlation between spontaneous agglutination in broth and other cultural features was not made, however, until two years later when a group of observations appeared correlating the agglutinative growth with peculiar colony forms. Among the first in this respect was Theobald Smith and Reagh [444] who in 1903 called attention to the "granular" colonies of B. icteroides on gelatin and to their agglutinating growth in broth culture. The colony variations in B. icteroides observed by Sanarelli,[423] and pictured by him in some detail in connection with his study of yellow fever in 1897, undoubtedly concerned the dissociation phenomenon.

From this time on, many observations have been made, bearing from one angle or another, on the serological aspects of this reaction. Since it is impossible within the brief scope of the present work to give due consideration to them all, I shall mention only those which seem to show most clearly the points at issue. The papers referred to divide themselves naturally into two main groups. First, those which were published before the phenomenon of microbic dissociation was clearly recognized and second, those which have appeared since that date, and which deal consciously with the chief dissociative types. This dividing date may be placed roughly in the year 1921 when the important contributions of Arkwright [16] and of de Kruif [119] materially assisted in focussing the attention of bacteriologists upon this important aspect of bacteriological study. It will be observed, however, that many contributions belonging in the first group have a publication date beyond 1921 by reason of the circumstance that the authors either were not aware of the publications of de Kruif and Arkwright, or failed to grasp the significance of the results reported by them. Although some may hesitate to accept the significance for dissociation of certain examples of the earlier work, I believe there are no instances mentioned in the following review which, though sometimes slightly obscure in themselves, are not validated by later results obtained under conditions involving full recognition of the chief types of dissociates. Some others will question the exactness of the results of early work because most of it was not performed with cultures based upon single cell isolations. I am of the opinion, however, that there is little need for apprehension on this point. While it is quite true that certain aspects of study of the dissociation phenomenon require utmost accuracy so far as starting with pure lines is concerned, in the more general aspects I fail to see any essential difference in results depending on whether single cell isolation or carefully repeated colony isolations were employed; and I might add that quite typical results have been secured when, so far as we are aware, only mass culture methods have been employed.

Intestinal Bacteria.—Most of the early work involving the relation of dissociation to serologic reactions concerns the members of the colon-typhoid-dysentery and the paratyphoid-enteritidis groups, together with some observations on the cholera vibrio. Although P. Th. Müller [355] in 1903 and several others had studied the phenomenon of spontaneous agglutination of B. typhosus in immune serum, Steinhardt [453] in 1904 was one of the first to point out the relation of such cultures to colony form and to describe with great accuracy the appearance of the R type. This form of culture not only occurred spontaneously in cultures but as she demonstrated by many tests could be produced artificially through the employment of typhoid immune serum. The resulting R type was inagglutinable. Somewhat similar results were obtained on B. typhosus by Walker [477] in 1904, by Loeffler [307] in 1906 (for B. coli) and by Porges and Prantschauff [392] in the same year. Moon [351] in 1911 obtained from a single cell strain two pure line cultures one of which was agglutinable, the other not. Both bred true for many generations but eventually became similar again. Gay and Claypole [187] in 1913 obtained from rabbits in which a carrier state had been produced, a culture which failed to agglutinate in a serum that agglutinated the stock culture at 20,000. The nonagglutinating culture was, however, agglutinated by serum from a rabbit immune to the same culture grown on rabbit blood agar. Bull and Pritchett [82] in 1916 were unable to confirm these results.

In 1911 Almquist [3] observed a typhoid culture which underwent certain morphological transformations with respect to the cellular elements. Out of long rods and filaments developed small coccus forms which he regarded as gonidia. These in turn transformed into minute granular bodies (Körnchen) which were filtrable through a Berkefeld candle. These bodies refused to grow on ordinary mediums at 37 C., but grew at room temperature and at 10 C., taking the form of a yellowish layer on agar. They were not pathogenic for rabbits or guinea-pigs. When used for the immunization of rabbits, the resulting serum agglutinated not only the granular bodies but also the original culture. Such serums also possessed immunizing properties against the original culture. Almquist called the culture Bact. antityphosum. According to Enderlein's [160] terminology, one would interpret this phenomenon as the transformation of normal rod forms and filaments, first into gonidia, then into gonites. Either the smaller gonidia (microgonidia) or the gonites might have been filtrable, according to much later work of Enderlein and others.

Eisenberg [158] in 1914 obtained, among other colony mutants of B. typhosus, his socalled "Zwergform" which gave an agglutinative growth and failed to agglutinate in typhoid immune serum. Teague and MacWilliams [458] in 1917 isolated from the blood of a rabbit that had been inoculated with typhoid culture two kinds of colonies, one small, smooth and opaque, the other large, irregular and transparent. The latter gave agglutinative growth.

The results obtained in 1921 on typhoid by Morishima [352] are of special interest, and related to similar results reported earlier by Park and Williams [382] for paradysentery. Morishima showed that typhoid organisms grown on normal serum did not become inagglutinable. When cultivated on specific immune serum (rabbit) they at first became inagglutinable but later (sometimes within a few days, but always by seventy-two days) they again became agglutinable. Here we observe that the reversion seems to have occurred "in spite of the continuous subjection to the unusual environment." He used 20% immune serum in extract broth and the same dilution of normal serum for controls. In a 1:10,000 dilution the immune serum no longer produced changes in agglutinability. The author pointed out that inagglutinability was accompanied by inability to absorb agglutinin. The inagglutinable strains, moreover, showed a modified reaction to varying acidity ("acid agglutination"). No other data than the circumstance of impaired agglutinability were presented to indicated a transformation from the S to the R type; but this transformation has occurred clearly in other cases involving a similar loss of agglutinating power under similar conditions, and thus leads us to suspect that Morishima's results concerned the dissociation phenomenon. In this connection, however, the reader's attention is directed to a curious feature of Morishima's results considered in section 12 of this paper.

Ishii [262] in 1922 in a study of spontaneous agglutination recognized the two chief types of B. typhosus and some other intestinal forms. One gave spontaneous agglutination while the other did not. Ishii did not, however, appreciate the significance of his work and failed to relate it to the dissociative process. The antigenic and serologic interrelation of the two types was not considered.

Regarding the serological aspects of dissociation among the paratyphoid-enteritidis forms fewer observations have been made. Although there may have been earlier reports in this field, Sobernheim and Seligmann [446] were among the first (1910) to demonstrate a change occurring in B. paratyphosus and involving antigenic, agglutinative and absorptive phenomena. In Baerthlein's [30] extensive report of 1918, he mentioned obtaining inagglutinable variants from all of the 14 species with which he worked. In particular he called attention to several agglutinative and absorptive differences in B. paratyphosus B and in B. dysenteriae Y. In the former case, of six colony variants, two varied in their flocculation reaction and failed to absorb the agglutinins of the other variants from specific sera. One of these two formed large, ragged colonies and was not agglutinable. The other variant was sometimes nonagglutinable and in addition possessed the ability to agglutinate spontaneously. Both of these we may now probably regard as R types. The former, Baerthlein states, after five and one-half months, gave a further variation which resembled B. typhosus in its reactions and agglutinated in high dilutions of typhoid serum. In 1918 van Loghem [308] also reported dissociation-like changes that occurred in a culture of B. paratyphosus after it had grown normally for many years. First it was observed to form indol. Later some of its variants agglutinated only slightly with the paratyphoid B serum but strongly with typhoid serum. An apparently similar transformation, but involving B. paratyphosus A, occurring under the influence of the lytic principle, was reported by Bachmann and de la Barrera.[26] In this instance a "mutant" was formed which came to manifest stronger serologic affiliation with B. typhosus than with the original paratyphoid A culture. D'Herelle,[248] in commenting on this case, gives us to understand that the results were caused by the bacteriophage and implies that they can be produced in no other way. Such a view, however, is erroneous as we shall see in section 14 when we consider in greater detail the dissociation-provoking power of the bacteriophage. The relation of the "mutations" described by van Loghem, by Bachmann and de la Barrera and by Baerthlein are apparently all on the same order and furnish support for the conception of bacterial convergence of the R types as first introduced by Schütze and dealt with more conclusively by Goyle,[206] as will be shown on a later page.

In January, 1924, appeared an abstract account of studies by Krumwiede, Cooper and Provost [289] dealing with the serological duality of paratyphoid cultures. The authors state that they were able to obtain two components from nearly every culture examined, and that these bred true. They differed from each other agglutinatively and in agglutinin absorption and are referred to as "group components" and "specific components." The group components among themselves showed marked cross agglutination and tendency to reciprocal absorption while the specific components showed little cross agglutination and only a slight tendency to reciprocal absorption. The antiserum of either component agglutinated the other antigen only to a slight degree. The authors did not, however, in their brief report relate the phenomena observed to the dissociative reaction, although they pointed out its significance in practical serologic identification of bacterial species. They did not correlate the two components with the usual colony types although they were doubtlessly observed. These results will be considered further with reference to a later publication by Krumwiede and his collaborators.[288]

Coming now to a consideration of the dysentery group, although earlier observations had been made pointing to the dissociation of these forms, Park and Williams [382] in 1917 reported one of the clearest of the early records dealing with certain serologic changes involving the dissociative reaction. A culture of the Flexner paradysentery bacillus was transferred daily for eleven consecutive days through broth containing homologous immune serum in amounts of 15, 4, and 1.5%. The agglutinative titer of this serum preceding passage was 800. After the eleven passages in the 15% serum broth the culture showed entire loss of agglutinating power, and was reported to have lost also the ability of absorbing the specific agglutinins. The cultures grown in 4, and 1.5% serum agglutinated at 60 and 100, respectively, and continued to absorb agglutinin. The most interesting aspect of this work, however, was the total or partial recovery of agglutinating power when the three cultures were returned to nonserum medium. The culture grown in 15% serum became agglutinable at only 200 after sixteen weeks of further cultivation while 4% serum culture returned to a 500 titer, and the 1.5% serum culture to 800 (original titer), in the same length of time. From a comparison of these results with those of Steinhardt it becomes clear that the changes observed by Park and Williams involved the S and R (or O) types of the dysentery bacillus. The reversion of the R (or O) type to the S is in accord with results reported several years later by Morishima. These have already been considered. Baerthlein [30] also has presented results on dysentery Y bearing out the work of Park and Williams, reported one year earlier.

Early observations dealing with the serologic aspects of dissociation have also been made on the cholera vibrio. Most of these deal with the modification in type determined by growth in homologous immune serum. As early as 1898 Ransom and Kitashima [501] observed that, when cultivated in immune serum, the vibrio lost its agglutinability and presented modified cultural characteristics. Similar observations were made by Hamburger [231] in 1903. The modified cultural features were such as Eisenberg [150] in 1912 clearly showed to be characteristic of what we now term the R type. The more detailed work of Balteanu [38] will be considered on a later page.

B. Pertussis.—One of the most important of the early group of studies was that of Bordet and Sleeswyk [64] in 1910 on dissociation in B. pertussis. More clearly than any other, this work demonstrated the form and the significance of the serologic changes that may accompany the dissociation reaction. Naturally, however, Bordet and Sleeswyk did not relate the phenomena observed to what we now call dissociation, or to the problem of spontaneously agglutinating cultures, which were just receiving their first serious study. This pioneer work, nevertheless, warrants presentation in some detail.

Bordet and Sleeswyk ascertained that a normal culture of B. pertussis growing on a blood medium could slowly be adapted to plain agar; but that in so doing the nature of the culture was distinctly changed. The blood culture was designated MS, the agar culture MG and their antigenic differences particularly were studied. We may omit the theoretical considerations which led to this work and turn to the results.

Rabbits were immunized against both MS and MG. The MS antiserum gave the following results: in intact form it agglutinated MS well and MG moderately; when absorbed with MS it agglutinated neither antigen; when absorbed with MG it failed to agglutinate MG but did agglutinate MS. The MG antiserum, intact, agglutinated MG but not MS. When absorbed with MS it still agglutinated MG; but when absorbed with MG, it failed. Thus the MS antiserum influenced both antigens while the MG antiserum influenced only its own. The serums did not behave symmetrically. It appeared that MS antiserum contained two agglutinins; moreover that, although MS culture absorbed two agglutinins, it was agglutinated by only one. The authors pointed out that the transformation MS → MG was a reversible one and very prompt. They concluded that these differences in pertussis were comparable with the "natural varieties" of B. dysenteriae. They shared the opinion of Grassberger and Schattenfroh [208] that, at least in agglutination, the immune serums do not exert their action on fundamental substances in the bacterial protoplasm, but on "accessory substances." They believed that such studies possessed important significance in that it would be impossible to effect a serologic diagnosis of pertussis if one made use of an antiserum obtained from injecting agar slant cultures and an antigen culture grown on blood medium. The MS culture of Bordet and Sleeswyk was unquestionably our present S type while their MG was either the R or O in course of transformation. Their cultures were probably to some extent mixtures of these forms, but with such a degree of prominence of one or the other that the serologic results were clearcut; more so, indeed, than we might expect under the circumstances. As we advance to a consideration of the later group of studies in which the serologic discrepancies are found attributable to antigenic differences definitely recognized as related to the S and R components, we shall see that the agglutinative and absorptive scheme first clearly pictured by Bordet and Sleeswyk may be regarded as forming the serologic type reaction or pattern for many of the agglutination results thus far described as involving the "double and single antigen" hypothesis. The underlying facts were anticipated at an earlier date by the studies of Theobald Smith and Reagh [445] on flagellar versus somatic antigens (to be considered later in detail), and were further elaborated at a later date in the work of Weil and Felix [483] on the H and O antigens of B. proteus X19.

In view of the importance of the work of Bordet and Sleeswyk I have reconstructed from their rather elaborate presentation a tabulation indicating the general trend of the serologic reactions involved. From these results slight departures may be expected, depending upon the purity of the types, the presence of intermediates, and the degree of "roughness" manifested by the R type in question; but, in general, later studies have shown a remarkable conformity to this scheme of agglutinative and absorptive reaction.

TABLE 2

SHOWING THE ESSENTIAL AGGLUTINATIVE AND ABSORPTIVE RESULTS OF THE SEROLOGIC STUDIES OF BORDET AND SLEESWYK ON THEIR TWO TYPES OF BACILLUS PERTUSSIS
(Reconstructed from their distributed data)

Antigen Culture Type	MS Antiserum Before Absorption	MG Antiserum Before Absorption
MS	Well*	0
MG	Moderate	Well
	Antiserums absorbed with MS	
MS	0	0
MG	0	Well
	Antiserums absorbed with MG	
MS	Moderate to well	0
MG	0	0

* The relative descriptive terms were used by Bordet and Sleeswyk.

In the above tabulation we may, as pointed out earlier, accept the MS antigen of Bordet and Sleeswyk as analogous to the S type ("double antigen"), while MG is analogous to the R or O type ("single antigen"). From the results presented in the table it will be clear why Bordet and Sleeswyk stated that the serological relations between MS and MG were not "symmetrical" but gave what they termed a "one-sided action."

The last paper in the older group of studies that I wish to consider is the contribution on serologic types of B. pertussis reported by Krumwiede, Mishulow and Oldenbusch [290] in 1923. On the basis of examinations of both old laboratory strains and freshly isolated strains by agglutinative and absorptive methods, they reported two distinct serologic types, A and B. Of these, most old laboratory cultures yielded both, while fresh isolations commonly gave only one. If I understand correctly the facts presented, and particularly the data incorporated in table 1 (p. 24), these two types were characterized serologically as follows: the A serum agglutinated the A antigen and also, to some extent, the B antigen. The B serum agglutinated the B antigen but had little or no effect on A antigen. When A serum was absorbed with A antigen it lost A agglutinins; also B agglutinins (?). When A serum was absorbed with B antigen the B agglutinins were lost and the A agglutinins much reduced. When B serum was absorbed with A antigen the A agglutinins were lost and the B agglutinins practically unchanged. When the B serum was absorbed with B antigen it lost B agglutinins; also A agglutinins (?). These results may be summarized as follows:

TABLE 3

SHOWING THE ESSENTIAL AGGLUTINATIVE AND ABSORPTIVE RESULTS OF THE SEROLOGIC STUDY OF KRUMWIEDE, MISHULOW AND OLDENBUSCH ON THEIR TWO SEROLOGIC TYPES OF BACILLUS PERTUSSIS

(Reconstructed from their tabular and distributed data.)

Antigen Culture Type	A Antiserum Before Absorption	B Antiserum Before Absorption
A	2,000	0 or slight
B	1,000	2,000
	Antiserums absorbed with A	
A	0	0
B	0 (?)	1,000
	Antiserums absorbed with B	
A	200	0 (?)
B	0	0

Although the above authors mention the older work of Bordet and Sleeswyk on B. pertussis, they state, "The results we have obtained do not in any way confirm the findings of Bordet." This conclusion, however, must be regarded as a misjudgment, probably based upon an imperfect understanding of what the results of Bordet and Sleeswyk actually were. If we compare the results as presented in tables 2 and 3 the similarity in both agglutinative and absorptive reactions is quite apparent. We may therefore conclude that, although denied by Krumwiede and his coworkers themselves, as also by Krumwiede, Cooper and Provost [288] in 1925, the study presented by them in reality confirms in a striking manner both the results and interpretation of Bordet and Sleeswyk [64] relating to the existence of two antigenic forms of B. pertussis, as first pointed out by them in 1910.

In connection with the older group of studies bearing on the relation of dissociation to serologic reactions one other should logically be introduced at this point; and this concerns the work of Weil and Felix and their followers on the X19 form of B. proteus. For reasons which will become apparent, however, it is desirable to postpone consideration of this subject until we have examined some of the contributions marking the first of the new series of studies in which the chief cultural types, S and R, were definitely recognized. The earliest records dealing with serologic reactions concerned with clearly recognized S and R forms as we know them today were those of Arkwright and of de Kruif, both reported in 1921, the former dealing with dissociative reactions in various members of the colon-typhoid-dysentery group, the latter with the dissociation of the animal pathogen, Bact. lepisepticum. For further reference to these studies, dealing with other aspects than the serological, other sections of this work must be consulted.

Bact. Lepisepticum.—The two cultural forms described by de Kruif [118, 119] were termed by him D (virulent) and G (nonvirulent). For our present purpose we may transpose these terms of reference to the S and R as more commonly employed. In de Kruif's experience rabbit antiserum for type S agglutinated both S and R at 2,000 after 16 hours at 55 C. The antiserum for R was less active, agglutinating S at 200 and R at 1,000. In some cases S did not agglutinate higher than 50. As de Kruif states, "the antigenic power, so far as production of agglutinin is concerned, appears to be decidedly stronger in the case of type D than of type G." Furthermore, absorption tests seemed to uphold the unity of the antigenic nature of S and R. Serum immune to the S form agglutinated S and R antigens at a titer of 1,500. But, after two hours of contact at 55 C. with the S antigen, the titer of this serum for type R had fallen to 200, and for type S to 80. The same serum, after absorption with type R, showed a reduction in titer to 40 for both S and R. De Kruif believed that R culture was not only more flocculable but also had greater binding power than S culture. He concluded that there was no "qualitative" difference in the antigenic nature of the S and R forms of culture. In the light of later work, this view demands reexamination. So far as de Kruif's results are concerned, they accord with the usual scheme at least insofar as they reveal a diversity of antigenic structure in S and R.

B. Avisepticus.—To the observations of de Kruif it may be added that Manniger [817] in 1919 had also reported evidence indicating the dissociation of B. avisepticus (Pasteurella avium), another Pasteurella closely related to, if not identical with, de Kruif's Bact. lepisepticum. He showed the production of a nonvirulent mutant from a highly virulent culture of normal type, as already indicated in section 8 of the present work. Among other points of interest he studied briefly the reciprocal serologic reactions of the virulent and nonvirulent cultures, respectively. The chief agglutinative results may be presented in the following tabulation.

Antiserums for	Antigen	Titer
Virulent	Virulent	320
Nonvirulent	Virulent	1280
Virulent	Nonvirulent	160
Nonvirulent	Nonvirulent	2560

From these results it is apparent that the agglutinability of the non-virulent form (R) in its homologous serum was greater than that of the virulent type (S) in either immune serum. Attention has often been called to this peculiar phenomenon in other bacterial species, namely, a hyperagglutinability of the R form. In my own experience, and in the experience of certain others, the agglutination titer shown by highly virulent fowl cholera cultures in their homologous serum is usually low, seldom above 400 to 600.

B. Typhosus and B. Dysenteriae.—Arkwright,[16] whose excellent report also appeared in 1921, examined nine cultures of Shiga, three of Flexner Y, four of B. typhosus, two of paratyphosus B and three of B. enteritidis and found that, among the Shiga cultures, all but one gave both smooth and rough colonies which afforded pure cultures of the S and R types. Among other cultures, some gave both S and R, some R only and some S only. Arkwright also pointed out that various R cultures may differ among themselves as well as from S. Although most of the R type agglutinated spontaneously in 0.85% salt solution, a concentration of 0.42 or 0.21% permitted stable suspensions. Shibley [440] has employed the same method for keeping in suspension R type cultures of streptococcus.

Arkwright inoculated rabbits with both culture types of typhoid and dysentery and studied the reciprocal agglutinative reactions. The S antiserums agglutinated S cultures but had little effect on R cultures. Similarly, the R antiserum agglutinated R antigens but had little effect on S. When a stock antidysentery serum was absorbed with S antigen the R agglutinins were left unchanged; but when absorbed with R antigen the R agglutinins were removed while the S agglutinins were left intact. In other tests with B. typhosus and Flexner Y homologous immune serums were employed. In order that these results may be compared with others, and particularly with the results of Bordet and Sleeswyk for pertussis, I have brought into tabular form the data relating to Arkwright's absorption tests on dysentery Flexner Y and its S and R forms.

From the results of the serologic tests presented above dealing with the R and S forms of dysentery Flexner, it can be observed that, although insufficient absorptions are manifested, the general picture resembles somewhat that given by Bordet and Sleeswyk, and by Krumwiede for pertussis; also as we shall see, by Griffith for the pneumococcus and by Smith and Reagh for the motile and nonmotile forms of the hog cholera bacillus. I believe that these examples are sufficient to demonstrate the fact that, serologically, the R type culture is not a random variant or mutant as many have supposed, but a definite form or stage of culture which stands in a definite serologic relation to the S form; moreover, that this serologic relation is not variable for different bacterial species, but highly constant, in whatever species the two dissociates may be discovered. We shall see later how this conception, complicated somewhat by the introduction of a third clearcut antigenic type, underlies the complex structure which Weil and Felix and their followers have raised about the diversified H and O forms of B. proteus X 19.

TABLE 4

SHOWING THE RESULTS OF ABSORPTION TESTS WITH SERUMS IMMUNE TO TYPES S AND R OF B. DYSENTERIAE FLEXNER Y.

(Compiled from data in Arkwright's table 7, p. 47.)

Antigen Culture	S Antiserum	R Antiserum
	Before Absorption	Before Absorption
S	20,480	160
R	320	2,560
	Antiserums absorbed with S	
S	1,280	0
R	160	1,280
	Antiserums absorbed with R	
S	10,240	0
R	160	160

Another contribution of interest referred to in greater detail in section 11 is that of Ørskov and Larsen,[513] mentioning certain serologic aspects of dissociation in a paradysentery culture. From the stock culture were isolated two chief types, first recognized by colony forms, which they termed "V" and "B." Both of these variants gave agglutinating serums while culture M (a further variant of V) gave a serum with no agglutinating power. The detailed serologic results of their study show that V antiserum agglutinated V well and M slightly; B antiserum agglutinated B and Bu (a further variant from B), but not V or M; and M antiserum agglutinated M only slightly and V, B and Bu not at all. Furthermore, absorbing the V antiserum with V antigen removed all V agglutinins, while absorbing with culture M left fair agglutinating capacity for V. Both the direct agglutination and the absorption tests are incomplete. They serve to indicate clearly the diversity of serologic type but without at the same time making clear the exact nature of the variants. What data are available, however, suggest that V was a type S, that B was an intermediate and that M was a full R and Bu an intermediate fairly close to B. The results of fermentation tests support these conclusions. On the whole, these results of Ørskov and Larsen, although lacking in analysis on the part of the authors, add evidence for the conception of the heterogeneous nature of the R type.

Salmonella and the Aertrycke Bacillus.—Andrewes [10] in 1922 obtained two different serologic forms from a single strain of a member of the group, Salmonella. One of these produced, and reacted with, the group agglutinins, while the other showed as dominant the specific agglutinogens and reacted to the specific agglutinins. In this case, however, the two colonies from which these cultures came were reported as alike in their morphologic characters. Griffith [24] described what was probably a similar case for the meningococcus and expressed his opinion that both of these instances concerned an incomplete S to R transformation.

The studies of Schütze [436] in 1922 on dissociation phenomena in a culture of the Aertrycke bacillus followed closely the lines established by Arkwright one year earlier. Here he found a variant whose agglutination and absorption relations to the original culture were lost. Serum immune to what we may regard as the S culture would not agglutinate the R form, nor the R serum the S culture. Such extreme diversities are not common in the literature. In certain tests the R culture was not agglutinated even by the R serum itself. White [487] made a similar observation for a certain R strain of Salmonella.

But perhaps the most important aspect of Schütze's work was the observation that the R type culture of one bacterial species may be agglutinated even to titer by the R immune serum of other species, even when the S forms of the same cultures reveal no or slight serologic relation. For instance Schütze states: "rough variants of Gaertner's bacillus, paratyphosus A and typhoid strains will agglutinate, sometimes to the titer limit, with rough sera from the paratyphoid B group, while smooth prototypes, from which they have been derived, remain quite unaffected." This phenomenon the author refers to as a sort of "serological cosmopolitanism" among the rough culturés. This observation, I believe, constitutes the first clear recognition of the relation of the dissociative reaction to the phenomenon of bacterial convergence. The phenomenon has parallels elsewhere in the literature. For example it appears in Laura Stryker's [454] and Yoshioka's [494] cases of pneumococcus dissociation (although the phenomenon was not recognized in either instance); also in Reimann's [405] case of the agglutination of the R type pneumococcus in heterologous pneumococcus immune serums; also presumably in Torrey and Buckell's [465] convergence of their types of gonococcus. It was also no doubt concerned in the divergent serologic reactions of the B. paratyphosus A of Bachmann and de la Barrera [26] as already reported. In all these instances, some of which remain to be considered, we can observe the phenomenon of bacterial convergence which manifestly depends on similarities in antigenic structure among the R forms; and which, according to the more recent study of Goyle,[206] may sometimes appear in relation to the intermediate (O) culture types, as will be shown in greater detail on a later page.

B. Paratyphosus and Related Organisms.—In 1925 Krumwiede, Cooper and Provost [288] presented in their comprehensive study of agglutination absorption various references to the serological behavior of certain variants, some of which were recognized by the authors themselves as R types. Still other variants, in the light of much earlier work, were manifestly intermediates, or "incomplete" R forms. Much of the work of these writers dealt with B. paratyphosus in which they could recognize three chief colony types, smooth, mucoid and rough (p. 135). The mucoids, as Fletcher [179] also had observed, were especially common in carriers and convalescents. Platings direct from feces often gave mixtures of S and mucoid, at other times mixtures of S and R. The rough forms were much more stable than the S or the mucoids, although all tended to breed true in plating as well as in passage through animals. The mucoid type might develop either S or R, a point to which we shall return in our consideration of the intermediate type. The R cultures showed greater agglutinability and grew as a sediment in broth.

Regarding the actual serologic characteristics of these types the authors present many tables giving the results of direct agglutination and absorption tests. With a few exceptions, however, the results are unusual and entirely out of harmony with nearly all other work on the serology of the S and R forms. They were able to detect no uniform difference in regard to the antigenic nature of the rough and smooth strains (p. 145) and concluded that colony variation was not a fundamental factor in agglutination results, with the possible exception of the mucoids. The state (p. 147), "In most of our work the rough and smooth varieties have been serologically similar." When their results were occasionally in agreement with those of other workers (Weil and Felix, Arkwright), they believed that "other variation factors" ("degradation," "vertical variation," "lateral variation") were involved in the process. They concluded that the rough colonies possessed a "duality" similar to that of the smooth (p. 171). They criticized Baerthlein's work on colony variation and correlated serological features as inconclusive and "difficult to understand" (p. 200). They doubted the reliability of the method of "aging" cultures for the production of colony variants as employed by Baerthlein and numerous other workers, and questioned whether this procedure "actually induced variation or permits a survival of existant variant individuals" (p. 198). They could not accept the results of Bordet and Sleeswyk,[64] on serologic variability of B. pertussis, as previously mentioned, and believed the differences observed were due to "pseudo-reactions" (p. 203). They considered Stryker's [454] detailed and thorough study of the pneumococcus variants (later fully confirmed by Griffith,[215] Yoshioka [494] and Reimann [405]) to be too brief to warrant interpretation "from our point of view" (p. 203). Regarding Griffith's work on pneumococcus serology as related to dissociation, these authors found it "impossible to tell how far some of the agglutination results were due to differences in reactivity and how far to degradation" (p. 203). Regarding the suppression on agglutinability determined by cultivation in homologous immune serum, as fully demonstrated in numerous cases in recent years (see later pages in "Incitants"), they state: "as far as we know the changes observed were not associated with proven antigenic change" (p. 204). They did not believe "that the rough colony variant is also a serologic variant" (p. 209).

A few points considered are however constructive, although they have been developed more clearly by other workers, both before and after the appearance of Krumwiede's study. They showed clearly that their "degraded" variants were characterized by an antigenic loss, such as also characterizes the O proteus type. They realized, moreover, that "degradation" might be followed by the acquisition of a quite new antigenic capacity not present in the original culture (p. 196). This was also pointed out by White [487] in England in 1925 and confirmed by Goyle [206] in 1926 with a great wealth of convincing experiment. Krumwiede and his collaborators also noted the parallelism between paratyphoid antigen heated, and the naturally "degraded" antigen (p. 197). As we shall observe a little later the explanation of this similarity (which is seen also in the relation between the H and O types of Weil and Felix) was presented by Goyle in 1926. Krumwiede and his collaborators also note the fact that the antigenic basis for common agglutination reactions lies in what they term the "group component" (probably the pure O or R antigen), and also in the antigen of the "degraded" variant. This point was definitely proved for typhosus and enteritidis by Goyle in his comprehensive work in 1926 dealing with the serologic aspects of bacterial convergence.

With these few exceptions, the work of Krumwiede and his collaborators dealing with the serologic aspects of dissociation shows a striking lack of conformity with the mass of recent studies. Although the authors failed to comprehend either the significance or the practical aspects of dissociative reactions, as indicated by the curious nature of their criticisms of other contributions which in reality attack the problem more directly, the points covered by them clearly dealt with the phenomenon of microbic dissociation. Their "degradation variants," "lateral variants" and "vertical variants" easily find a place in the system, although this sort of terminology is both confusing and misleading.

In concluding this consideration of the work of Krumwiede and his collaborators I may add that the chief obstacle that probably prevented them from more completely recognizing the underlying phenomenon of antigenic dissociation was the circumstance that they did not work with well stabilized R types and failed to recognize the wide latitude that always seems to characterize the cultural, morphologic and serologic expression of these forms, depending on the manner of their production and the degree of their stability. In addition, Krumwiede's work serves as an excellent example of the future necessity of providing a preliminary groundwork in the simple phenomena of dissociation before attempting the difficult antigenic analyses relating to this reaction.

B. Diphtheriae and Hofmann's Bacillus.—Although several investigators have studied the serologic relation of clinical diphtheria cultures to the types of Hofmann, the results have been variable. In the main there has been observed an apparent lack of relationship. Among the latest workers in this field are Bull and McKee [81] who in 1924 studied the subject by complement-fixation methods. They observed that cultures of B. diphtheriae and Hofmann strains could be clearly distinguished by means of such tests and concluded (p. 107) that "the common parentage of all strains of B. hofmanni and B. diphtheriae cannot be established by means of the complement-fixation reaction." This and similar instances serve to illustrate a fallacy which has been very common in dealing with the serologic relationships of bacteria. It involves the erroneous assumption that different colony strains coming from the same pure line must show a high degree of serologic homogeneity. This is well illustrated by the statement of Eberson,[145] who became involved in difficulty over this point (see Mellon [327]). Eberson states (p. 28): "Bacteria are unicellular organisms which are produced from other like organisms by the process of fission—an act in which the entire parent substance may be said to divide itself equally into two fractions, both of which constitute the resulting pair of bacteria. Under such conditions the offspring cannot possess what was not present in the parent." Thus Eberson, as also many others, is led to assume that, if two cultures, descended from the same assumed pure line, do not manifest symmetrical serologic reactions, a conclusion that the two forms may have a common parentage is scarcely justified. If there is one significant point in the serologic aspects of microbic dissociation it is the fact that there is no necessary symmetrical serologic relationship between different daughter colony strains. Indeed, when we deal with the R and S forms of bacteria, as will be shown in future pages of this section, the expectation is that the serologic reactions of the rough variant, as also certain other variants, will be quite unlike those of the parent form; and may be quite unlike the reactions of sister colony cultures taken from the same plate. The time is past when lack of serologic symmetry must necessarily be interpreted as indicating lack of "common parentage"; and this holds true whether the tests involve agglutination, precipitation or complement-fixation reactions. In other pages of this paper is presented evidence favoring the view that the Hofmann types represent an R form of the clinical diphtheria culture type. Under these circumstances a symmetrical serologic relation with the clinical form would never be expected. The criterion of relationship or nonrelationship which Bull and McKee seek to employ is therefore of no value for eliciting the information which they demand regarding the position of these two bacterial types.

Pneumococcus.—In the pneumococcus serologic disparities have been reported many times, beginning perhaps with Neufeld [363] in 1902. In most cases these have related to "two different types" of the organism, differing not only in cultural and colonial features (though these were recognized clearly only at a later date), but also in serologic characters and virulence. Neufeld reported a certain pneumococcus strain, agglutinating well in immune serum, which suddenly became non-agglutinative and at the same time nonvirulent. He also observed that serum immune to a certain virulent strain agglutinated foreign virulent strains, but not a foreign strain which was from the beginning non-virulent. Strains that agglutinated well lost this ability when grown for some time on agar.

In 1916 Laura Stryker[454] carried out an investigation of fundamental significance dealing with pneumococcus variability determined by growth of the virulent organism in homologous immune serum. Although she did not recognize the fact of dissociation, merely observing that "some biologic change must have taken place" in her cultures, the data presented in her paper clearly indicate that she was concerned with the dissociative phenomenon as more clearly recognized by Griffith, Reimann and Amoss some years later. Her results merit report in some detail.

Stryker grew virulent pneumococci in 10% homologous immune serum broth in continuous series, transferring every two days or every week. Control cultures were carried in normal serum broth. The first change noticed in the normal culture involved a sedimentary form of growth in the immune serum medium. The sediment was in the form of a hard mass and was broken up only with difficulty. After five or six transplants in the same medium, however, the sedimentary growth became flocculent and was easily broken up by shaking. Type 2 pneumococcus culture required a longer time for this change to occur. In both types the form of the individual cells was changed to threads or to swollen cells, often clumped. In this modified culture state some of the characteristics of the normal culture remained: gram-positive staining, solubility in bile, and fermentation of inulin. Other characteristics, however, changed: capsules were lost, virulence was lost and the antigenic and opsonic reactions were modified; the culture agglutinated with greater difficulty and phagocytability was increased.

Regarding the culture growth, this lost the moist, confluent and greenish appearance of the normal pneumococcus culture and grew in the form of dry, brownish colonies which showed a tendency to produce hemolysis. In broth they gave a sediment which still persisted after 25 passages in a plain medium. Regarding agglutination in immune serums, there was observed a tendency to develop nonspecific agglutinins. Serum immune to the modified culture (type 2) agglutinated the modified strains of both types 1 and 2, but failed to agglutinate normal strains. Serum immune to the normal type 2 agglutinated the homologous normal strain and also, though to a slighter degree, the modified cultures of types 1 and 2.

In absorption tests, Stryker showed that when antiserum for a normal type 1 or 2 culture was absorbed with type 1 or 2 modified strains, the agglutinins for both homologous and heterologous modified strains were removed, while agglutinins for homologous normal strains remained. When, on the other hand, antiserum for a normal type 1 or 2 culture was absorbed with normal homologous strains the agglutinins for both normal and modified strains were removed. Here again we have a striking similarity to the early case of Bordet and Sleeswyk involving B. pertussis. These serologic results, Stryker assumed, were due to a suppression of certain receptors, as Cole earlier believed to be the case in certain serologic disparities with B. typhosus. We can see clearly, however, that they are in line with many other instances of the same sort dealing with the socalled "double" and "single" antigens. Other interesting points in Stryker's paper are considered further in sections 11, 12 and 13 of the present work.

As stated earlier, although the splendid work of Stryker has given one of the clearest pictures of the serologic aspects of the dissociative reaction in the pneumococcus, she did not recognize the biological significance of the phenomena observed. We shall see presently, however, that our view of dissociative changes in this case is amply justified by the results obtained somewhat later by Griffith, by Reimann and by Amoss, working with recognized R and S forms of the pneumococcus.

In 1919 Mildred Clough [102] examined many strains of the same organism and observed nine which precipitated with types 1, 2 and 3 serums. These atypical strains were slightly virulent and highly susceptible to phagocytosis. None of the nine absorbed precipitin from type 1 or 2 serums, and serums immune to these strains agglutinated these strains only. Nicolle and Debains [371] according to Eastwood [143] found among many typical pneumococcus cultures about 30% which agglutinated with several of the type serums, but mainly with type 2 serum. Yoshioka [494] in 1923 presented a very clear picture of pneumococcus serologic variability involving dissociative aspects. He showed that whenever typical and virulent pneumococci were placed under such conditions as determined a loss of virulence (as drying, unfavorable medium, incubating at 39 C., etc.), such cultures also experienced marked antigenic changes. These changes involved not only loss of agglutinating power in their homologous, immune serums

(i. e., homologous to the unmodified culture type), but newly acquired agglutinability in heterologous serums. What is perhaps more important, however, they had also gained a susceptibility to equal agglutination with serums immune to certain streptococcus forms (Aronson virulent). Assuming, as I believe we justifiably may, that these pneumococcus variants represented either R types, or cultures in the course of transformation toward the R type, we have here another interesting case of the "serological cosmopolitanism" of the R forms as developed by Schütze [436] for certain intestinal bacteria. In the light of this and other work, there can be small doubt that we shall soon find means to transform any of the standard pneumococcus types into quite different serologic forms. Indeed, Berger and Engelmann [48] have already reported such a transformation of a type 3 pneumococcus into type 2. What we mainly lack at the present time with reference to the serology of the pneumococcus is a thorough study of the relation of the various types to the distribution of the fundamental heat stable and heat labile antigens.

The several studies just reported have served to give us an excellent picture of some of the results of dissociation in the pneumococcus; but it remained for Griffith [215] (1923) in England, following soon the important lines established by Arkwright and de Kruif, to correlate many of the curious points in the behavior of this organism with the definitely recognized products of dissociation; namely, the S and R culture forms. The serologic characteristics of these variants, although according well with the findings in other bacterial species, merit presentation in some detail. Other aspects of Griffith's work are presented in other sections of this work.

To state the matter briefly, Griffith produced the R form of the pneumococcus by growing normal, virulent cultures in their homologous immune serum. These modified strains were noncapsulated and were lacking in virulence. It is their serologic characteristics, however, that are of present interest. Rabbit serum immune to the S form agglutinated S and R about equally, but only at a titer of about 160. The R antiserum agglutinated R at 640 but S scarcely at all (10). When S antiserum was absorbed with S antigen it lost agglutinins for both S and R culture. When S antiserum was absorbed with R antigen it failed to agglutinate R but still showed some agglutinating power for S. When R antiserum was absorbed with R antigen (twice) it gave no agglutination of R antigen (it always lacked agglutinative power for S). When R antiserum was absorbed with S antigen its titer for R dropped from 640 to 160. Here we have clearly the same "one-sided action" described so well by Bordet and Sleeswyk for B. pertussis in 1910. In order that the comparison may be facilitated, I have brought together in tabular form Griffith's essential serologic results.

TABLE 5

SHOWING THE ESSENTIAL AGGLUTINATIVE AND ABSORPTIVE RESULTS OF SEROLOGIC STUDIES OF GRIFFITH ON THE RECOGNIZED S AND R TYPES OF THE PNEUMOCOCCUS

(Reconstructed from Griffith's distributed data.)

Antigen Culture Type	Type S Antiserum Before Absorption	Type R Antiserum Before Absorption
S	160	0
R	160	640
	Antiserums absorbed with S	
S	0	0
R	0	160
	Antiserums absorbed with R	
S	"Moderate"*	0
R	0	0

* Griffith does not state exact titer.

Griffith also presented interesting results of precipitin tests which other reports lack. For antigens, Berkefeld filtrates of broth cultures were employed. When S antiserum was added to S filtrate there formed a copious precipitate characterized by firm masses which could not be easily broken. When the S antiserum was added to R filtrate no precipitation occurred. When the R antiserum was added to either S or R filtrate no precipitation occurred in either case. These results seem to indicate that the precipitinogen was absent in the R cultures, and that the respective serum failed to possess a rough precipitin. Thus, the precipitin seems to be related to those antigenic groups which distinguish the S type of culture and also underlie the characteristic of virulence. The discrepancies between precipitin and agglutinin reactions in general are, on the strength of such findings if confirmed, likely to afford an interesting field for further study.

Although we owe to Griffith the first clear recognition of the S and R types of the pneumococcus and the demonstration of the association of the "specific soluble substances" with the S form, as shown in the section on "Dissociation and Virulence," other interesting features of pneumococcus dissociation were added by Reimann[405] in 1925 supporting the view of "serological cosmopolitanism" of the R types already introduced into the field of pneumococcus dissociation by Stryker and by Yoshioka. Reimann ascertained that, when serums of types 1, 2 and 3 were applied to the S and R forms of a type 1 pneumococcus strain, the S culture agglutinated in type 1 serum only, while the R culture agglutinated in all three standard type serums. The "pure" S form was thus "type-specific" while the R form was not. According to Reimann's interpretation, and as Griffith had earlier stated, the R form of culture had lost the specific soluble substances. Reimann's results thus confirmed a number of preceding observations pointing to the heterologous nature of the R type in its serologic behavior. The serologic interrelation of the S and R cultures was not considered by Reimann, nor was it mentioned in the paper by Amoss,[8] although the latter presents certain brief immunologic features of pneumococcus dissociation to be mentioned on a later page.*

* Since the writing of the foregoing pages there have appeared two valuable papers by Julianelle[360, 360a] dealing with the antigenic structure of Friedländer's pneumobacillus and including interesting data on the serological relations of the S (capsulated) and the R (noncapsulated) forms of culture. Julianelle observed that it is only the S type which produces the specific soluble substance, analogous to that noted by Griffith, Heidelberger and Avery and others in the pneumococcus. This substance, lacking nitrogen and appearing to be of the nature of a polysaccharide, seemed to endow the culture with the virulence characteristic of the S type. This carbohydrate when separated from the cells was nonantigenic; but in the form in which it existed in the cells was strongly antigenic, and upon injection gave rise to an antibody specific for the type (A, B, C or X) of the pneumobacillus. In this manner it seemed to condition the "type-specificity." This carbohydrate was found to be lacking in R type cultures and in the substance of S type cultures the cells of which had been decapsulated by the acid method of Porges. The remaining antigen, as also the dominant antigen of the type R cells, appeared to be a nucleoprotein which was highly antigenic, giving rise to distinct antiprotein antibodies. The serologic reactions involving these two different antigens (carbohydrate and nucleoprotein) were as follows:

Serum immune to the S type culture contained antibodies which "cause the type-specific agglutination of encapsulated cells, precipitate the carbohydrate derived from organisms of the homologous type, and afford passive protection in white mice against infection with bacilli of the same type." S antiserums were found to contain negligible amounts of antiprotein and agglutinated only irregularly cells of the R type.

Serum immune to type R culture, or to the nucleoprotein constituting its chief antigen, agglutinated or precipitated R culture or R protein, but did not react with the type S cells or with the polysaccharide. Such R serums reacted not only with the homologous R culture, but also with decapsulated cell substance and with any heterologous type R strains; also with the nucleoprotein of B. coli, B. aerogenes and the granuloma bacillus, thus manifesting the "serological cosmopolitanism" noted elsewhere in this work. It may be added here that Julianelle's conception of the nature of the type R culture is that of a "degraded" form of the organism, and in this respect his view coincides with that of Krumwiede. Although neither entertains notions of microbic dissociation in the biological sense, the work of Julianelle indicates in an especially clear manner the extent to which thorough-going antigenic analyses may lead us closer to the chemical realities underlying the dissociative reaction.

Meningococcus and Gonococcus.—In the section dealing with dissociation and virulence, I have pointed out the interesting data secured by Atkin [21, 22] in the colony dissociations of these diplococci. Although we have little knowledge of the serologic interreactions between the clearly recognized types, either in meningococcus or in gonococcus, it is desirable to mention one other matter bearing on the often observed relationship between certain strains of these two species. Early observations dealing with this subject were introduced in 1906 by Bruckner and Cristéanu [74] and by Vannod,[471] followed by those of Wollstein [491] in 1907, Teague and Torrey [459] in 1907, Gurd [219] in 1908, Dopter and Koch [135, 136] in 1908, Arkwright,[12] also Elser and Huntoon [157] in 1909, Colombo [107] in 1911 and by Arkwright [13] in 1912. In a general way, as Arkwright has pointed out, the records of these investigators have demonstrated that a marked similarity exists between certain strains of the two species from the point of view of agglutination, complement fixation and the generation of antibodies. He stated: "The differences between the individual races of gonococcus, or of meningococcus as tested by these methods, appear to be almost as great as between the two species." And again: "It is therefore unlikely that well-marked constant differences along these lines will be found between the groups and subdivisions of the whole class to which the Gonococcus and Meningococcus both belong" (l.c., pp. 116 and 117). Wollstein [491] who worked with monovalent serums was able to observe no difference between many meningococcus and gonococcus strains. Also Colombo, who studied the relation particularly by complement-fixation methods, was unable to discover specific differences. Arkwright,[13] who considered the subject further in 1912, came to the following conclusions: "Meningococcal sera produce complement fixation as readily with some gonococcal extracts as with the extracts of some strains of meningococcus; whereas no reaction is obtained with some heterologous meningococcal extracts. A monovalent serum usually reacts better with an extract of its homologous coccus than with extracts of other strains of Meningococcus or Gonococcus, but a gonococcal extract sometimes gives a better reaction with a meningococcal serum than the homologous extract does." . . . "No satisfactory distinction between Meningococci and Gonococci can be demonstrated by means of complement fixation tests."

From the facts presented above it becomes clear that the serologic relation existing between the meningococcus and the gonococcus parallels to a degree the serologic affinities manifested by the R forms of culture of several members of the colon-typhoid-dysentery group as also of the pneumococcus-streptococcus group; and thus seems to suggest the existence of a sort of bacterial convergence—at least with respect to the serologic reactions. Atkin has already presented data which indicate the facility with which freshly isolated meningococcus and gonococcus cultures dissociate into the intermediate or into the R type; also the probable frequency of this reaction, after they have been grown for a time on artificial culture mediums. From this it seems probable that it is through the medium of the O or R forms of culture that the somewhat noteworthy serologic affinities between the meningococcus and gonococcus are maintained. With our present knowledge of the fact of dissociation in these species, and the means by which it may be brought about, the problem is now open to experimental attack. Such a study is likely to reveal more closely than heretofore the nature and cause of the serologic "discrepancies" so frequently observed; and perhaps in this way to establish on a firmer basis our methods of serologic diagnosis of meningococcal and gonococcal infections.

B. Cholerae Suis and the Hypothesis Relating to the "Flagellar" and "Somatic" Agglutinative Antigens.—Before continuing further the serological aspects of dissociation, it is desirable to consider its possible relation to one of the theories of agglutination current in earlier years, and actively revived in recent time in connection with the "double" and "single" antigen hypothesis. As we shall see later, this problem has a bearing on the H and O forms of B. proteus first described by Weil and Felix while in this country it had been elaborated much earlier at the hands of Theobald Smith and Reagh [445] in 1903 and by Beyer and Reagh [50] in 1904. Already in 1897 Malvoz,[316] and one year later Dineur,[128] had called attention to the possible rôle played by the flagella in agglutination, a conception also furthered by Nicolle and Trenl [370] in 1902. Smith and Reagh accordingly undertook to demonstrate the separate nature of flagellar and somatic antigen, using for this purpose certain cultures of the hog cholera bacillus. Although all common strains of this organism are motile, they had at their disposal a nonmotile strain. They stated, however, that between this strain and their other cultures there was no difference except in motility. First they showed that the action of rabbit serum immune to the motile type (M) on the type M antigen was different from the action of the serum immune to the nonmotile culture (NM) on the nonmotile form. These differences were as follows: the reaction was delayed; the clumps were in the form of small, compact granules; loose clumping, characteristic of the motile form, was absent; to obtain an active serum against the NM form required a long and intense immunizing procedure. All of these points, as the combined results of many investigations have shown, are characteristic of intermediate or R type cultures. The results of cross agglutination tests with the two immune serums are tabulated.

TABLE 6

SHOWING THE RESULT OF CROSS AGGLUTINATION TESTS BETWEEN SERUMS IMMUNE, RESPECTIVELY, TO THE MOTILE AND NONMOTILE TYPES OF SMITH AND REAGH'S HOG CHOLERA CULTURES

(Assembled from Smith and Reagh's distributed data)

Antiserums for	Antigen	Agglutination Titers
Motile form (S ?)	Motile	10,000
	Nonmotile	500
Nonmotile form (R or O*)	Motile	20-200
	Nonmotile	200-500

* See "O" under proteus dissociation, p. 144.

These results show a distinctly characteristic "one-sided action." Next, an incomplete series of absorption tests was performed with the serums absorbed with M and NM antigens as shown below.

TABLE 7

SHOWING THE RESULTS OF ABSORPTION TESTS WITH SERUMS IMMUNE, RESPECTIVELY, TO THE MOTILE AND NONMOTILE STRAINS OF SMITH AND REAGH'S HOG CHOLERA CULTURES

(Assembled from Smith and Reagh's distributed data)

Antiserums for	Antigen Absorption	Titers after Absorption	
		Motile	Nonmotile
Motile form	(Unabsorbed)	20,000	200
	Nonmotile	20,000	100
	Motile	"decided loss"	"decided loss"
Nonmotile form	(Unabsorbed)	500	500*
	Nonmotile	(No data)	(No data)
	Motile	40	40

* More nearly typical results are shown in table 6.

From these results the authors conclude that the agglutinins are differentiated into "flagellar" and "body" (somatic) types, and that therefore the motile bacteria carry a double antigen while the non-motile carry a single. This view has been reinforced by the further study of Marion Orcutt,[377] in 1924, in which she reported having separated the actual flagellar antigen in pure form. Although the experimental data on agglutination and absorption presented by Smith and Reagh and their immediate followers are too incomplete to make possible a satisfactory comparison with other work involving the "double" and "single" antigens, it is nevertheless apparent that there exist similarities with the data presented by Bordet and Sleeswyk and others. We know, moreover, from the statement of Smith and Reagh that the agglutination of the nonmotile culture was in the form of small, compact granules; and this phenomenon, as has been shown by Goyle for B. typhosus and B. enteritidis, as described by Weil and Felix for B. proteus X19 (to be mentioned presently), and by Balteanu [38] for Vibrio cholerae, is characteristic of the intermediate or O type of culture—perhaps sometimes of the R. It therefore seems possible that the nonmotile hog cholera culture of Smith and Reagh represented such an O form, or a mixture of O with the variant, R. The motile form, on the other hand, was unquestionably the normal S culture. The true R type of the hog cholera bacillus is probably that described by Orcutt [376] at a later date.

While the view of flagellar as opposed to somatic antigens and agglutinins elaborated by Theobald Smith and Reagh and their followers on the primary postulates of Malvoz is sufficiently alluring and has been maintained even to the present time by many workers (including Felix,[172] Braun and Schaeffer [66] and others for B. proteus and by Feiler [167] for B. typhosus, as well as by Furth [188, 190] for typhoid, paratyphoid and dysentery bacteria), certain modifications have been required. Since it was found that certain nonflagellated bacteria could be split into two forms, R and S, giving exactly the same order of serologic reactions as those observed in the R and S types of motile bacteria, the flagellar hypothesis could scarcely stand. When the new results concerned capsulated bacteria such as the pneumococcus, the S type of which is capsulated while the R form is not, it was naturally logical to modify the flagellar hypothesis into a capsular hypothesis; and this was quickly done. But, if we go a step further we encounter exactly the same sort of serologic disparities in the behavior of organisms that are not noteworthy by reason of possessing capsules, namely, B. pertussis, the streptococcus, and Vib. cholerae. Among these we can also observe the same sort of "one sided action." We are therefore driven to dispense, not only with the old flagellar hypothesis, but also with the capsular hypothesis. When, as is certainly now the case, we are pushed to the morphologic extremity of our arguments, we may perhaps fall back upon the recognition of ectoplasmic versus endoplasmic substances to explain the phenomenon. It is possible that here we may yet make a stand. I regard it more probable, however, that we shall eventually come to disregard all distinctly morphological elements in respect to differences in antigenic behavior and begin to study serologic differences in terms of the presence or absence of certain specific soluble substances wherever they may have their point of origin in the bacterial cell, and with whatever cell structures they may seem to be related. Griffith has already made a beginning in this direction in his observations relating to the specific soluble substance of the pneumococcus in cultures of the S type.*

Micrococcus (Brucella) Melitensis.—That microbic dissociation is operative in cultures of the organism of Malta or Mediterranean fever is strongly suggested by certain recent work of Et. Burnet.[85] It is a well known fact that many typical, clinical cases of this disease yield blood cultures with which the agglutination test is negative or slight. Burnet states that fully one-quarter of all cases fall within this class. Often the organism obtained agglutinates only in its homologous serum and sometimes agglutination is quite absent. As a result of examining many cultures Burnet points out that nearly all strains of melitensis, when tested against a type serum, divide themselves sharply into two groups: those which agglutinate well (titer, 1,000-2,000); those which agglutinate poorly (titer, 100-200) or not at all. In some instances there are "intermediates," but these are few in number. These circumstances hold for cultures obtained either from natural infections in goats or from human cases. Some serums from goats or human cases agglutinate the second type of culture only. Out of 35 strains examined by Burnet, 14 belonged to the second group. On the basis of these observations this investigator sets up two distinct types of the Malta fever organism; and these he refers to as melitensis I and II. He points out that type II is identical with the so-called para-melitensis of various workers, notably of Bassett-Smith;[30] also with organisms of the "group 6" of Evans, and the "group 4" of Fusier and Meyer.[192] Certain "intermediate" forms are also mentioned.

* Since the present work went to press there has appeared (November) the splendid antigenic analysis of the capsulated (S) and noncapsulated (R) forms of Friedländer's bacillus by Julianelle.[606, 608] There is not now opportunity of considering in suitable detail the interesting bearing of this paper on the serologic aspects of dissociation among the capsulated bacteria. It may only be remarked that the results seem to indicate a correlation between the presence of capsules and the antigenic basis of the specific soluble substances of the pneumobacillus; also to indicate that the S type organisms, deprived of their capsules by the acid method of Porges, behave antigenically like the R forms. The author believes that the sort of reactions observed are analogous to those noted by Smith and Reagh and others in the case of their "flagellar" and "somatic" antigens. He believes, furthermore, that the dissociation observed in the pneumobacillus is due to "a cleavage of the specific antigen complex in the animal body rather than a condition of cultural development." Although this conception is somewhat at variance with the conclusions that I have drawn from earlier studies, Julianelle's work introduces certain new problems of considerable interest in the field of dissociative behavior.

When rabbits were immunized with melitensis I and II respectively, serum immune to melitensis I agglutinated I well and II poorly. On the other hand, serum immune to melitensis II agglutinated II well, but I not at all. It is of interest to compare these results with the classical case of Bordet and Sleeswyk [64] in their study of the modified cultures of B. pertussis, as well as with some of the other instances mentioned in this section dealing with the chief culture dissociates, S and R, or in some cases, O. It is also of interest that Burnet states the rabbits given the type II culture were difficult to immunize, but that the agglutination of type II strains usually reached a higher titer than the type I strains. The results of the absorption tests presented by Burnet agree in all respects with the results of direct agglutination. In addition, he showed that several strains of B. abortus (which he as well as Evans regarded as identical with M. melitensis) all fell into the type I group; no abortus strains revealed the type II antigen.

Burnet's type II strains, moreover, were found to correspond with the "para-melitensis" cultures of Bassett-Smith.[39] The latter were found by Bassett-Smith to be characterized by a form having longer rods than the "normal" culture type. This para type, morever, was more easily agglutinated by nonspecific serums and, in addition, showed a tendency to undergo spontaneous agglutination. Bassett-Smith also found rabbits difficult to immunize with this strain.

In discussing the appropriateness of the older term, "para-melitensis," Burnet suggests that when used in this sense, the prefix "para" does not accord the same significance as obtains in the case of the terms, paratyphoid or paradysentery. He therefore suggests the use of "para" for bacterial species which are distinguished by differences in pathogenic quality as well as in antigenic function. For species which depart from the "normal" in antigenic function only (as in the case of his Melitensis II) he prefers the use of the numerals, I, II and III, as in the pneumococcus types. He therefore terms his melitensis variant "Melitensis II."

In addition to the data mentioned above, Burnet presents the results of many tests involving the influence of heat on the antigens of melitensis I and II. These observations also, although not adequately supported by data on the cultural and biochemical characteristics of the two types, are sufficient to suggest strongly that Burnet's type II cultures, as well as paramelitensis of many other workers, represent the R form of culture, or at least a culture (such as O) in the course of transformation to the R state. The serologic tests alone are nearly sufficient to warrant this conclusion. Unfortunately the subject of differences in virulence or in immunizing power were not studied and further work in this field is likely to add other interesting data.

Although the rôle played by microbic dissociation in the phenomena described above is strongly suggested, the details become more clear in a second contribution by Burnet dealing with the phenomenon which he terms "Entrainement" (influence of one culture on another which association) and which is considered in detail in section 11 of the present work.

B. proteus.—In 1916 and 1917 Weil and Felix [483] awakened new interest in B. proteus by reporting certain observations on variation in connection with their study of its possible rôle in typhus fever. Today we can see that these variational phenomena are concerned with dissociative reactions of a fairly typical sort, although they also present some points of interest not commonly found in the usual reaction. Briefly, Weil and Felix reported the splitting of their culture of B. proteus X19 into two chief types to which they gave the terms, H and O. The H form, which appeared to be the normal, grew on solid culture media with a broadly spreading film ("Hauch"), while the O variant grew in isolated and discrete colonies without spreading ("ohne Hauch"). In agglutinating serum the H form clumped in large loose ("grob") flocculi, easily dispersed by shaking. The O form, on the other hand, agglutinated in small, granular ("fein") clumps. As was shown by Sachs and Schlossberger in 1918, when the H antigen was heated at 80 to 100 C. for one hour, flocculent agglutination (in presence of immune serum) no longer occurred although some fine clumps were observed. By a similar treatment the O antigen was not altered; it still agglutinated in fine clumps. These antigenic modifications were soon confirmed by many others. In view of these results Weil and Felix concluded that normal proteus cultures possessed two antigenic groups or receptors, H (heat labile) and O (heat stable), while the O variant had but one—the heat stable. Similar antigenic groups were observed in B. typhosus and paratyphosus in 1918 and 1920 by Weil and Felix;[484] also in B. typhosus by Furth [191] in 1922 and in B. enteritidis by Grushka [217] the same year. Subsequently they were observed in many other species. All of these studies, many of them supplemented by data on cultural variation as well as serologic, have done much to lend new emphasis to the "double antigen hypothesis" and the "double" as opposed to the "single" antigen type of organism. In casting about for the explanation of these reactions, however, it is unfortunate that most of the workers have fallen back on the old flagellar and somatic antigen hypothesis of Malvoz, rather than on the newer conceptions of bacterial variation embraced by the phenomenon of microbic dissociation.

In 1919 Braun and Schaeffer [66] attempted the production of "artificial" O types of B. proteus by growth on phenol agar and by "starvation" upon an impoverished agar medium. They made use of the proteus X19 strain of Weil and Felix and succeeded in obtaining types which they believed agreed with Weil and Felix's OX19 in serologic tests. Both types maintained their group antigen (O) but lost the individual antigen (H). Felix and Mitzenmacher,[173] however, held that the phenol and the starvation antigens differed from the "natural" O antigen, although Braun and Schaeffer believed that they had disproved this contention and had shown the identity. Whether the starvation antigens actually did differ from the phenol antigens is not so clear, since the matter was not thoroughly tested. A difference seems probable, however, in the light of the distinct difference in the cell types of the two cultures, as amply demonstrated by Braun and Schaeffer.

Relation of Proteus H and O to S and R Types.—In view of the apparent resemblance in several respects between the H and O forms of Weil and Felix and the S and R forms of many other intestinal bacteria as developed by Arkwright, it was of importance to ascertain to what extent, if at all, the analogy holds true. Such a comparative study was made by Arkwright and Goyle [18] in 1924, followed by a more comprehensive treatment of the subject by Goyle [206] alone in 1926. We may first consider the former report. Arkwright and Goyle observed in B. typhosus, B. enteritidis (Goyle) and B. dysenteriae (Arkwright) antigens regarded as analogous to H and O. They stated that the presence of the two agglutinins could be proved by demonstrating that the type of agglutination is changed after heating an emulsion at 100 C. for 10 minutes; and that "the heated emulsion will absorb S agglutinins but not H, and can be used for obtaining a pure agglutinating serum." They also stated that variants could be obtained containing only one antigen which might be "either R (H) or the S (O)," according to their view.

The conclusions of Arkwright and Goyle are in opposition to the trend of much other evidence. They are, for instance, open to criticism on this ground. In the experiments on B. typhosus and B. enteritidis (conducted by Goyle) it is stated that "in order to obtain 'smooth' variants containing as small a share of the 'rough' factor as possible, the normal cultures were grown on nutrient agar to which 1 in 1,000 phenol had been added. On this medium daily subcultures were made. . . ." And again, "these 'S' (smooth) variants have not been obtained by selecting colonies and consequently may not be so pure nor so persistent as the 'R' (rough) variants." It is thus (in the case of typhoid and enteritidis, at least) cultures which arose from phenol treatment, and which were assumed to be of the S type (in the sense that S is the "normal," as intimated by Arkwright on p. 110 of the same paper) that the authors found identical with the O form of Weil and Felix. It should be added that in the case of B. dysenteriae (in the portion of the study conducted by Arkwright) it was the "normal" culture without phenol treatment that was assumed to be the S form. It thus appears that Goyle, in his portion of the work, modified a normal culture (S) by use of phenol and then compared the result with O. Braun and Schaeffer had already clearly shown that one way to change H to O in proteus was to grow on phenol agar.

Throughout the experiments, moreover, much use was made of "stock" immune serums, the exact nature of whose antigenic basis is naturally a matter of doubt. Experimental data based upon such a manner of scientific approach to a problem of this sort are difficult to accept as furnishing grounds for satisfactory conclusions. It is natural that they should have been called into question. In the opinion of White [487] the conclusion of Arkwright and Goyle that the O and H antigens were essentially those of smoothness and roughness respectively, was without justification. He believed that their conclusions were based on two unwarranted assumptions: "that R cultures grown on phenol agar would not also become pure O without at the same time becoming smooth;" and "that any O antigen in the rough cultures was either identical with that of the smooth cultures or was at least demonstrable by the sera employed." White was fortunately able to show that the rough variants contained a heat stable antigen peculiar to the R forms. For this antigen agglutinins were present in serums made with the R form, but were absent in serums made with the smooth form. The following conclusions of White may be put more or less in his own words. The smooth culture normally exhibits both H and O antigens, of which the former is more obvious in the tests with Salmonella. In some freshly isolated Salmonella cultures, and in those grown on phenol agar, the O property comes to the fore as in the case of B. dysenteriae. Roughening is therefore concerned with a fundamental change in the O antigen, O smooth becomes O rough, substances which are quite distinct. Often rough cultures show a sufficient amount of "smooth O-antigen" to absorb all or nearly all of the "smooth O-agglutinin" from a smooth serum. But a truly "smooth" culture has no "rough O-antigen." The absorption test therefore tends to be "one-sided." White stated further that the process of roughening is often associated with the reduction or the total loss of the H antigen and that the change is permanent. In this manner White brought the serologic changes involved in the H → O transformation into closer agreement with those of the S → R change, but left many points still to be clarified.

In 1926 the proteus dissociation was again considered by Goyle [206] independently. And we may note several important modifications from the results and conclusions previously reported with Arkwright. His work involved mainly a very detailed and clearly presented study, by direct agglutination and absorption methods, of the three forms of B. enteritidis and B. typhosus. These three types were first indicated by Arkwright and Goyle, but characterized more fully by Goyle independently.

Smooth Form Normal: Containing two antigens (similar to those of the H culture of Weil and Felix), the heat labile H and the heat stable O.

Smooth Form Variant: Containing ordinarily only a single antigen, O (heat stable), but sometimes also the R, as below.

Rough Variant: Containing mainly the heat stable R antigen but often some of the heat labile, smooth H antigen found in normal smooth cultures and sometimes, according to White, some O antigen.*

The stock cultures employed by Goyle were all normal cultures serologically, and in agglutination formed flocculent clumps. The R variants were produced by plating out old broth cultures. The smooth variant

* For purposes of effecting greater clearness at this point, attention may be called to the relation between the mode of designating a given culture type and the mode of designating the antigenic configuration or constitution of that type. Although we may refer to one of the chief culture forms by a symbol indicating its most strongly revealed, or so-called dominant, antigen, it must be borne in mind that such a culture may contain one or more other antigens which occur here as minor antigens (quantitatively speaking), but which may also occur as dominant antigens in other culture types. Thus the smooth form of culture may be designated by the symbol, S, because the S antigen is dominant in this culture form. But, at least in many bacterial species, and perhaps in all, the smooth culture also contains a larger or smaller fraction of the antigen O; or sometimes a fraction of the antigen R. This circumstance naturally affords the experimental basis for the recognition of the culture type characterized by the "double," or perhaps we might term it the bivalent, antigen. Whether there exists any culture type possessing the smooth antigen (S) alone, cannot at present be stated, but I am not aware that such a type of culture has ever been described. Thus the smooth culture which we designate S may be said to have the antigenic configuration SO; or perhaps in some cases (dependant upon the species concerned, or the stage of development) the configuration SR. On the other hand, when we consider the antigenic configuration of the "pure" intermediate (O), and the "pure" rough (R), we find that these often seem to possess, respectively, a monovalent antigenic constitution, O and R,—thus agreeing with the symbols designating the intermediate and the rough culture types. At the same time, depending on the stage of development of the culture, it seems that the O culture type may sometimes contain a residual fraction of S antigen; and similarly that the R culture type may sometimes contain a residual fraction of O antigen. Whether any single culture may contain all three of the chief antigens we do not know, but under certain conditions of growth, or on aging, it seems possible. The probability, moreover, that these few bivalent combinations of the chief antigenic units (S, O and R), as recognized at present, by no means represent all of the possible antigenic configurations will receive more detailed consideration on a future page.

O, was produced by growth on phenol agar. All these were employed unheated for the production of immune rabbit serum. All the immune serums were tested against all of the antigens by agglutination and absorption methods. Regarding the agglutination tests the type of precipitate is of interest, and may be found to possess significance in the diagnosis of types among other bacterial species involved in dissociative reactions. For this reason I present Goyle's results in some detail.

TABLE 8

SHOWING THE TYPE OF PRECIPITATE PRODUCED IN GOYLE'S CASES OF SIMPLE AGGLUTINATION INVOLVING B. ENTERITIDIS SERUM IMMUNE TO CULTURES H (SMOOTH NORMAL), O (SMOOTH VARIANT) AND R (ROUGH VARIANT) AGAINST THEIR RESPECTIVE ANTIGENS AND IN CROSS TESTS *

(Data selected from Goyle's tables 2 to 7)

Antigen Culture	Agglutination with Antiserums for		
	Normal S	Variant O	Rough R
Normal S	Flocculent-granular	Flocculent-granular	Flocculent
Variant O	Granular	Granular	Granular
Rough R	Flocculent	Flocculent	Flocculent

* Analogous results obtained with B. typhosus.

Table 8 indicates clearly the form of the agglutination precipitate determined by the nature of the antigen or of the immune serum employed, according to Goyle. The normal, smooth antigen (S) precipitates in a form which is a mixture of floccular and granular components in accordance with the double antigenic nature of the culture. The smooth variant, which Goyle takes as identical with the O form of Weil and Felix, always gives a granular ("fein") precipitate, incapable of resuspension. Finally, the rough antigen gives, like the normal smooth, a purely flocculent precipitate but without the granular admixture. All of these precipitate-forms have been described frequently for agglutinating or nonagglutinating cultures, but their relation to distinct antigenic states or forms of culture has not commonly been detected. It would be a matter of considerable importance to be able to correlate the form of flocculation with a definite serologic type, as seems to have been done by Goyle for B. enteritidis and B. typhosus.

It is of importance to note, however, that some others, as Jordan,[274] have found the R type culture characterized by granular antigens rather than floccular. Jordan found that a smooth strain serum of B. paratyphosus B agglutinated R antigen in higher dilution than the smooth itself. Also that the precipitate formed was finely granular. Jordan's rough serum did not agglutinate the smooth strains at all, but agglutinated the R antigens in a granular precipitate. A flocculent precipitate under these conditions was never observed. There thus appears to be a discrepancy between the results of Jordan on paratyphosus and Goyle on enteritidis and typhosus. We may recall, however, that White had shown that rough cultures of Salmonella often contain considerable amounts of O antigen (which gives, according to both White and Goyle, a granular precipitate). It seems possible that Jordan's R cultures may have contained enough of the O antigen to have given the results observed. In any case Jordan's report raises an interesting question relative to the antigenic transition from O to R culture, and one which is in need of further study. His observations are supported by those of Cowan [111] for her R types of streptococcus, by those of Griffith [215] for the R pneumococcus, and by others.

Turning again to the work of Goyle, it is the absorption tests presented by him that possess greater significance in indicating the antigenic constitution of the three culture types already mentioned. While it is impossible to present in a brief statement the results of the many tests necessary to demonstrate the points at issue, the most essential findings may be summarized for B. enteritidis. B. typhosus, it may be noted, gave almost identical results. In order to make the presentation of this matter more concise I have brought together Goyle's more essential data in the following tabulation:

TABLE 9

SHOWING GOYLE'S ESSENTIAL RESULTS ON ABSORPTION TESTS INVOLVING THE THREE TYPES OF B. ENTERITIDIS CULTURE: NORMAL (N OR S), SMOOTH VARIANT (O), AND ROUGH VARIANT (R)*

(Compiled from Goyle's tables 2 to 7 inclusive)

Antigen Culture	Agglutination Titers with Antiserums for		
	Normal S (N)	Variant O	Rough R
	Serums Unabsorbed		
N	12,800	6,400	12,800
O	3,200	800-6,400	0-1,600
R	3,200	1,600	6,400-12,800
	Serums Absorbed with N Culture		
N	No data	0	0
O		0	0
R		0	6,400
	Serums Absorbed with O Culture		
N	12,800	3,200	12,800
O	0	0	0
R	12,800	1,600	6,400
	Serums Absorbed with R Culture		
N	800	400	100
O	1,600-3,200	800	0
R	0	0	100

* Results essentially the same for B. typhosus.

The data presented in table 9 indicate by antigenic analysis the existence of the three chief types of enteritidis culture designated as normal smooth (S), smooth variant (O) and rough variant (R). The antigenic constitution of these may be stated as follows, making use of Goyle's terms of reference.

Normal N (or S): Contains N (= S) antigen (heat labile) mainly; also some O (heat stable) but no rough antigen (R).

O type (= O). Contains O antigen (heat stable) mainly, but also some N (= S); less often some R antigen.

Rough type (= R). Contains the R antigen (heat stable) mainly and usually some N (= S); often, according to White, some O antigen.

These results therefore clearly relate the H antigen of Weil and Felix with the normal or S form, and not with the R as first suggested in the earlier publication of Arkwright and Goyle. Moreover, the O type of Weil and Felix is shown not to be related to the S as first stated by Arkwright and Goyle, but to be an independent antigenic type essentially different from both R and S. For this culture type Goyle lets stand the symbol (O) first used by Weil and Felix. To summarize therefore, we have one heat labile antigen, S (H), and two heat stable antigens, O and R. These may enter into various combinations. These combinations typify the culture types, and show a rather orderly manner of appearance, one being lost as another is gained. The order in which the antigens are gained is, moreover, the same as the order of transition from one culture form to another: $S \rightarrow O \rightarrow R \rightarrow S$. The direction of these changes, which was not pointed out or commented upon by Goyle, is a point of considerable interest which will be referred to later.

The establishing of the O antigen of Weil and Felix as a new antigenic type, apart from S and R, raises the question of the nature of the O culture type, characterized by the possession of this (O) antigen in greatest abundance. I may say directly that I believe that it is the antigen of the transitional or intermediate culture type, or at least one of these types. It arises from S and passes into R. Although this conclusion is not yet proved by evidence which we can regard as conclusive, many observations point to its truth. In the first place it may be pointed out that, although culture types may exist lying outside of the S, the R and the intermediates or transitionals, such have not been reported so far as I am aware, unless we regard the filtrable form of bacteria as a distinct culture type. The expectation is, therefore, that any culture that is not S or R is one of the intermediates, which, as we have seen, embrace a considerable range of variability. In the second place, Braun and

Schaeffer have shown that, when proteus cultures are grown on phenol agar, there is produced, after a few generations, a new form of culture which differs from the S (H) in the morphology of the individual cells. They show all the peculiar shapes which have been observed in the recognized transitional types of other bacterial species. We know, moreover, that this form of culture reverts rather easily to the normal H form when returned to nonphenol mediums. This also is characteristic of the transitional as we have seen in several instances. We also know that a more permanent variant than the phenol O can be obtained by growth of proteus on starvation agar; and that a still more permanent variant can be produced under the influence of immune serum. But I doubt that either of these were the "extreme R" form of proteus. Indeed, it is doubtful if the extreme R of this organism was ever observed by Weil and Felix or others of their school, although it may have been observed by Bronislawa Fejgin.[169] It is only recently that Mr. Weaver in our laboratories has obtained, in the dissociation of a proteus culture on phenol agar, a new chromogenic form, differing markedly from both S (H) and O which, though not yet examined fully, may represent the R type of proteus. It is not until this form has been studied serologically in relation to the known H and O types that we shall have the full answer to the problem of proteus dissociation. The evidence supplied by Goyle on typhosus and enteritidis is strongly suggestive, but in such cases we are scarcely justified in drawing final conclusions from analogies. Proteus X19 itself must be further studied.

To the foregoing exposition a further word must be added regarding terms of reference to the proteus dissociates, in view of the fact that, as we now see, Weil and Felix introduced a new antigenic type, the parallel existence of which in enteritidis and typhoid has been demonstrated by Goyle. Since a part of the undesirable complication in the earlier study of Arkwright and Goyle was occasioned by their use of the symbol S (smooth) in reference to cultures grown on phenol agar, and which as I have intimated were undoubtedly not true S but transitional forms, it seems desirable to return to the precedent set earlier by Arkwright himself (and maintained by him in his section of the paper alluded to)—namely, to let S stand for the normal culture type. Furthermore, since Weil and Felix were the first clearly to point out the distinct serologic characteristics of the type which they termed the O, it seems appropriate to continue the use of this symbol in referring to cultures possessing the characteristics of the transitional form. In this group, however, we must expect a greater variability than in S or R because of the great instability of the intermediate type. In proteus Weil and Felix have already called attention to the serological differences existing between OX19 and OX2.

Vibrio Comma.—The serologic aspects of the dissociative reaction in relation to the cholera vibrio have been studied recently by Balteanu,[38] largely following the lines established by Weil and Felix for B. proteus and by Goyle for B. typhosus and B. enteritidis. The results concern partly the two culture types regarded by the author as analogous to the H and O forms of proteus—the former characterized by possession of the double antigen, HO, the latter by the single antigen, O. As has been noted, Weil and Felix extended their early study of the antigens of B. proteus to cover Vib. comma and these authors had in 1920 concluded that in this organism the heat-labile antigen (H) was absent, only the heat-stable (O) being present; that is to say, the heated assumed normal culture agglutinated only in fine granules, without evidence of the flocculent precipitates ("grob") characteristiic of the heat labile antigen. H. Brutsaert [76] also reported in 1924 that he could detect only the heat-stable antigen in the cultures that he examined. As Balteanu states, if the view of Weil and Felix, supported by Brutsaert, is correct, and Vib. comma does not contain the heat-labile antigen, then the case is different from that existing in B. proteus; and, to use Balteanu's own words, "the theory of the labile, flagellar and the stable, somatic antigen is clearly limited in its applicability."

Balteanu made use of two cholera immune serums: one, a Lister Institute polyvalent serum which possesses little interest for us because of the uncertain nature of the antigens by means of which it was produced; the other a monovalent serum immune to the socalled Pottevin strain of cholera vibrio, possessing more significance in our analysis of the reactions. Only the latter will therefore be further considered. One Pottevin serum was produced by the injection of antigen heated at 58 C. for 30 minutes; another by the same antigen heated at 100 C. for two hours. Both serums were employed on a living culture, and on a culture heated at 100 C. Both direct agglutination and cross-absorption tests were performed, and the essential results may be summarized as follows.

The receptor complex of V. comma manifests the same sort of differentiation as that observed in B. proteus and in cultures of the typhoid-paratyphoid group. To state the facts more specifically, there are present in cholera cultures two antigenic substances, heat-labile H and heat-stable O, corresponding to the "flagellar" and "somatic" antigens of Smith and Reagh in B. cholerae suis. The heat-stable component resists heating at 100 C. for two hours or more, while the heat-labile H is destroyed. The agglutination of the H antigen occurs in the form of loose floccules while the O antigen agglutinates in fine granules. Whole cultures (unheated) give a flocculo-granular or mixed precipitate, except at the upper titer limits where the flocculent precipitate is dominant. Balteanu stated, however, that the differences were not so sharp as in the case of B. proteus, and attributed this circumstance to the fact that a monotrichous organism like the cholera vibrio possesses relatively less flagellar substance than bacteria like B. typhosus and B. proteus.

The data presented above are sufficient to suggest the presence of the dissociative reaction in Balteanu's Pottevin strain of the cholera vibrio but confirmatory evidence is derived from his further study (l.c., part 2) of the serologic behavior of certain "variants" which he obtained from various cholera stocks. As I have already noted in other parts of this work, "variants" or "mutants" of the cholera vibrio have often been described, the modified culture usually manifesting itself by differences in morphology, colony form and biochemical reaction. According to Balteanu, Bordet [57] in 1896 reported inagglutinability of the cholera vibrio as a result of animal passage. Baerthlein,[27] moreover, in 1912, Eisenberg [150] in 1912 and Baerthlein [30] again in 1918 described such colony variants. Shousa [515] also in 1924, according to Balteanu, described an R type of the vibrio differing serologically from the normal form. Cantacuzène [90] in 1925 derived an inagglutinable variant of a vibrio obtained from the marine invertebrate, Phascolosoma, by maintaining a normal culture in contact with immunized tissue.

The work of Balteanu, however, affords more concise information regarding the antigenic structure of the cyclogenic variants occurring in various strains of this species. When old cultures in nutrient broth, peptone solution or on agar were plated this investigator observed among the "normal," round, translucent colonies three variants. These were as follows:

Circumvallate, Rugose Colonies: These were small and yellowish and possessed thickened borders; also radially arranged ridges. These colonies could not easily be suspended in salt solution or distilled water and were firmly adherent to the medium. In broth they formed on the surface a thick, wrinkled film which eventually fell to the bottom of the tube leaving the liquid clear. This type was not constant for, after several days, colonies on agar showed regeneration fringes from which, by subculture, the original, transparent form could be obtained. The same culture type was observed on old agar cultures after they had begun to become dry.

White Ring-Colonies: These were observed on old agar plates poured from old broth cultures of two of the strains (Pfeiffer and Kedah). "The colonies were whitish and semitranslucent. Sometimes they had an opaque center and more translucent margin, resembling the ring forms of Baerthlein." This culture type was inconstant and reverted to the normal on subculture.

Opaque Colonies: These were obtained from the Pottevin strain. They were usually "prominent, opaque colonies, sometimes with a slightly irregular surface. They appeared dense and white by transmitted light, were very firm in consistency and were adherent to the agar. The growth was extremely difficult to emulsify but by careful rubbing up with distilled water a stable suspension was eventually obtained. Such an emulsion was partially precipitated in salt solution (NaCl 0.85 per cent) after 24 hours at room temperature. Heating to 100 C. somewhat increased the stability in the presence of salt." The absence of any R receptor in this culture led Balteanu to believe it was not similar to the "rough" forms described by Arkwright in 1921. From this type there were sometimes obtained "intermediate colonies" which later reverted to the normal form. This "opaque variant" as a rule, however, was constant on agar, although reversion occurred after repeated subculture in broth. Plating cultures that were 10 to 20 days old yielded cultures with translucent margins; and further subculture from these gave the normal type. This was the only variant that was fairly stable on agar.

The morphology of the third variant revealed an organism somewhat smaller than the normal and surrounded by "a thick layer of pink staining material" . . . . "a sort of slimy exudate simulating a capsule." Sometimes the slimy covering enclosed two or more organisms in a common matrix. Motility in this culture type was absent and flagella were lacking. The range of acid agglutination (ascertained by following the technic of Beiniasch) was higher for the variant (5.5×10^{-4} to 4.4×10^{-8}), than for the original culture (1.38×10^{-4} to 1.1×10^{-8}).

In studying the serologic characteristics of this variant Balteanu prepared four immune serums: one against normal culture heated at 58 C. for 30 minutes; one against normal culture heated at 100 C. for two hours; one against the opaque variant heated at 58 C. and the fourth against the opaque variant heated at 100 C. All the agglutination tests were performed in a medium containing only 0.22% NaCl — in which the variant remained fairly stable. The heating at 100 C. was done in a Koch sterilizer.

The results of the serological tests, designed to reveal the serologic and antigenic relationships between the variant and the normal culture, were as follows: The opaque variant behaved like the O form of B. proteus; that is, only the heat-stable O antigen was found — no heat-labile H antigen was detected. The immune serum produced against the variant, however, contained some H agglutinins, thus demonstrating that some H antigen remained in the O culture. Conplement-fixation reactions showed that the O form was as efficient a receptor as the normal, as had also been demonstrated in the case of B. proteus.

One other highly interesting feature of the work of Balteanu was the demonstration that the immune serum derived from the heat-labile antigen H was heat-stable; while the immune serum obtained from the heat-stable antigen O was heat-labile. This important observation links for the first time serologic aspects of dissociation with the interesting observations made by Joos [272] in 1903, in which he called attention to the inverse relation existing, as regards heat-stability, in the so-called "α" and "β" agglutinogens of B. typhosus and their homologous agglutinins in immune serums. It thus transpires that Joos was working with cultures containing the "double" and the "single" antigen. This subject is treated in greater detail on another page of this work.

The splendid antigenic analysis of V. comma carried out by Balteanu was purely objective and he did not speculate regarding the significance either of his own or of similar findings in other bacterial species, although he was inclined to relate the differences between the H and O antigens to the presence or absence of flagella, and thus to bring his results into apparent relation with the "flagellar" and "somatic" antigenic hypothesis. Indeed he performed certain experiments with "flagellar antigen" which seemed to support this view. Although I have discussed the subject of the flagellar and somatic antigens at some length in connection with the studies of Smith and Reagh on B. cholerae suis, I may say once more at this point that I believe neither in the work of Balteanu nor in earlier investigations is there any real evidence that such a thing as flagellar antigens, as distinct from somatic antigens, actually exist. It is true that the centrifugal methods, as employed by Balteanu and others to secure "pure flagellar antigen," probably do represent a higher concentration of flagellar substance than is present in the "body substance" (centrifugate); but such "flagellar substance" has never to my knowledge been separated from the soluble or extractive substances also present in the same suspensions. I believe this may be a justifiable criticism of the work of Marion Orcutt [377] on flagellar antigens in 1924. It seems to me, therefore, that we are scarcely safe in regarding serological results that seem to indicate the presence of a heat-labile antigen as due to a flagellar fraction. These heat-labile bodies might equally be regarded as representing soluble substances from the bacteria themselves. Furthermore, we do know, particularly from Griffith's study of the relation between the specific soluble substances of the pneumococcus and the S type of that organism, that the S antigen is more likely to be present in this form of culture as a soluble product than in the heat-stable culture. Moreover, as indicated elsewhere in this work, the phenomenon of the "double" and "single" antigen is by no means limited to those bacterial species characterized by the possession of flagella.

Concluding the matter of the relation of microbic dissociation to the serologic reactions observed in V. cholerae, I believe that the subject has been made sufficiently clear through the facts already presented. The socalled "normal" type of V. comma is the S form (the "helle" of Eisenberg, and the H type of Balteanu) and carries the double antigen, just as in the case of B. proteus. It passes over during aging (B. did not force the dissociation by the use of phenol) to the O form, characterized by the single antigen—and perhaps in some cases to the R type. Around the O type are grouped intermediates or transitional forms, the existence of which Balteanu's descriptions make quite clear. Manifestly, however, this writer did not study, nor attempt to produce artificially, the well-stabilized R form of Vib. comma; at least an R form comparable with that obtained by Firtsch while working with the spirillum of Finkler-Prior in 1888. This possible culmination of his work, following the lines of antigenic analysis already laid down by Goyle, is much to be desired. Our knowledge of the antigenic structure of these forms demands study of the clearly recognized R types as well as of the S and the O cultures.

To the foregoing consideration of Vibrio cholerae it may be added that Shousha [515] in 1924 succeeded in recognizing what he believed to be the S and the R forms of culture and pointed out distinct cultural, biochemical and serological differences existing between these types. It is not established, however, that he worked with well stabilized type R cultures.

Conclusions.—In concluding this section it may be noted first that there exists a far-reaching truth in Baerthlein's statement that, among bacteria in general, colony differences are correlated, not only with striking biochemical differences, but also with fundamental serologic and antigenic disparities of considerable importance. And to this we may add that similar striking variations in antigenic constitution may occur even when the colonial and cultural features appear to be almost identical. These differences seem to involve in most cases a simplified antigenic structure of the R type (and also of the O type, when it appears) as compared with the more complex and variable antigenic structure of the S culture. In other words it is more commonly the O and R antigens that are found in apparent purity, while in the S type culture there is a combination—usually of S and O. On the other hand, the culture characterized morphologically as the R (rough) may contain not only R but also S antigen; and, according to White, some O as well. The nature of the various antigenic configurations produced by these basic antigens, either alone or in certain combinations, is therefore sometimes such as to warrant the use of the terms "double" and "single" antigen cultures as proposed by Bordet and Sleeswyk for B. pertussis in 1910 and as revived by Weil and Felix for their H and O strains of B. proteus X19, although in other cases "double" antigens may be present throughout. The data already presented, however, make it clear that there exist at least two possible combinations of the "double antigen" type: One is the combination of S with O (undoubtedly typified by the H type of Weil and Felix); the other is the combination of S with R, as observed in all those cases where R is undergoing what we have termed "reversion," but what, as we may eventually find, is not a reversion at all. In addition to these, White has demonstrated apparently the OR combination. It thus becomes an important and interesting question as to methods by which these types of double antigen culture may be differentiated. From the experimental data supplied by Schütze on typhoid, paratyphoid and enteritidis, by Arkwright and Goyle on typhoid, enteritidis and dysentery, by Goyle alone on enteritidis and typhoid and by Weil and Felix, Braun and Salomon, Braun and Schaeffer and a host of other workers on proteus, it becomes apparent that, in the configuration supplied both by the SO culture and by the SR culture, we are dealing with a combination between a heat labile antigen (S) and a heat stable antigen (O or R). It would therefore seem that the differentiation might be effected on the basis of the heat stable component; and here possibly, we can observe a difference. As first shown by Weil and Felix, and as confirmed by most of their followers, the O type of proteus X19 is characterized by the "fein" as contrasted with the "grob" flocculation. It is manifestly the "fein" element (O antigen) that gives the slightly granular appearance to flocculating "normal" culture. This point was confirmed by Arkwright and Goyle in fact though not in actual interpretation; and later by Goyle (in both fact and interpretation) for both typhoid and enteritidis. The heat stable R antigen, however, as shown by Arkwright and Goyle and by Goyle alone, also by some others, undergoes flocculent precipitation and in such cases, in combination with antigen S (as in the rough SR cultures, to be mentioned later), does not introduce the mixed, granular element. Such cultures, therefore, according to Goyle give only normal flocculation. In addition, we know that the rough culture type is characterized by spontaneous flocculation, in which it is likely to differ from the pure O. Outside of these means of detecting the nature of the double antigen cultures, SO and SR, there remains the possibility of plating, which will usually reveal the R type colony with its characteristic form and structure. The O colony form may sometimes differ little from the normal S, or it may be quite characteristic, as already noted in section 6 of this paper.

In all these distinctions we should bear in mind, however, Jordan's [274] recent report on the R type of B. paratyphosus B giving granular agglutinations. As stated earlier, the details of the O → R transformation are in need of further study.

The possible relation of these observations to the interesting studies of Joos [272] on the dual nature of the typhoid antigen, although not altogether clear, is sufficient to demand their mention. Joos postulated, on the basis of certain elaborate tests involving typhoid immune horse serum, produced by both heated and unheated antigens, that these antigens contained two different agglutinogens, α and β. The former, he believed, was present in living bacteria, but was easily destroyed by heating. The β agglutinogen was also found in living bacteria but was more resistant to heat. Injecting the living bacteria into animals, therefore, gave rise to α and β agglutinins, while the injection of heated culture gave rise to the β agglutinin only.

Thus far we can follow Joos with the analogy of the normal typhoid culture, containing its heat labile S (or Joos α antigen and its heat stable O antigen, producing upon injection the "double agglutinin" immune serum, while the O type culture (or Joos β) produces the "single agglutinin" serum. Beyond this point we lack data for adequate comparison. Joos goes on to state that the influence of heat on the agglutinins is the reverse of its influence on the agglutinogens. The α agglutinin is heat resistant while the β agglutinin loses its activity when heated. The α agglutinin, moreover, is wholly incapable of uniting with the β agglutinogen, while the β agglutinin can combine with either the α or β fractions of the antigen. These results of Joos were confirmed in a measure by both Kraus and Joachim [287] and by Scheller.[429] The second phase of his complicated study goes beyond any available data we possess on the relation of the S, O and R types to the "double antigen" reaction, except for the following instance. Balteanu [38] in his recent, excellent antigenic analysis of the cholera vibrio has shown the presence of the two types of antigen, H and O, the former being heat-labile, the latter heat-stable. When these antigens were employed for the immunization of rabbits, serums were obtained whose reactions have already been noted. But this observed fact is of significance. The serum immune to the heat-labile antigen was heat stable, while the serum immune to the heat-stable antigen was heat-labile. It is thus suggested from these limited data that the serologic system with which Joos worked involved the presence of the H and O antigens, if not the R. This problem demands further study.

That the issues involved in these serologic disparities between different subcultures from the same pure-line strains possess great significance for pathology and for serologic diagnosis, as first pointed out years ago by Bordet and Sleeswyk [64] for B. pertussis, by Bernhardt [46] for B. diphtheriae and B. typhosus, and as again called attention to more recently by Mellon [325], cannot easily be denied. For the evidence reported makes it clear that the antigenic relationship between the S, the O and the R forms arising from the same pure line may often be less marked than the relation shown between some of these types and unrelated bacterial species, when judged by agglutination, precipitation or complement-fixation tests and results. As indicated on a foregoing page, Schütze [436] was perhaps the first to point out this circumstance. He showed that serums immune to certain Salmonella strains agglutinated, not only the R form of quite different strains of Salmonella, but also the R forms of B. coli and B. dysenteriae. That these cross agglutinations between different species was due to the heat stable antigen mainly was shown by White [487] in 1925. Moreover, in his positive cross agglutination tests between B. typhosus and B. enteritidis, Goyle [18] showed in 1926 that the reaction may occur with S, O or R forms of culture; but that it ordinarily concerns the heat-stable antigens only, and therefore shows most strongly in the O and R types; indeed he concludes that the heat-stable R antigen is common to both bacterial species. These results seem to go far in explaining not only Schütze's cases of "serological cosmopolitanism" of the rough forms, but also other cases of bacterial convergence already noted. They moreover give us a clue to the assumed enteritidis-typhoid transformations of Sobernheim and Seligmann and of van Loghem; to the assumed paratyphoid-typhoid transformations of Baerthlein and of Bachmann and de la Barrera; to the colon-typhoid mutation of Malvoz; to the convergence of the gonococcus strains observed by Torrey and Buckell; to Esther Stearn's case of convergence of water-borne bacteria under the influence of gentian violet; to the serologic heterogeneity of Reimann's and Stryker's pneumococcus R types, as well as to the pneumococcus-streptococcus serologic affiliations described by Yoshioka, the changes noted by Atkin in the old stock strains of Gordon's meningococci, and the pneumobacillus-colon-aerogenes serological affiliations reported by Julianelle.[505, 506] Although Reimann did not specifically refer to the matter in his paper on pneumococcus dissociation, the agglutinative results which he presents for the R type antigen against standard type 1, 2 and 3 pneumococcus serums afford, so far as they go, an interesting confirmation of this convergence taking place within the species. In all these instances, one of the most interesting features thus concerns, not only the loss of type specificity through dissociation, but also the gain of a sort of antigenic heterogeneity in which the antigenic structure of fairly diverse species comes into apparent relation. In this connection, it may be of interest, in passing, to recall the bacteriophagic cosmopolitanism in which a similar diversified action is observable. It might make an interesting study to ascertain to what extent, if at all, the distribution of bacteriophagic action parallels the "serologic cosmopolitanism" in the case of some of the intestinal bacterial types. In another paper [227] I have briefly touched upon this point.

All of these observations combine, moreover, in suggesting to us that the S → O → R transformation is not always so simple a thing as the continuous loss of antigenic substance, as some writers would have us believe in the S → R transformation. It involves also the gain of new antigenic substance to replace the old. Here, there exist problems of much serological interest. But, whatever may come out of it, the diversity of the reactions observed in relation to microbic dissociation can leave us only with a stronger impression of the inadequacy of flocculation and absorption tests, as usually performed, as a means of differential bacteriologic diagnosis—at least aside from a few well known instances in which their worth has been, in a measure at least, proved. I believe that many of the discrepancies in serologic reactions, as at present employed for diagnosis, will find their solution in the results of the dissociative reaction, operative either in the cultures employed for immunization, or in the test cultures themselves.

It is also apparent, however, that these discrepancies will not be fully explained until we embark on a new trend of study in our search for the meaning of the varied serologic types of this and that pathogenic species. The modern business of collecting new serologic types, applying one or another test, and then constructing artificial pigeon-holes for their safe-keeping, really gets us nowhere; it has not advanced in any appreciable degree our knowledge of the intimate nature of these forms. It seems that what we need most at the present moment is to learn how to create these various types. Then we shall know both their ancestry and their progeny, and all their interrelations. Then, indeed, we may come to understand their position in what we may sometime recognize as the *antigenic cycle*. Then, and then only, I believe, shall we be in a position to take full advantage of what such new knowledge may make possible in the development of more effective methods of serologic diagnosis, and perhaps of immunologic procedure.

10. DISSOCIATION AND IMMUNOLOGIC RESPONSE

Closely related to the serologic and antigenic reactions of the S, the O and the R types is the question of immunologic response which it may be possible to elicit by the injection of the O or the R type culture, as opposed to the "normal" or S type which has undoubtedly been employed in the greater part of immunization practice. Indeed, the question of the relation of the culture type has seldom arisen, except perhaps in the case of anthrax and a few other infections in which an attenuated virus has been employed in prophylaxis. We have already considered the relation between attenuation and the R type with reference to several pathogenic species; and we have seen that, in many of them at least, the type R culture was non-virulent or only slightly virulent. If the quality of virulence is localized in one or more of the distinct culture types it might seem possible that the immunizing power would also be so restricted to a certain culture form. Bearing on the specific immunizing value of inoculation with clearly recognized type R culture, the first evidence of significance was supplied by the work of de Kruif [118-121] in 1921-22.

The Pasteurella Types.—De Kruif [118] had already demonstrated the virulent character of his type D (S) of Bact. lepisepticum and also the harmless nature of his type G (R) culture (section 8). Two rabbits had been given injections intrapleurally 14 days previously with 0.1 and 0.5 cc. respectively of type G culture, and had survived the inoculation. These animals, along with three controls, then received injections with one cc. of 10^{-4} to 10^{-6} dilutions of type D culture. The principals survived the infection while the controls died in four to nine days.

As de Kruif pointed out, it seemed remarkable that such solid immunity should have been produced by a single, small injection of the type G culture. But this calls to mind another instance. Many years ago I [222, 223] was able to demonstrate for a strain (52) of the organism of fowl cholera (B. avisepticus), which is closely related to, or perhaps identical with, de Kruif's organism, a similar remarkable immunizing power. Less than 1×10^{-8} cc. given subcutaneously to rabbits produced a local abscess. Even 3.0 cc. of this culture was easily tolerated. Animals so treated were immune to at least 2.0 cc. of a virulent culture (48) of which 1×10^{-8} cc. was fatal for unprotected rabbits within 14 to 36 hours. Culture 48, when killed and used as a vaccine, showed slight immunizing ability. The immunizing power of culture 52 was unique, but its mode of action was never explained.

The results obtained by de Kruif have led me to reexamine this culture which has been maintained on plain infusion agar for the 12 years or more since the tests mentioned. I find many of the cultural features now characteristic of the R type as outlined by de Kruif for Bact. lepisepticum type R. That de Kruif has given the true explanation for my earlier results seems highly probable.

As remarkable as were de Kruif's results in producing a solid immunity against fatal infection with virulent cultures of the bacillus of rabbit septicemia by the injection of living cultures of the avirulent R type, they were not new in the field of Pasteurella infections. In 1919, in Budapest, Manniger [317] obtained somewhat similar, though less striking, results by the injection in the living state of what we may now recognize as the R form of the closely related B. avisepticus. The origin of this avirulent culture has been described in the section dealing with "dissociation and virulence" and does not need to be repeated here, since we are now concerned purely with the immunologic results. Suffice it to say that the original culture form (S) was highly virulent for hens, pigeons, rabbits and mice. The "mutant" was practically lacking in virulence and produced death only when injected in tremendous amounts, as for instance, a complete agar slant in suspension (larger animals). When gray mice were injected once or twice with one-half to one oese of the nonvirulent form they were able to resist 0.000,02 oese of the virulent culture, which circumstance actually represented a marked protection, considering the high virulence of the normal culture. When pigeons were injected once or twice intramuscularly with 2 oese, protection resulted; 0.5 to 1.0 oese, however, did not protect. When hens, six to eight weeks of age, were given 1.0 to 2.0 oese considerable resistance resulted. With the immunization of rabbits Manniger was not so successful. But it is to be noted that his protective doses were administered intraperitoneally, and I was able to show many years ago that in immunization with an avirulent strain of the same organism the intraperitoneal route was of slighter value. A distinct reaction (abscess and drainage) was necessary before immunity could be produced.

Manniger manifestly worked with a culture of the R type—or at least one in the course of transformation toward the R form. It was manifestly not the extreme R, since reversion to the virulent type seems to have been easily possible. When taken into consideration with the results obtained by de Kruif, and with my own earlier results, the record is of much interest since it reveals the fact that the immunizing power of the R type of Pasteurella may not be an unusual occurrence, but perhaps the general rule. These results naturally raise the question as to the possible immunizing power of R type cultures of the closely related human form of Pasteurella, B. pestis. So far as I am aware this question has never been studied. It is of interest, however, that by the use of "bacteriophage suspensions" d'Herelle [246] has been able to immunize with much success against both barbone (Pasteurella bovis) and human plague (B. pestis). It seems clear, from d'Herelle's analysis of the mechanism of the protective reactions here involved, that the bacteriophage is not directly concerned, but that the immunity is due to the generation of protective antibodies possessing opsonic significance. The question arises regarding the extent to which the remarkable results reported may be due to the R type antigen present in the bacteriophage suspensions, i. e., the Pasteurella culture after lysis.

Streptococcus.—Further interest in the relation between dissociation and immunologic response attaches to Cowan's [112,113] work with the streptococcus. Among other points she studied the immunizing value of the less virulent R type when injected into mice and demonstrated that definite resistance was produced against virulent S cultures when subsequently administered by another route. When the R injection was given intraperitoneally it tended to produce abscesses which healed by fibrosis. There seemed to exist a correlation between the formation of these abscesses and the production of immunity. The results obtained by Cowan, however, do not reveal such definite protection as that observed by de Kruif in resistance to Bact. lepisepticum. In the first place the R inoculations of Cowan were themselves sometimes fatal, especially when larger amounts of culture were injected. Moreover, the protective value of R streptococcus was not manifested in all cases. Cross protection (i. e., immunization with an R form of one culture and infection with the S form of another culture) gave some evidences of success, but the number of cases are too few to serve as a basis for final conclusions. Discussing the "local" versus the "general" nature of the protection afforded by inoculation of R culture, Cowan concluded that it was not of the former type. In some cases heat-killed R culture, as well as S, produced resistance against living S culture.

Pneumococcus.—Regarding the immunologic significance of the pneumococcus dissociates little can be said at present. The only workers to consider the matter in relation to the recognized types are Griffith and Amoss, and neither have dealt with the subject of active immunity as was done by de Kruif and Mary Cowan. Griffith [215] immunized rabbits by many repeated injections of his types S and R, some of the rabbits receiving as many as 26 injections of the latter culture. Protection tests with the resulting serums were performed on mice infected with 0.1 cc. of the virulent type 1 S strain whose MLD was 10^{-8} cc. When injections of 0.2 cc. of S serum were administered with the infecting culture protection resulted. Under the same conditions of infection, however, the R antiserum showed no protective power. No tests involving the production of active immunity by injection of R cultures (which showed so strikingly in de Kruif's case with Bact. lepisepticum) were reported.

Amoss [8] studied the protective value of rabbit S (Amoss' C) and R (Amoss' Z) antiserums for mice infected simultaneously with type S culture. The serums were produced by injection of heat-killed antigens. Two-tenths cc. of serum was mixed with varying amounts of the virulent culture and injected intraperitoneally. Of 15 mice which received each type of serum, one died in the group protected with S serum, while all died in the group receiving R serum. All who have studied the point have reported the greater difficulty in building up a strong antiserum against type R. No report is given of attempts to infect mice which have tolerated inoculations with type R culture, so that we are unable as yet to compare results of pneumococcus and streptococcus immunology.

Salmonella.—White [487] has also made some tests of the immunizing power of R type cultures of Salmonella and observed that the injection of rabbits with the "roughest of the rough strains" (showing no flocculation and no absorptive action on flocculation) gave perfect immunization against the more virulent, smooth strains. A rabbit was immunized with the rough "Kral" strain and was given a final immunizing dose of the living organism. Twenty days later the animal was given one-fourth of an agar slant culture of "Sweet" (smooth), while a control animal received one-fortieth of a smooth slant. The control died in two days while the rabbit immunized with R was unaffected. White thus concluded that "the total loss of flocculating power during roughening does not appear to influence the immunizing value of the strain." But, since White had already shown that "the destruction of flocculating antigen by heat coincides with almost total disappearance of immunizing activity," he concluded that the true immunizing antigens were "thermolabile bodies, quite distinct from the agglutinin-stimulation antigens. . . ." White also called attention to the point that, not only for the R types of Salmonella but for Shiga and Flexner strains as well, the lack of flocculating power in an antigen (White's R antigen gave granular agglutinations) is associated with slight power to invade the tissues, and therefore concluded that "the antigens which flocculate are essentially substances connected with invasive activity, and that the corresponding agglutinins—probably agglutinins as a whole—are essentially anti-invasive agents." As we may see later, this is no doubt partly true, but the situation will be found to involve still other factors.

The Anthrax Vaccines.—A consideration of the relation of dissociation, and particularly of the R type of culture, to immunologic response would not be complete without mention of the Pasteurian method of vaccination against anthrax, which appears to involve this phenomenon. As is well known, Pasteur [383] found it possible to weaken progressively the virulence of active anthrax cultures by growing them at temperatures above optimum (42 C.) for 15 to 20 days (Vaccin I), or for 10 to 12 days (Vaccin II). Chamberland and Roux [94] produced similar attenuation by potassium bichromate, Chauveau by increasing the atmospheric pressure and Arloing by direct sunlight. Pasteur's cultures lost largely the power of spore formation and virulence and produced some degree of resistance when inoculated into animals. Regarding the actual nature of the modification produced in Pasteur's cultures there exist few original data on which to base an opinion since he leaves us with slight record of the colony characteristics of his cultures. We do know, however, from definite experiments performed in subsequent times, and particularly from the work of Preisz,[394] that one effect of growing B. anthracis at high temperatures is such as to produce two new forms of culture—the "white form" (S ?) and the slimy transitional (O?) which at times seems to exist in great abundance. We know, moreover, that these culture types possess characteristics quite different from the original, virulent form. The slimy, transitional type carries a reduced virulence (which, however, easily regenerates) while the "white" culture is often without virulence. Although it can scarcely be doubted that the immunizing value of Pasteur's vaccines is in some way connected with these modified culture types, strange as it may seem, there appear to be no records as to the behavior of pure cultures of these widely different forms in immunologic or serologic reactions. It seems probable, however, that the Pasteur vaccines are virulent cultures which have been partially attenuated by heat-modification in the direction of the S form; and that the degree of this modification is greater in the case of Vaccin I than in Vaccin II. The ultimate understanding of these matters can be reached, however, only when more definite experiments have been concluded making use of the clearly recognized forms (R, S, and transitional) of the anthrax bacillus produced in controlled laboratory dissociations. Mr. Nungester in our own laboratories has definitely isolated these chief types by means of the normal dissociative reaction and colony selection and has studied them sufficiently to validate many of the observations of peculiar colony variations presented in earlier pages of this paper.

The Cholera Vibrio.—Balteanu [38] in his excellent study of the H and O antigens of this organism isolated a socalled opaque variant, which was probably the intermediate or O form; also a rugose variant, probably the R. Although the latter was not studied fully, the normal form was contrasted with the opaque variant with reference to immunizing power. The actual tests were, however, somewhat limited, representing only a comparison of the cross reactions. Twelve guinea-pigs were given three successive doses. Six received culture of the normal form and six received culture of the opaque variant. All were tested 10 days after the last dose. The six receiving the normal form were tested with the opaque variant, while the six receiving preparatory doses with the opaque variant were tested with the normal culture. In the words of the author, "All lived except one, which had been vaccinated with the original form and this one died 2 days after the test dose of the opaque variant." Unprepared controls injected with one-half the test dose of the two forms respectively died in 10 to 12 hours. The author's conclusion was that the opaque variant gave "as good protection as the normal." He had previously shown that the opaque variant was "rather less virulent." Unfortunately the immunizing power of the form in which we are at present particularly interested, namely the R, was not tested. An interesting field of study thus lies open in respect to the immunologic reactions of the cholera dissociates.

Nature of the Immunologic Response.—To the above one further point may be added regarding the possible nature of the immunologic response. In numerous instances as will be shown, (section 11, Immune Serum), contrary to the mass of textbook information on the subject, one influence of homologous immune serum upon normal, virulent culture is to accelerate a transformation into the commonly less virulent and more easily phagocytable R type culture. Such observations, involving largely experiments in vitro, but also many somewhat more obscure observations of reactions in vivo, might appear to bring into view a new immunologic principle relative to one aspect of the body defense against virulent, invading bacteria, involving the opsonins and their actual mechanism in promoting phagocytosis. The matter may be stated briefly as follows:

We know that normal serum, unless perchance germicidal for the bacterial species in question, usually has the ability to hold the S form of culture up to type; and perhaps, in some cases, to enforce the R → S reversion. We also know that homologous immune serum in suitable strength has, on the contrary, the ability (in vitro, and to some extent certainly in vivo) to effect a transformation from S to R. We know, moreover, that immune serum contains antibodies in the form of opsonins or bacteriotropins and that these often have the marked ability of facilitating phagocytosis both in vitro and in vivo, by reason of a still unexplained influence on the bacteria.*

Finally, we know that, in certain cases at least (pneumococcus, Pasteurella, pneumobacillus, for example), the R type of cell is easily and quickly phagocytosed in vivo while the S type is not.

The interesting question therefore arises: Are the immune bodies that we call opsonins and bacteriotropins agents by whose action the transformation in type of the infecting organism in immune serum (in vivo) is accomplished? In other words, is the "preparing action" (on the bacteria), commonly attributed to the opsonins, merely the enforcing upon the virulent forms of a dissociation by virtue of which reaction they become amenable to phagocytosis? I may say in conclusion that many details of the opsonic theory of immunity (which is surely convincing in its broader aspects) are not so clearly established but that renewed study might be profitably undertaken along the lines of this conception. In such a case we might be led to consider with greater interest than heretofore the obscure "antiblastic theory" of immunity clearly suggested by Charrin and Roger [96] in 1892 and elaborated by Ascoli [20] in 1908 in opposition to Bail's [31] aggressin theory. But in this connection, if the transformation from S to R can be demonstrated to occur in vivo as readily as in vitro, then these two at present opposed

* Bull [89] has expressed the view that the agglutinins play a part in the protective reaction.

theories will be brought into unity through their mutual relation to the dissociative reaction. I believe that a conception of immunity based upon such a unification of the antiblastic and aggressin hypotheses, and embracing our present notions of the trophic antibodies, may yield the most effective basis for new lines of immunological study lying outside of the toxin-antitoxin reactions.

Finally I wish to call attention to one further point of immunological significance dealing with active and passive immunity in two species in whose mode of action in infections there are many similarities—Bact. lepisepticum and the pneumococcus. De Kruif [119] has pointed out for the R type of the former (as I also noted [223] many years ago for an unrecognized type of the Pasteurella, B. avisepticus*) a most remarkable immunizing power through the agency of living R type cultures. He did not test the protective value of the serum immune to the avirulent R type. On the other hand, Griffith, and after him Amoss, showed the high degree of protection resulting from the use of serum immune to the S type of pneumococcus and the absence of protective power in the R immune serum. They did not test the immunizing action of injections with living R type culture. In view, however, of the results of de Kruif (and it may be added of Cowan on streptococcus immunity, and of P. B. White on immunity from an R type Salmonella) the important question arises: In the case of the pneumococcus can a durable immunity be produced by the injection of living R type cultures, comparable in effectiveness with the passive immunity afforded by the injection of S immune serum? Notwithstanding the important results already reported on pneumococcus dissociation, these immunologic aspects of the problem have been omitted by Stryker and Griffith as well as by Reimann and Amoss. Also in the case of Bact. lepisepticum we still await information regarding the presumable lack of protective power in the serum immune to the R type. But in all the results that are thus far clear we find the inherent suggestion that there may exist in the body a type of immunity whose potency is not appreciably manifested by antibody phenomena, and which is not transferable by means of socalled immune serums. It may be that interesting results

* In 1914 I wrote (l. c., p. 401) the following: "May we regard this as a hint that in other pathogenic species there exist, among the many avirulent strains, one or two perhaps that may be characterized as immunizing cultures? . . . One might inquire whether, among the many avirulent cultures of the pneumococcus . . . there exist strains which, by some form of inoculation or through some strictly local reaction, would call forth the immune response to more virulent culture material. . . . In this disease . . . might it not be worth while to make a systematic study of the reciprocal relations existing between various virulent and avirulent types?"

will come from a further study of the immunizing properties of certain R type cultures of pathogenic bacterial species.*

Conclusion.—In the section dealing with the relation of dissociation to virulence it has been shown that in one and the same bacterial culture there may exist, side by side, cells or colonies, some of which represent organisms of the highest virulence while others represent organisms possessing no virulence whatever. From the data presented in this section it begins to appear in a similar manner that the diversity of biologic activity manifested by different cells in the same culture may be carried even further. In one and the same bacterial culture there may exist side by side cells or colonies some of which represent organisms that are of no or of slight value for immunizing purposes; others which represent organisms of remarkable immunizing power but destitute of virulence. In 1914 I [223] pointed out that certain cultures were not necessarily of immunizing value merely because they were virulent or otherwise "typical." Apparently this conception is not limited to different cultures occurring independently in nature, as I then assumed, but may even concern different organisms in the same culture, and even in the same pure-line. I believe that these later observations open a field of considerable significance for immunological theory and practice. There now may certainly be said to exist, not only in nature but also at times in the laboratory, cultures of pathogenic bacteria that may properly be spoken of as "immunizing cultures," because they possess this power far in excess of other cultures of the same species. They are not ordinarily, I believe, to be found associated with acute disease; nor are they to be found commonly in pure form in laboratory stocks. But the point of special interest is that such cultures can be produced artificially by resort to appropriate methods of cultivation. Some of these methods are already known and others will undoubtedly be discovered. It is needless to say that they may not be the same for all bacterial species (section 11).

As to the extent of applicability of this conception—that must be left for the future to ascertain. At present we can safely say, however, that the fact is established for Bact. lepisepticum and B. avisepticus

* In quite recent studies on the immunologic aspects of the dissociative reaction in Friedländer's bacillus (pneumobacillus) Julianelle [505, 506] has reported some facts of interest. Serum immune to the capsulated S type (and therefore containing the type specific immune bodies) was found to protect white mice against the virulent S culture. Serum immune to the noncapsulated R type (and therefore lacking the type specific immune bodies) had no protective influence in 0.2 cc. amounts. This immune serum likewise failed to agglutinate the capsulated cells or to precipitate the soluble specific substance. In a similar manner serum immune to the R nucleoprotein (acetic acid precipitable material) failed to protect when given in similar amounts. The protective value of immunizations by means of the R cultures was not reported, and therefore this question still remains open for the pneumococcus.

among the Pasteurella, and will probably be shown for B. pestis, another member of the same group. Also, though to a lesser degree, for certain strains of the hemolytic and the greening streptococci; and probably for Salmonella. The situation with reference to the pneumococcus remains to be ascertained. In this case and in many others the artificial production of cyclogenic variants, and a careful comparison of the relative immunizing power of these dissociates, is at present one of the most important fields of investigation in vaccine therapy. It may not be found that it is always the R type of culture that is of chief immunizing significance (and the situation may not be found the same with toxic forms as with those characterized by marked aggressiveness) any more than it may always be observed that it is the S type culture that carries virulence. Different bacterial species may vary in this regard. The important fact to bear in mind, however, is that we are coming to regard not only virulence, but also immunizing power, as a quality not distributed equally through the elements of the culture, but localized in certain cells existing in a definite, but doubtlessly transitory, cyclogenic stage.

11. THE INCITANTS TO ACTIVE MICROBIC DISSOCIATION

Of the cases of microbic dissociation thus far mentioned perhaps the majority have been observed as occurring "spontaneously" or have been the incidental or accidental result of endeavors having another purpose in view. In the majority of instances, moreover, the striking phenomena observed have not been recognized at the time as possessing significance other than that they were "mutations," or that they demonstrated the possibility of adaptation to a new environment, or that they manifested the nature of "impressed variations." Major interest has nearly always centered in the end result of the process and seldom in the nature of the process or reaction itself. Indeed, if any investigator has considered the process at all, it has usually been in terms of some simple sort of mechanism by virtue of which certain pre-formed bacteria, better fitted to survive, did survive; while others, unfitted, perished. It was as simple a reaction as the "survival of the fittest"—and one possessing about the same general significance. For this reason microbic dissociation has not been studied as a significant, independent biologic phenomenon and few experimental attempts have been made, either to prevent its appearance or to enforce it when it might be of service; moreover parallel dissociative trends among bacteria at large have seldom been noted.

The earliest methods consciously employed for securing that sort of microbic variation now termed dissociation involved the use of old broth or agar cultures, or old gelatin cultures as employed by Firtsch[178] in 1888, and emphasized anew by Baerthlein[30] in his study of colony variation and correlated features in 1918. Baerthlein plated on agar, incubated for 24 hours, then permitted the plates to stand at room temperature for some days or weeks. Similar methods had been employed by Preisz for B. anthracis, by Feiler for B. proteus, by Eisenberg for many species and by Penfold and others for intestinal bacteria. Under these conditions daughter colonies of a new type often arose within, or at the margin of, the old colonies. These new colonies represented the "mutant" type and could be isolated in pure culture. In other cases Baerthlein merely plated old broth cultures and obtained directly two or more colony types with their characteristic culture growths. Arkwright, Weil, Felix, Eisenberg, Wreschner, Manniger and many others have also employed this method.

Although in the case of many bacterial species one merely needs to be on the watch for dissociation to make its appearance in old cultures on solid mediums, or in old broth tubes, there can be no doubt that dissociations similar to those described in previous pages can be produced in the laboratory easily and quickly as soon as we have discovered the critical environmental conditions required to incite the process; or to develop it to macroscopically observable proportions.* At least I believe we are justified in searching first for the "critical environment," rather than for things more intangible and mysterious.

That the critical environment or condition will be the same for all species of bacteria seems unlikely, although that all conditions might impress the physiologic mechanism of the organisms in a somewhat similar manner can scarcely be doubted. Different substances are required for growth of different bacteria, but they all grow; different substances may possess different lethal or bacteriostatic effects for different bacteria, but death or bacteriostasis results in all cases. Similarly, I believe, we may be justified in assuming that different incitants to dissociation may be required for different bacterial species, although the same one might be sufficiently effective for several members of the same bacterial group, and perhaps for members of some other groups as well.

* For reasons which will subsequently appear, I believe it doubtful that, in "enforcing dissociation" on a culture, we impress upon it any reaction that is foreign to its established physiology. "Enforcing a dissociation" is more probably bringing into the foreground, and making observable macroscopically, a type of physiological behavior already inherent, and presumably obscurely operative, in the culture mass. We must recall that normal, smooth cultures always contain some O antigen.

With the exceptions to be mentioned later, we have little concise information regarding the influence of various incitants to dissociation. On the strength of numerous incidental observations, however, it is suggested that the category of incitants may include such diverse stimuli as operate through temperature, food substances, physical state of the medium, volume of the medium, oxygen tension, desiccation, antiseptics, foreign protein, products of bacterial metabolism, body secretions, normal serum, specific antibodies, products of growth (including microbic associations); and to these may perhaps be added the influence of certain colony types. Certain of the more conspicuous instances in which some of these incitants to dissociation are found to be operative will be reviewed in the following pages.

Temperature.—That temperatures higher than the optimum are able to stimulate variation in cultures has long been known; and practical use has often been made of the fact. It was observed clearly in pigment production (Laurent [297]) although here the phenomenon may concern only fluctuating variations. Also in fermentative reactions (Wilson,[488] Coplans, Adami, according to Gurney-Dixon [220]) and in virulence, as early exemplified in the first immunization experiments of Pasteur with fowl cholera and anthrax, and in many other cases in more recent years. In Pasteur's work with anthrax protection it scarcely need be said that his cultures, modified by growth at 42.5 C., were relatively nonvirulent in proper doses and were ordinarily nonsporogenic. This was demonstrated with special clearness by the studies of Pasteur, Chamberland and Roux [383] in 1881. Bail [32] later demonstrated the effect of heat in modifying the abilty to form capsules and the closely related character of virulence. Preisz [394] showed the effect of heat in causing the development of peculiar mucoid colonies of the intermediate or transitional type between the S and the R forms. The studies of Katzu,[276] also on anthrax, have again quite recently revealed the influence of high temperatures in producing peculiar secondary colonies. In all of these instances we can recognize the modification of the culture toward a "mutant" form of growth.

The transformatory effect of temperatures higher than normal on virulent clinical types of the diphtheria bacillus was first recorded by Roux and Yersin [416] in 1890. The results clearly dealt with the production of O and R forms which were nonvirulent. Cultivation of virulent cultures at 39.5 to 40 C. yielded nonvirulent cultures resembling the pseudodiphtheria bacillus and characterized by different colony formation, modified growth in broth, changed cell morphology and fermentation features. There can be little doubt that Roux and Yersin produced the dissociation of the diphtheria bacillus by heat. Hewlett and Knight [251] manifestly accomplished the same result seven years later by heating their virulent cultures at 45 C. for 17 hours. With reference to the colon-typhoid-dysentery group, it is scarcely necessary to mention the long list of workers who, beginning with Rodet in 1894, have observed heat derived variations, some of which were manifestly what we now regard as the R type. Further, we have the dissociation of B. proteus X19 from the H into the O form under the influence of heat (42 C.) as shown by Weil and Felix, Hirschfeld and Zajdel [252] and others. Similar results were obtained by Braun and Schaeffer [66] by the use of phenol agar and by "starvation" agar. The derived O forms were apparently the same serologically, and were similar to the "natural" O type of Weil and Felix, although further study of this subject is needed. Hirschfeld and Zajdel, also Sachs and Schlossberger,[420] studied the same problem in B. typhosus, B. paratyphosus B, and B. dysenteriae.

Heat has also been employed for the purpose of producing serological variations in pneumococcus and streptococcus. In 1922 Yoshioka [494] reported obtaining from virulent pneumococcus cultures of the three standard types serological and cultural variants as a result of growing in successive passages at a temperature of 39 C. Similar results were obtained with Aronson's virulent streptococcus. When we compare the sort of variants thus obtained with those produced by Griffith in 1923 and by Reimann and Amoss in 1925 under the influence of immune serum, it is clear that Yoshioka's variants represented the R types, or transitional forms moving in this direction of variation.

In the section dealing with "dissociation and virulence" I have mentioned the curious instance reported by Reddish [402] from Rettger's laboratory dealing with the socalled B. sporogenes which was believed by Reddish to be a necessary contaminant in nearly all of his botulinus cultures from various sources. I have also suggested that this form might have been an R type of the normal B. botulinus, produced by the action of heat as commonly employed by American workers in preparing the cultural material for dilution and plating in the process of isolation. If later work should demonstrate this to be the case, here also we would have another instance of the influence of heating on the production of dissociation in an anaerobic species.

Quite recently Soule [450] has shown the influence of temperatures higher than optimal in forcing the dissociation of B. subtilis. Whereas in broth at 26 and 28 C. the reaction occurred slowly, raising the temperature to 37 C. greatly accelerated the reaction; and when the broth cultures were grown at 45 C. a large percentage of dissociation occurred within a few days. When agar slant cultures, which ordinarily showed slight or no dissociation in the earlier hours of growth at room temperature, were maintained at 37 C. dissociation progressed rapidly.

Food Substances.—In this connection the work of Reiner Müller, Thaysen and of Penfold are of special interest. Penfold[386] in 1912 reported a "mutational" change in colonies of bacteria of the intestinal group characterized by the formation of "papillated colonies" within the mother colonies after a period of five to nine days when the organisms were grown on agar containing various sugars; or, in one instance, sodium mono-chlor-acetate. The daughter colonies, similar to those seen earlier by Neisser, Massini, R. Müller and others, were regarded as mutants, although the process was not spoken of as dissociation. Perhaps the clearest case among those reported by Penfold is that of B. typhosus seeded on isodulcite agar. Here the daughter colonies appeared red, giving indication of fermenting the isodulcite while the original culture was unable to do so. It may be added that these results were obtained with cultures arising from single cell isolations. The new form was spoken of as "an isodulcite mutant" and occurred uniformly in the case of all typhoid strains studied. Reiner Müller[357] had already shown in 1909 a similar mutational capacity on isodulcite, and in addition that B. coli colonies did not respond in this manner.

Similar results were obtained by Penfold by growing B. typhosus on lactose agar plates containing neutral red. He was able to demonstrate also "raffinose mutations" in the case of B. paratyphosus B, but not in the Aertryckè bacillus. This difference Penfold regarded as the only cultural distinction and one worthy of use as a common test. In addition he demonstrated the influence of sodium mono-chlor-acetate in causing many intestinal bacteria to form papillated colonies. The "mutants" in this case were more resistant to the medium, but were in turn able to give rise to other papillated colonies when the strength of the sodium acetate was further increased. This reaction was observed in B. coli, B. enteritidis, B. paratyphosus A and B, and in B. grundthal, but not in B. typhosus. Six "sucrose mutants" were mentioned by Thaysen[460] in 1911.

Eisenberg[140] has also shown the readiness with which the anthrax bacillus could be transformed from the spore to the sporeless state by growth on glycerol agar. This result Eisenberg attributed to the acidity produced from glycerol fermentation, although others assigned it to the special germicidal influence of glycerin. If acid substances stimulate this change, the case is at variance with most instances in which it appears that dissociation occurs most readily in an alkaline medium. Glucose mediums produced similar though less uniform results. De Kruif [120] has reported that high concentrations of peptone favored dissociation in Bact. lepisepticum.

Wilson [488] in 1906 obtained, through the addition of urea to culture mediums, results in culture modification which we may regard as incipient dissociation. From B. typhosus, B. coli, B. pyocyaneus, B. enteritidis and Friedländer's pneumobacillus he thus produced filamentous and leptothrix forms. Adami, Abbott and Nicholson [1] observed similar effects from the addition of saliva to medium in which B. coli was grown. According to Gurney-Dixon [220] Connal [108] has shown that the spinal fluid, in cases of spinal meningitis, may contain as much as 0.5 % of urea, and it is thus interesting to consider the relation of such an urea content upon the dissociation of the meningococcus in vitro, as also of the influenza bacillus in the spinal fluid as reported by Ritchie [406] in 1910 in cases examined by him.

Under the heading of the influence of food substances (although the circumstance probably involves osmotic and other influences) the study of Matzuschita [321] in 1900 on the effect of high salt content may be mentioned. Matzuschita grew various species on agar medium containing from 5 to 10% of sodium chloride. The organisms are reported to have assumed unusual forms and to have given modified cultural features. Cocci gave off rod or filament forms. In addition, many rod forms produced, besides long filaments, simple cocci and giant coccoid bodies such as have been described numerous times in cultures manifestly undergoing the early stages of the dissociative reaction.

In concluding this subject a word may be added regarding the effect of sewage in producing dissociative changes in many species of bacteria. According to Wilson (1910) (See Gurney-Dixon [220]) Almquist noted that the effect of sewage on B. typhosus was the production of leptothrix forms, such as are observed at the beginning of the dissociative reaction in many species of bacteria. The effect of sewage in producing mucoid forms of culture has been noted since by many investigators. Almost any sample of sewage when plated on Endo agar reveals many socalled mutants of intestinal bacteria, especially of B. coli. Similar results have been obtained from the application of sewage filtrates to pure cultures of B. typhosus and related organisms. In addition, it is now a well-known fact that sewage filtrates represent the most prolific source of lytic principle, active against many species of bacteria. From sewage-contaminated river (Huron) water I have isolated lytic principle for all of the following bacterial species: Bact. aerogenes, B. alkaligenes, B. cavisepticus, B. coli, B. cholerae suis, B. dysenteriae, B. enteritidis, Bact. gallinarum, B. icteroides, Bact. ozenae, B. paratyphosus A and B, Bact. pneumoniae, B. psittacosis, Bact. pullorum, B. pyocyaneus, Bact. rhinoscleromatis, B. typhosus, B. typhi murium, M. albus. Students working in my laboratory have obtained lytic principle for M. citreus, Streptococcus fecalis and Streptococcus lacticus. Also Koser [285] has obtained one for a true thermophile. Sewage filtrates are thus highly provocative not only of microbic dissociation, but also of transmissible autolysis.

Starvation.—"Starvation methods" have also been employed for the purpose of enforcing dissociation. These are exemplified by the work of Braun and Schaeffer [66] on Proteus X19 and of Feiler [167] on the typhoid bacillus. Braun and Schaeffer employed an agar medium containing a minimum quantity of nutrient broth and observed that after a few transfers on this medium the X19 strain was transformed to one giving the cultural and serologic characteristics of the OX19 of Weil and Felix, a type which we can recognize as possessing some characteristics of the R, and which corresponded also with the form obtained from phenol agar ("single antigen" type). Following the method of Braun and Schaeffer, Feiler in 1921 produced a similar transformation in the typhoid bacillus. The "permanence" of the starvation culture was less marked than that of the phenol dissociate; and much less than that of the dissociate produced by the action of homologous immune serum. The types of Proteus X19 produced by starvation were found by Braun and Schaeffer to have lost all motility and all flagella. Feiler found the starvation R type of B. typhosus composed mainly of short plump forms with no motility. According to his observations phenol worked a more thorough dissociation in respect to producing nonflagellated forms. Moreover, the reversion to the normal type of culture on normal mediums was effected most quickly with the starvation dissociates, more slowly with phenol cultures, and most slowly with the R types produced by immune serum.

Physical State of the Medium.—There exists universal agreement among all investigators who have studied the dissociation problem, knowingly or unknowingly, that a liquid medium favors or accelerates the reaction, while a solid medium delays it. This circumstance was pointed out early by Penfold and later (with full recognition of the dissociative process) by Arkwright,[16] de Kruif [120] and by Webster.[481] Soule [450] has more recently found that, while the type S subtilis is highly stable on agar, the dissociation proceeds rapidly in cultures maintained in plain broth. I have observed the same phenomenon in many cultures of members of the colon-typhoid-dysentery group. Even the presence of an undue amount of water of syneresis at the bottom of an agar slant culture may be an important factor in many cultures. Eisler and Silberstern [156] were thus able to show that cultivation of B. typhosus on moist and dry agar gave respectively two antigenically different types of culture. On the other hand, rapid transfers through plain broth may apparently increase the S type or even cause a reversion of "pure" R strains (Jordan [273]). Feiler [167] found that such rapid passage of an immune serum R strain of B. typhosus effected a transformation to the S type after some 18 passages; while 25 passages on slant agar determined no change in the R culture. Apparently a dry medium is best calculated to produce the greatest stability in both R and S cultures, and such observations are confirmatory of the older observation of Beijerinck [42] to the effect that microbic stability is always most marked on solid mediums. This view was also upheld by Eisenberg in all of his many and varied studies on bacterial variation.

When we turn to the effect of liquid culture mediums, more specifically upon some of the curious intermediate (O) types, we observe a particularly unfavorable action on the stability of the culture so far as continued dissociation is concerned. Here we note the rapid disappearance of the O type as narrated by Firtsch [178] for Vibrio proteus, the disappearing O form of B. typhosus in broth, as depicted by Bernhardt,[49] and the rapid disappearance of certain presumably intermediate types of B. coli in broth as described by Bordet [61] in connection with his studies on transmissible autolysis. Many other observers have noted colony forms of various bacterial species which would grow on agar, though perhaps faintly, but not in broth. It can scarcely be doubted that the dissociative reaction is the explanation of many of these instances. As we shall see later, these curious phenomena may have a close bearing upon the significance of the bacteriophage and its mode of action.

Volume of the Medium.—Certain experimental data, particularly those of Soule [450] on dissociation in B. subtilis, have shown that dissociation does not occur under conditions in which there are no opportunities for growth, as in physiologic salt solution for example; or in a tube of broth medium when the active stage of growth has been passed. It is thus suggested that if two different volumes of medium, one small the other large, were employed, the degree of dissociation, after a given time, would be greater in the large volume than in the small one. Such a test was actually performed by Soule, the amounts being about 5 cc. of broth in a tube and about 300 cc. in a flask. Each was seeded very lightly with B. subtilis type S. After 16 to 18 hours active growth in the tube ceased and the degree of dissociation, as ascertained by plating tests, had remained at a low point. In the 300 cc. flask, however, growth was still proceeding actively after 60 hours and the percentage of dissociation had at this point reached a much higher value. We may thus conclude that, if the medium is one in which dissociation can occur at all, the results will be more pronounced as the volume of medium is increased.

In this connection it is of interest to note that the volume of the medium employed may also have an influence on the time required for the reversion of the R type to the S. This point which was studied by Feiler [167] is considered in the following section. From these facts one might predict that a procedure for raising virulence by increasing the relative number of type S organisms would succeed as well through the inoculation of a single large volume of medium as by repeated transfers through several small volumes.

Oxygen.—Few data are found bearing on this point. Novy and Soule [499] observed that when the available oxygen surrounding agar slant cultures of B. malleus was reduced to 0.1% (but in large volumes of gas) peculiar erosive changes in the culture, strongly suggestive of dissociation took place; and that these were accompanied by the generation of a new form of culture growth quite different from the original. These changes have not been fully studied. On the other hand, it can be observed that daughter colonies of B. typhosus, as reported by Morishima [352] (four strains), appeared within three to five days under anaerobic conditions while, under aerobic conditions, they required eight days. Webster [481] has stated that the dissociation of Bact. lepisepticum is concerned with the oxygen relations and can be prevented in broth cultures by the addition of hemoglobin, serum or certain inorganic catalysts. Soule has thus far not been able to detect an appreciable influence of modifications in oxygen content upon dissociation in B. subtilis although some of his experiments suggested the influence of 10% CO_2 (in either air or nitrogen) in maintaining or "protecting" the S type against dissociation on agar plates. In this instance the criterion was the "marginal" dissociation which often affects colonies that have started as S type culture. In connection with the reversion of the R form of M. tetragenus (noncapsulated) to the S type (capsulated) Wreschner [492] has reported that, in tubes of serum-broth exposed to air, a 25% dissocation occurred in 24 hours, while in similar tubes in which the medium was covered with paraffin, no reversion whatever had occurred within the same period. This is in accord with my own experience that secondary R colony formation is less common in agar slant tubes that are rubber capped or sealed with wax (after a brief period of growth in air) than in tubes left open to the air permanently. We know, however, from the study of Schattenfroh and Grassburger [427] that the dissociation of the butyric acid bacillus (anaerobic) takes place under anaerobic conditions of growth. In this case, however, the cultures do not grow well under aerobic conditions. The relation of oxygen to the inciting of dissociation is thus in need of further study. It may be pointed out, however, that there is some evidence to indicate that the R type culture grows best under at least partial anaerobic conditions; it is more commonly found at or toward the bottom of the tube. Firsch [178] in 1888 clearly showed the difference in the top and bottom areas of gelatin tubes as concerned the distribution of the chief types of Vibrio proteus. These and similar observations raise the question regarding the nature of the modified, nonvirulent culture of B. tuberculosis obtained by Vaudremer [472] in deep, nonglycerolated potato culture. The manner of growth and the morphology of the cells, taken in conjunction with the absence of virulence, is suggestive of the R type of culture in general, although of the fact of dissociation in B. tuberculosis we at present know little.

Antiseptic Substances.—The influence of antiseptics in producing dissociations is detectable in many older contributions to the subject of bacterial variation and mutation. Phenol was the first of these substances commonly employed. In 1890 Chamberland and Roux [94] reported the influence of phenol in modifying the virulence of B. anthracis. In this case a different type of culture was produced. Phenol, among other means, was employed by Malvoz in 1892 in his alleged (Villinger [475]) transformation of B. coli into B. typhosus, "or near to it." In this case he grew his colon cultures in phenol broth at 42 C., and it is therefore difficult to differentiate between the effect of the phenol and the effect of the heating. Unquestionably, however, this treatment served to produce R types of B. coli, which, as we now know from the work of Schütze,[486] may show analogies with other members of the colon-typhoid-dysentery group and paratyphoid-enteritidis group as well. Indeed, we may perhaps recognize in the report of Malvoz one of the earliest instances dealing with what we now term "bacterial convergence." Villinger [475] in 1894 studied again the influence of phenol on B. coli and although recognizing its modifying effect on cell morphology and motility, as well as on other characteristics, could not agree with Malvoz to the extent of believing that a transmutation to B. typhosus had occurred. A good review of many of these old experiments dealing with the colon-typhoid transformation is to be found in the work of Kiessling [277] in 1893. Fraenkel [183] also in 1889 in his study of the effect of phenol on bacteria has presented data bearing on the dissociation problem.

In studies on B. proteus X19 phenol has often been used to enforce the dissociation from the H type to O, which as a result of the studies of Arkwright and Goyle,[18] White,[487] and Goyle [206] alone, we have come to regard as an intermediate form. Braun and Salomon,[67] as later Braun and Schaeffer,[66] employed phenol agar (0.17%) for the production of their OX19. Feiler [167] accomplished a similar dissociation of B. typhosus by a similar means. In the early work of Arkwright and Goyle on B. enteritidis, B. typhosus and B. dysenteriae these workers at first believed that they had increased the S fraction of mixed S-R cultures by the use of phenol. This result, however, was quite opposed to those of Feiler's detailed and careful experiments and cannot be accepted in the light of much contradictory evidence. Indeed, Goyle has come to a quite different conclusion in his later paper. This point has been considered further in section 9 of this work. In addition to the above instances, Jeanne Lommel [511] has most recently reported the influence of phenol and of formol as modifying the fermentative characteristics of B. coli, resulting in the production of a strain that fermented saccharose and thus resembled B. communior.

Regarding other antiseptic substances, Penfold [385, 386] has shown the influence of sodium mono-chlor-acetate in enforcing dissociations of B. typhosus, and causing the formation of secondary cultures. F. M. Burnet,[87] while observing the $S \rightarrow R$ transformation of B. typhosus in plain broth, also noted the higher proportion of R forms produced in oxylate broth. Eisenberg [148] showed the origin of secondary colonies of B. typhosus when cultures were submitted to the action of egg albumen, which Rettger has regarded as possessing germicidal power for several bacterial species. Chamberland and Roux [94] showed the partial dissociation, accompanied by loss of spore formation, of B.

anthracis when cultures were submitted to the action of a 1:2,000 dilution of potassium bichromate. Möhler and Washburn, according to Gurney-Dixon,[220] produced variants of B. diphtheriae by the use of iodin terchloride.

The action of antiseptic substances has also been studied with reference to producing in chromogenic bacteria changes which are presumably on the order of dissociations although certain aspects of the matter are not clear. Among these contributions we find of particular interest those of Franz Wolf [490] (1909) and of Baerthlein [28] (1912). Wolf studied B. prodigiosus, Sarcina lutea and Staphylococcus albus and aureus. Serial growth of B. prodigiosus in mediums containing traces of potassium permanganate, cadmium nitrate, mercuric chloride or potassium bichromate produced some dark red races and some white. Such changes often seemed to be permanent. It is of special interest that the same antiseptics (for example, mercuric chloride, cadmium nitrate and potassium permanganate) could produce both dark red and also white races. The dark red forms produced in this way were permanent but the white often reverted. Dobell [130] believed that these changes were purely a matter of chance and might be "permanent, partly permanent or impairment." In a few instances Wolf observed white strains, simulating M. albus, arising from M. aureus.

Baerthlein plated out old broth or agar cultures of B. prodigiosus and obtained some dark red colonies and some white. Others were white with red spots, or red with white sectors showing the so-called "sector-mutation." According to Baerthlein these "mutant races" bred true in most cases but might show reversions if left for a considerable time in the same culture tube.

Dyes.—The forcing of dissociation by dyes has received considerable attention under the guise of securing the adaptation of bacteria to antiseptic or bacteriostatic agents. Thus Ainley-Walker and Murray [478] in 1904 noted that the action of methyl violet on B. typhosus, B. coli and the cholera vibrio caused the development of variants. Also, in 1906 Loeffler [307] reported that by the use of malachite green he had produced four new types of B. coli which remained constant in their new characteristics. Two of these we can easily recognize as R types. In 1912 and 1913 Cecil Revis [404] studied the adaptation of B. coli to malachite green and to brilliant green. He found the culture, through "training," could be made to grow in 0.05% of brilliant green but in plates containing this amount two types of colony arose. The two forms differed in cultural features and in fermentation reactions, and undoubtedly represented the S and R type.

Referring again to B. coli, a report by Esther Stearn [452] on adaptation phenomena in some water-borne bacteria is of special interest in showing certain dissociative changes. This investigator was able to demonstrate successive biochemical alterations in B. coli-communis A and B, B. aerogenes B and in the lactic acid bacillus, under the influence of gentian violet (1:200,000). The degree of change varied with the time of exposure (48 hours, 120 hours, 5 months) but interest centers especially in the fact that, whatever the organisms were at the start, all of them came through the experiment at the end of five months manifesting the type reactions of B. coli communior A. Even the lactic acid bacillus ultimately reached this endpoint. It seems probable that the results in this case were due to the dissociative reaction, determined by the dye and by aging. If this is so they possess the significance of demonstrating that the changes which occurred were cumulative over a considerable time, although some of the cultures showed change at the end of 120 hours. So long as it is able, the organism apparently responds to meet the changing exigencies of its environment. Here we apparently have a case of bacterial convergence as indicated by the fermentative reactions of the endforms, which may be regarded as the R types of the respective cultures employed, and we may bear in mind the serologic convergence of the R forms of Schütze (typhosus, paratyphosus, dysenteriae and coli); also the apparent serologic convergence of the gonococcus cultures whose "common type" was pointed out by Torrey and Buckell. If the work of Esther Stearn can be confirmed, the observations will mark one of the significant findings in the field of dissociation and bacterial convergence. It seems probable, however, that serologic tests of the end forms might have revealed differences not shown in the fermentation reactions.

The foregoing observations for the most part made no mention of the nature of the phenomena involved except for the occasional statement that they concerned "mutations." The characteristics of the mutating types were seldom followed up, and most of the descriptions of them are meagre. So far as I am aware, the only investigators up to 1926 to report actual experimental studies involving in any way the conditions determining dissociation (with the exception of those concerned with the H and O forms of proteus, in which dissociation was naturally not recognized) were Cowan, de Kruif, Webster, Griffith, Reimann and Amoss. Cowan worked with the streptococcus, de Kruif and Webster both with Bact. lepisepticum; Griffith, Reimann and Amoss with the pneumococcus. Other dissociations have been reported as such, but the interest has centered in the results rather than in the process.

Even Cowan's work at the beginning was not so much a matter of enforced dissociation as the taking advantage of slow and natural variations of cultures on blood agar, whose value in affording R cultures she estimated by testing the ability of certain colonies to yield spontaneous agglutinating growth (indicative of the R type) in broth. It is by no means necessary, however, to perform such a tedious selection in order to produce dissociation in the streptococci, for variants often arise on blood agar plates when seeded with old cultures. There is little doubt that a still more rapid dissociation of streptococcus can be incited by some such simple means as those employed by Stryker, Yoshioka, Griffith, Reimann and by Amoss, to be considered subsequently; or such as employed by Atkin (deep, Gordon "trypagar" plates grown for a few weeks) for colony dissociation of the gonococcus and the meningococcus. Faith Hadley [499] has secured striking dissociation in colonies of Streptococcus fecalis, mitis, salivarius and hemolyticus by cultivation for several weeks (first at 37 C., then at room temperature) on deep "trypagar." Under these conditions the colonies may often attain a diameter of 12 to 18 mm. in the case of fecalis, and reveal a great wealth of daughter-colony and other structural variation. It is only by studying microscopically as well as macroscopically the remarkable features of such giant colonies that one can properly appreciate the undoubted complexity of morphologic and physiologic behavior occurring within an isolated streptococcus colony. To conclude this phase of the subject, however, it may be said that probably any bacterial species may reveal marked colony variations provided it can be made to grow for a sufficiently long time on solid mediums. The chief reason why such phenomena have not been observed more frequently in the past is that colonies seldom have time to mature before the medium dries and growth is suspended. Dissociation, we must not forget, is dependent on continued growth.

Reaction of Medium.—In Bact. lepisepticum de Kruif [120] has shown dissociation in cultures freshly isolated from rabbits. The reaction occurred rapidly in broth and peptone solutions but more slowly on agar and only to a slight extent in mediums containing normal horse or rabbit blood or serum. An acid medium (P_H 6) retarded the tendency while an alkaline reaction (8.5) often accelerated it. Attention may here be called to the fact that Gratia [209] pointed out that 8.5 was most favorable for transmissible autolysis, and that d'Herelle [248] recommends 7.8. Both Reimann [405] and Amoss [8] reported 7.8 as the most favorable reaction for pneumococcus dissociation, while acid mediums hindered the reaction. The older observations of Preisz [393] involving what we can now certainly recognize as dissociation in B. anthracis were made on cultures grown on "alkaline agar." In dissociations of many members of the colon-typhoid-dysentery group, also in the capsulated bacteria, I have found that a reaction of P_H 7.8 favors the dissociation. Apparently B. pyocyaneus does not dissociate in an acid medium.

Animal Passage.—Although animal passage has always been recognized as the most effective means of exalting the virulence of pathogenic bacteria it has seldom been employed for the recognized purpose of modifying the morphologic or cultural type. It was employed unsuccessfully by Eisenberg [149] in 1912 in the attempt to cause a reversion of a nonsporogenic race of B. anthracis to the sporogenic type. Many passages through guinea-pigs showed the culture to possess some residual virulence but its nonsporogenic character was not changed. Preisz had already stated that when the sporogenic function had once been lost it could not be regained. Bail and Flaumenhaft,[34] however, who obtained the two colony variants, presumably S and R, of which only the latter (their β form) was virulent, found that the inoculation of guinea-pigs with R gave typical anthrax death and only R organisms could be found on autopsy. The inoculation of the S type (their α) resulted in a greatly delayed death; and then also only the R form could be obtained at necropsy. From examinations made at intervals before the S-inoculated animals died, they noted that the R form began to arise at about the 90th hour of the infection and increased from that time onward.

Baerthlein [30] in 1918 showed that passage through normal animals caused the heightened development of the S type (capsulated) of the pneumobacillus from mixed S and R cultures; but that, when the "pure" R form (noncapsulated) was injected, it did not revert to the S state. The same thing was shown by Wreschner [492] for his "absolute" noncapsulated variant (R) of M. tetragenus when passed through mice. The influence of passage in stimulating the overgrowth of the S form of Bact. lepisepticum was demonstrated by de Kruif; also by Mary Cowan for streptococcus and by Griffith, Reimann and by Amoss for the pneumococcus. Curiously, Schmitz [434] obtained different results in his injections of guinea-pigs with pure line strains of B. diphtheriae. By intraperitoneal injections, as well as by the use of collodion sacs, he was able to demonstrate that, in the course of time, the virulent form was robbed of all of its chief characteristics and became nonvirulent. The effect of passage of S cultures through immunized animals usually gives this result, so far as present data indicate. Whether the S → R transformation occurs under natural conditions in the living animal as soon as the bacteriotropic antibodies have adequately developed is a highly interesting and important point because of its bearing upon a possible mechanism of body defense to which attention has already been called by Griffith with reference to pneumococcus infection, and which has been considered in detail earlier in this work. Griffith was unable to detect evidence of the S → R transformation in the living animal (immunized) although the S form of the pneumococcus, after being injected, rapidly disappeared from the body. It is logical to expect that further study will reveal this transformation occurring in vivo.

With respect to the effect of passage of virulent culture through the normal animal Yoshioka [494] has recently obtained modified, avirulent forms of the virulent, hemolytic, Aronson streptococcus by following the method of Morgenroth. A mouse was given injections with the virulent type. Four hours after the injection a culture was taken from the subcutis and plated. Among the mass of hemolytic colonies could be found a few of the greening variety. These were nonvirulent and gave quite different serologic reactions. Naturally these methods involving animal inoculation for the production of variants are open to some criticism, but support for their significance is added by the circumstance that the modified culture was serologically similar to the dissociate produced by growing the virulent type at 39 C., and by other methods. Unfortunately, however, Yoshioka did not present data on the correlated cultural features of the variants, a matter which always lends valuable support to any conclusion regarding such assumed modifications, so far as the production of the R form is concerned.

Regarding the influence of passage of virulent cultures through immunized animals, several instances bearing on this point are presented in the following pages dealing with the influence of continued growth in immune serum. We shall see that the most common effect is the production of a modified culture which has lost all, or at least the greater part, of its original virulence.

Normal Serum and Ascitic Fluid.—Although it has often been recognized that normal blood produces marked changes in the form of bacteria, whether these changes are always of the nature of dissociations is not so clear. Observations of this sort begin to appear in the literature soon after the description by Pfeiffer of the phenomenon which bears his name, and have concerned largely the typhoid bacillus and the cholera vibrio. Regarding the latter there is evidence that, not only immune serum but also normal serum, may cause a granulation of the vibrio and that at least some of the granules are viable when transplanted into a fresh medium. This has been demonstrated more convincingly, however, for the action of normal serum on the typhoid bacillus for which many normal serums are known to be germicidal. Knowledge of these matters was first presented by the various works of Pfeiffer, Kolle and Fränkel, and Radziewsky;[400] and further by the work of Eisenberg[147] who presents an excellent review of the influence of both normal and immune serum up to the time of his own work in 1903. It would seem, however, that a dissociative influence of normal serum is not clearly established for B. typhosus unless it also possesses some natural germicidal power. Danysz[115] showed that when "first vaccine" anthrax culture was grown in rat serum the organisms were modified. Since, however, the first vaccine is itself a modified culture, the final results obtained by Danysz will not be further considered. Hess[250] in 1921 showed that when virulent anthrax cultures were grown continuously in normal horse serum (12 hour passages) they lost their capsules in from 14 to 31 generations. The loss was permanent unless the organisms were again grown in animals. It occurred more quickly at 41 C than at 37 C. It is of interest that one of the cultures (Stamm III) with which Hess worked possessed the characteristics of what we have termed the S type. This was transformed to a noncapsulated state in the shortest time of all—namely, seven generations on normal serum. Many other workers have produced modifications of cultures by growth in normal serum, but the changes reported are not such that we can ascertain that the transformation to the extreme, or even to an intermediate R form, was attendant. In the case of most pathogenic organisms the result has been, rather, an increase in the S type. When the foreign serum has intrinsic germicidal influence the result is manifestly different.

To the above it should be added that aging an S type culture in contact with normal serum (as in a 10% serum-broth) may not result in producing an increase in the number of S organisms in the culture, and in this respect is quite different from actual passage of organisms in series through serum-broth cultures. Nungester,[499] for instance, has observed that aging a culture of B. anthracis in such a medium for some weeks resulted in an increase of R type organisms. The same reaction was observed not only in the case of the S type culture, but also in cultures of several other types (including the slimy variety) which Nungester succeeded in isolating from his original pure line of B. anthracis. The R type culture seemed to be the common goal of all the transformations. Similar results were obtained by aging in plain broth. Results from serial passage through serum-broth are not yet available from Nungester's experiments.

The modifying effect of normal serum on bacteria submitted to its action in the course of culture has to do, in part at least, with the phenomenon (often mentioned in the literature) of "serum-fastness," as treated by Braun and Feiler,[65] Rosenthal [413] and others. Such an adaptation to normal serum is manifestly accompanied by a modification of the organisms in which they become more resistant to whatever slight germicidal power the normal serum possesses. Rosenthal showed that such serumfastness of B. typhosus to normal rabbit serum occurred after three or four passages. Such resistance of the bacteria was not correlated with increased resistance to chemotherapeutic agents and was not "type-specific." It was still detectable after 20 passages in broth without serum. On agar the reversion to normal was accomplished more quickly—sometimes after a single passage. It is perhaps important to note that such cultures possessing maximum fastness for normal serum showed almost normal agglutinability, which indicates that they had undergone no significant beginning of the transformation toward the R type. The strains grown in inactive serum, on the other hand, showed, together with normal susceptibility (i. e., no serumfastness), a strong agglutinin-fastness. It is also important to note that "fastness" to the bactericidal power of the serums did not lend to avirulent bacteria the quality of virulence. With reference to this point Rosenthal states: "Die Festigkeit gegen die bakteriziden Substanzen des Kaninchenserums verleiht somit den Typhus Bazillen nicht die Eigenschaft der Virulenz." A similar result was obtained by Braun and Feiler. Although the accomplishment of a "fastness" to bactericidal influences did not serve to increase virulence, Rosenthal regarded such an adaptation as a necessary preliminary to the establishment of an infection.

The dissociative influence of ascitic fluid in culture mediums is well known to all who have worked with problems of mutation in pathogenic forms; and it may be that a part of the aversion to its use is due to its influence in effecting variations of such magnitude that they have been regarded as contaminations and assigned to the ascitic fluid as a source. The dissociative action of peritoneal fluid is to be seen in the reference by Crowell [114] to the transformation of the virulent, granular type (S) of the diphtheria bacillus into a nonvirulent, solidly staining coccoid form in the peritoneal cavity of the guinea-pig. In the case of normal serum, whether the organisms are killed or permitted to grow in a modified form, seems to depend on the strength of the serum, much as we shall see is also the case in immune serum. The modified type has usually been reported as a nonagglutinating culture.

Immune Serum and Immune Blood.—The use of normal blood or of blood serum for increasing virulence is too well known to demand further consideration; and the manner of its action in accomplishing such an effect has been shown in the experiments of de Kruif, Cowan, Griffith, Reimann and Amoss, to mention only more recent studies, as well as suggested by the reports of numerous earlier writers (Eisenberg, bibliography). The influence of immune serum, however, demands further attention. A survey of the effects of such serum on the "normal," virulent type (in the cases to be reported it seems justifiable to assume the S type or largely so) or on the definitely recognized S type, seems to reveal something of a discrepancy. Some works indicate the transformation of the virulent form into one possessing little or no virulence, and often characterized by a spontaneously agglutinating or non-agglutinating (in immune serum) manner of growth. Other reports, on the other hand, indicate a heightening of the virulent character by much the same method, indicating a sort of immunization of the bacteria against the destructive working of the immune serum. This last conception is theoretically plausible; and, on the strength of a few questionable instances combined with certain speculations of Welch, has descended through some generations of textbooks to such an extent that it may be said to represent at the present time the most common view of the matter. Zinsser (p. 249) [496] for instance states, "This lessened susceptibility to antibodies is noticeable not only in strains cultivated from the body in disease, but can be produced artificially by cultivating bacteria in inactivated homologous immune serum." After citing the observations of Walker [477] on B. typhosus and of P. Th. Müller [355] on B. typhosus and Vibrio comma, both of which indicated increased virulence of the respective organisms in immune serum (and both of which have been amply contradicted by much later evidence), Zinsser says, "Such strains not only increase in virulence but lose in both agglutinability and susceptibility to bactericidal effects." Again he states (p. 13), "but additional evidence pointing in this direction has been brought by experiments in which it was shown that bacteria cultivated in the serum of immune animals not only gained in resistance to destruction by the serum constituents, but at the same time were rendered more pathogenic." And again (p. 13), "It is interesting, moreover, to look upon this process of adaptation as a sort of immunization of the bacteria against the defensive powers of the host, a conception early suggested by Welch." In the following considerations we shall see that a view of the influence of immune serum in producing non-agglutinating forms of bacteria is amply justified by an abundance of evidence; but that this phenomenon is accompanied, as a rule, by greater virulence, as most commonly reported in the textbooks, is a conception against which a great mass of evidence is directed. In much of this evidence we shall see that many instances support the view that the fundamental phenomenon determining the observed changes is microbic dissociation.

Metchnikoff [345] in 1887 reported that the virulence of B. anthracis was diminished when the organisms were brought into contact with anthrax immune serum. Anthrax organisms which developed in the blood of sheep that had undergone vaccination lost the power of killing rabbits. Roger,[409, 410] also Charrin and Roger,[95] demonstrated the same phenomenon when they injected virulent streptococci, pneumococci and B. pyocyaneus into rabbits that had been immunized with these organisms, respectively. Indeed these authors [96] in 1892 formulated a theory of immunity based on the observed attenuating power of immune blood, according to which virulent organisms were not killed but lost their pathogenic power when coming into contact with the blood of immune or vaccinated animals. In these cases we unfortunately know little regarding the cultural or serologic modifications impressed upon the organisms concerned.

In 1893 Sanarelli [422] studied the influence of serum immune to Vibrio metchnikovi against the homologous organism. He inoculated virulent cultures into animals that had been vaccinated, and found that, although the bacteria had not been killed in the body, they had lost their virulence and were noninfective. He ascertained, moreover, that the same result appeared when serum from the vaccinated animals was brought into contact with the vibrios in vitro. In these cases, however, the inoculum containing the assumedly attenuated microbes also contained some of the immune serum. When the organisms were removed from this serum by subculturing in broth or by washing, and then injected, Sanarelli found that they possessed as much virulence as the original unmodified culture; sometimes more. He therefore came to the conclusion that the influence of the immune serum was, not to produce an attenuation, but sometimes even to increase virulence. In interpreting these results, however, we must bear in mind that the vibrio was not cultivated for any length of time in the immune serum (only 48 hours), and we are not aware that the immune serum possessed any unusual strength. We are not therefore justified in assuming from the results of these tests of Sanarelli, or others reported by Metchnikoff [347] in 1892 on attenuation of the hog cholera bacillus, that the growth of pathogenic bacteria in immune serum had the effect of increasing their virulence, although this has been the conclusion usually drawn from these cases, as well shown by the comment of Stryker [454] on these and similar instances. We shall be able a little later to compare these results with others in which we can be assured that the bacteria have actually been grown (continuously) in immune serum. And we shall see moreover why growing bacteria in immune serum for 24 hours, or even 96 hours, may not be sufficient to eliminate their virulent features when they are again placed on a normal culture medium.

Hamburger [231] in 1903 demonstrated that the cholera vibrio gained virulence when cultivated in normal guinea-pig serum but lost it and appeared as another type of organism when cultivated in guinea-pig cholera antiserum. Another report bearing on the same organism is that of Ransom and Kitashima.[401a]

Nicolle [366] in 1898 demonstrated the influence of typhoid immune serum in transforming the normal type into a self-agglutinating form, clearly the R type, but which was still virulent. The sojourn of the culture in the serum was, however, brief and immediate reversion occurred. Saquépeé [421] in 1901 accomplished somewhat similar results by growing B. typhosus in collodion sacs in immune rats. After a period of five months the culture became nonagglutinative, simulating the so-called "eberthiformis" type obtained from natural sources. In vitro the results were not so clear. Typhoid culture was placed in a tube containing 5 drops of broth together with 15 drops of immune serum. At the end of 45 days no change in the agglutinative power had occurred. When, however, the medium was changed every few days, a slight reduction in agglutinability occurred. The question of virulence was not considered. Steinhardt [453] in 1904 reported that when grown in strong immune serum typhoid and dysentery bacteria lost their virulence and became spontaneously agglutinative, then yielding cultures which we can now recognize from her description of the colony form as belonging to the R type. Somewhat similar results were reported by Reiner Müller [358] in 1911. Moon [351] in the same year showed that two distinct types of B. typhosus arose from a single cell—one agglutinable, the other not. Zinnser and Dwyer, according to Morishima [352] made observations in 1918 on the loss of agglutination by B. typhosus. Especially convincing studies, however, so far as their bearing on dissociation was concerned, were reported by Feiler[167] in 1920; others by Morishima in 1921.

Feiler studied in considerable detail the influence of phenol, starvation and immune serums on strains of B. typhosus, and B. paratyphosus B, following the same general lines as those of Braun and Schaeffer for B. proteus. By all the means mentioned Feiler was able to produce from normal, motile cultures variations characterized by modified cultural and serologic features; also by loss of motility and flagella. While phenol and starvation, according to Feiler's view, caused a loss of certain agglutinogens, continued cultivation in homologous immune serum caused the loss of all agglutinogens. Feiler thus held that the loss of agglutinability involved the loss of receptors—partial in the case of phenol but complete in the case of typhoid immune serum. He also held to the distinction between ectoplasmic (flagellar) and endoplasmic (somatic) agglutinogens, as earlier treated by Malvoz, Nicolle, Smith and Reagh and others. The permanence of the changes produced was also discussed, as considered further in section 12 of the present work. Feiler's results agree fairly well with the earlier reports of H. Sachs[419] and of Weil,[482] Felix and Mitzenmacher[178] on B. proteus; and he accepts a similar explanation, based on the "double" and "single" antigen (receptor) hypothesis. He did not present data on the virulence of the modified types.

The work of Morishima in 1921 containing data on the influence of immune serum on B. typhosus followed the trend of Feiler's study. Since this has already been mentioned in the preceding section, and since one important aspect will be referred to later in the present section, it is sufficient to say here that, like Feiler and others, Morishima was able to produce a different and nonagglutinating form of the typhoid organism by serial cultivation in typhoid immune serum. One may perhaps conclude from his results that the organisms passed through the intermediate (O) stage, in which they were inagglutinable, to the R form in which they were again agglutinable, although the report lacks data on the cultural features of the variants. In general Morishima's results confirm the results of Feiler's more detailed and comprehensive presentation of the serologic aspects.

Regarding the diphtheria bacillus, we have no exception to the general rule that a strong immune serum is able to effect a transformation of the virulent S type into the nonvirulent R form. Although this has been accomplished by several workers, the clearest picture is given by

Bernhardt[49] in 1915 in his valuable study of the various diphtheria types. He made use of immune human and guinea-pig serums, both active and inactive. To the immune serum in one cc. amounts he added varying quantities of 24 hour virulent culture and sampled by plating at 24, 48 and 72 hours. From such plates he obtained atypical colonies giving short rods of the Hofmann type (diphtheroid), which for the most part were nonvirulent. His further description of these forms indicates that they were an R type of the diphtheria bacillus as outlined in a previous section, but probably not well stabilized.

With reference to the pneumococcus, many older reports have indicated a transformation in the type of culture from a virulent to a nonvirulent by the use of immune serum. The latter type usually gave a spontaneous agglutination and was noncapsulated. Older observations dealing with this point are found in the works of Friel, Laura Stryker and Yoshioka. Much more conclusive experiments on the variation in this species, and with full recognition of the chief dissociative types, S and R, have been reported in more recent time by Griffith, Reimann and Amoss. Jacobson and Falk[266] have also reported obtaining rough colonies by growing cultures in broth containing immune serum. These reports will be mentioned presently. First we may review the older studies.

Friel[185] in 1915 demonstrated that when virulent cultures were grown in homologous immune serum, they underwent definite cultural changes, lost their virulence and became phagocytable even in normal rabbit serum. Laura Stryker[454] in 1916 reported growing virulent pneumococci for many generations in a broth with 10% homologous, immune serum. She observed that in all cases (types 1, 2 and 3) a new form of culture was produced lacking both capsules and virulence. The original culture was modified by degrees but eventually came to resemble what we now recognize (mainly through the work of Griffith) as the R type pneumococcus. The essential characteristics of Stryker's modified types have been presented in detail in section 9. The modified strains remained permanent even when grown on normal mediums without immune serum, but reverted in all cases after several passages through mice. Stryker believed that the degree of permanence was correlated with the number of passages through immune serum broth and the truth of this view was shown later by Griffith. The work of Yoshioka[494] in 1923 was largely confirmatory of the significant results presented by Stryker seven years earlier. He obtained similar modifications of his types 1, 2 and 3 pneumococcus cultures by drying and also by growing at 39 C. The variant culture showed itself by loss of virulence, decrease in agglutinability in immune (homologous) serum, and increase of agglutinability in heterologous sera, thus indicating bacterial convergence. These changes did not appear at the same time in all of the cells of the culture but only in certain colonies. The detailed serologic aspects of Yoshioka's work have been considered in section 9.

Since Griffith [215] (1923) is the leader in this field of study on the pneumococcus in which attention was focused on the R and S types in particular, his results are worthy of special notice. This is also true because he worked with standard type 1, 2 and 3 pneumococcus cultures. Virulent cultures (type S) were grown in series in concentrated homologous immune serum, heterologous immune serum and normal serum being used as controls. While at the beginning the cultures commonly showed only S colonies, after the first immune serum transfer some R colonies appeared on plating. The second serum transfer yielded a larger number of R colonies and, after the third transfer, R culture had entirely replaced S. These results, obtained with concentrated immune serum, could be duplicated in some measure by the use of immune serum in a dilution as high as 1:256. Heterologous and normal serums generally had no power to produce R colonies; once a type 2 antiserum gave R colonies in a type 1 culture. Griffith reported one experiment that is almost unique in dissociation studies: he tested the dissociation-provoking power of R and S antiserums on S type culture. A virulent S culture was grown in R antiserum; three passages gave no change in the organisms concerned. On the contrary, when S culture was grown for one generation in the S antiserum, a mixture of S and R colonies resulted. A colony of each type was picked from these plates and the subsequent cultures injected into mice. The S culture was fatal in a dilution of 10^{-7} cc. while the R culture was harmless in 0.2 cc. amounts.

The only instance, so far as I am aware, in which both S and R immune serum has been used against both R and S forms of culture, is found in the work of Soule [450] on B. subtilis. Soule made use of these immune (rabbit) serums in various dilutions but found that 10% was as effective in accomplishing the results reported as any higher concentration. When R and S forms of culture were placed in the S immune serum, no change occurred in R, but the S culture gave 80% dissociation into R at the end of 24 hours. This was the most complete and rapid dissociation observed by Soule under any conditions. When, on the other hand, R and S cultures were placed in R immune serum, the S type culture underwent no appreciable change, while the R culture experienced a 40 per cent reversion to the S type. So far as I am aware, this is the only instance reported indicating the reversion-producing effect of an R immune serum on R culture. It naturally opens several interesting lines of inquiry.

To the instances mentioned above may be added a reference to the tubercle bacillus, although we know little regarding dissociation in this species. Karwacki [508] has recently observed that by cultivating the tubercle bacillus in the blood serum obtained from tubercular cases he could obtain a morphologic dissociation of the inoculated organism. There arose in the culture coccus or rod forms that lacked the usual acidfastness. The colonies of these nonacidfast organisms grew more rapidly and became larger than the normal colonies containing the typical acidfast bacilli. Although many important details are lacking in this study, the presence of the dissociative reaction is strongly suggested. Apparently the tubercle bacillus is no exception to the general rule that growth of a culture in immune serum of immunized or partially immunized animals, is one of the surest methods of initiating those cultural changes which concern the dissociative reaction.*

Although the majority of evidence, as indicated briefly above, demonstrates the fact that virulent bacteria, grown continuously in immune serum, lose their agglutinative reaction and usually their virulence, and sometimes become spontaneously agglutinative—all of which suggests a change to the R type of culture—there are on the contrary some instances in which a heightened virulence and no appreciable change in the type of culture have been reported. These include the observations on B. typhosus by Eisenberg; [147] also Steinhardt [453] found that in weak immune serum the virulence of dysentery and typhoid cultures was increased and that the S type of culture was maintained. Ainley-Walker [477] grew typhoid organisms in typhoid immune serum and concluded that this procedure effected a diminution in agglutinability, a heightening of its virulence and an increase in its resistance to serum protection. Bordet [56] pointed out that the inoculation of Metchnikoff's vibrio into animals already immunized caused a change in the organism in favor of greater virulence.

From these last mentioned results one may gain the view that there exists little unity in the reports on the influence of immune serum on

* To the above it may be added that Julianelle [505, 506] has very recently shown that growth of the S type (capsulated) Friedländer's bacillus for 6 or 10 passages in broth containing 10% serum immune to the S type served to produce colonies of the R type (noncapsulated) culture. While the S culture was fully virulent, the R culture was without virulence.

the "normal" bacteral type, especially when the text books commonly intimate that the chief result of growing bacteria in immune serum is to increase their virulence. If one reviews these various reports with some care, however, it appears that these discrepancies may be explained by the following circumstance, which is shown with special clearness in the work of Steinhardt. Whether the normal S type will develop in its homologous serum and become more virulent, or whether it will develop in the serum and become less virulent (the latter reaction being accompanied by the S → R transformation) will depend on three circumstances: the actual titer of the immune serum, the concentration of the immune serum, and the length of time that the microbes are in contact with the immune serum (which must always be sufficiently fresh to permit continuous growth, preferably by means of successive passages). If the titer of the serum is too low, modification of the bacteria does not occur, except that they may become more virulent. If the serum is employed in too great a dilution modification does not occur (Park and Williams, Morishima), or perhaps the culture may become more virulent. If the time of contact of the bacteria with the immune serum is not sufficient, or if the number of passages through the immune serum (serum broth), is not sufficient, modification will not occur. This last result is best exemplified by the work of Metchnikoff [346, 347] and of Sanarelli [422] which led them erroneously to conclude that immune serum had no influence in diminishing the virulence of their treated cultures, but in some cases actually served to increase it. It is this old notion that has found its way into modern conceptions regarding the influence of immune serum on bacteria, as expressed by Zinsser in his textbook, by Stryker in her work on the pneumococcus and by many others. It has, moreover, doubtless been maintained in part by the somewhat natural expectation that there should exist a correlation between (immune) "serumfastness" and virulence in pathogenic organisms. That such a situation commonly exists for the bacteria at least seems very far from the truth, as shown by a great mass of evidence when it is correctly interpreted; although in the case of the pathogenic protozoa and particularly in the trypanosomes the situation is not so clear. Even in the action of immune serum against the cholera vibrio (the classical instance of bacteriolysis), although the vibrios disappear, it is by no means so certain that the culture is destroyed. We must come to the conclusion that there is considerable difference between the apparent lysis of living cells and their actual death; for, in such instances, there may be involved merely a transformation into another living form with which we are only slightly acquainted. I suspect that this circumstance may have an important bearing upon certain problems in serum therapy.

We may thus conclude that, among the bacteria at least, wherever the question has been adequately studied and the results of the experiments adequately analyzed, the influence of a strong, homologous immune serum in continued application is such as to incite microbic dissociation, and thus to modify the virulent culture type in the direction of the R form; furthermore, that this resulting modification involves, in its more extreme aspects, far-reaching cultural changes, antigenic and serologic alterations of great significance and, finally, a loss of virulence which in some instances may be complete.

Pleuritic Fluid (in Tuberculosis).—Although we at present have little knowledge regarding dissociative reactions in the tubercle bacillus, it cannot be doubted that here, as in other types of infection, the eventual discovery of different forms of culture will sometime play an important part in problems of prophylaxis and therapeutics. So far as definite observations on the cultural behavior of B. tuberculosis itself are concerned, the chief support for this view lies at present in highly significant findings recently reported by Karwacki.[508] These concern peculiar cell reactions observed to occur in the tubercle bacillus when in contact, either in vitro or in vivo, with the pleuritic fluid of certain tubercular patients. Karwacki studied the reactions of 22 strains of the tubercle bacillus in 20 samples of pleuritic fluid drawn from tubercular patients. In 16 of these fluids the organisms underwent a "mutation" revealing giant forms, acidfast streptococcus forms and small cocci which might be either acidfast or cyanophilic. In many instances the transformation was accompanied by complete dissolution or lysis, and the author wonders if the now commonly observed filtrable forms of the tubercle bacillus may not have their origin in this reaction. In any case, the cultures often became "sterile" in the old morphologic sense.

The author also points out the very significant fact that there existed a distinct relation between the clinical character of the pleurisy and the transforming power of the pleuritic fluid obtained from the same case. The "mutation" reaction was observed most often in the fluid from the rapidly-terminating cases while, in the severe and chronic cases the action of the fluid was slight or absent. In such instances, however, the transforming action increased gradually and in the stage of resorption gave distinct transforming power. In view of these results

Karwacki believes that the transformation reaction possesses prognostic value, since the retrogression of the pathologic state depends in some measure upon "mutation" of the tubercle organisms into a nonacidfast form in which they are more readily disintegrated or destroyed.

While admitting the unknown nature of the transformatory and often bacteriolytic reaction, the author suspects that it may play a part in immunity to tuberculosis. Calmette has expressed the view that the lysins are incapable of modifying either the morphology or the physical properties of the tubercule bacillus; but with this view the findings of Karwacki are not in agreement, since they demonstrate the existence in the pleuritic fluid of a distinctly bacteriolytic agent, capable of accomplishing marked transformations in the type, and perhaps in the physiological reactions, of the tubercle bacillus in vivo as well as in vitro.

To the foregoing exposition it scarcely needs to be added that the important observations made by Karwacki are in harmony with much other work indicating the dissociation-furthering power of immune blood and of the body fluids of infected or partly immunized animals. If later study confirms the fact that Karwacki was dealing with the actual dissociation of the tubercle bacillus, as now seems most probable, and that this reaction involved distinctly protective aspects, we have here one more piece of evidence that at least one important defensive mechanism of the body against infection is the precipitation of microbic dissociation in vivo. In this connection I believe that Karwacki has opened the door to a new and significant field of investigation in the study of tubercular infections.

Products of Growth or of Dissociation.—It has long been recognized by bacteriologists that metabolic products might exert a strong repressive influence on the further development of the same, or on the growth of another, bacterial species. This subject has received interesting treatment by Eijkmann,[146] Rahn,[401] Hajós [229] and others, to say nothing of the more unusual features concerned with the filtrates of B. anthracis and B. pyocyaneus as studied by Emmerich and Löw,[158] Gamaleia,[194] Malfitano [315] and others. The effect of these growth products in modifying the form of culture has received less attention although it is undoubtedly of primary significance in many of the peculiar effects observed when different cultures are grown together in association, as in certain instances to be mentioned shortly. Indeed, our present knowledge of symbiosis and its effects in the case of pathogenic bacteria leaves much to be desired. It is the purpose of the present section, however, merely to indicate some of the examples suggesting the influence of products of growth in promoting dissociations of the typical sort in sensitive cultures of several common bacterial species.

Quite recently Ørskov and Larsen [513] reported a peculiar instance definitely related to dissociation in a member of the paradysentery group, in which the filtrates of growth coming from a colony of a certain type ("B") showed definite inhibiting action on growth of the original culture. The subject is of such significance as to merit reporting in some detail. Plating a 14 day extract-agar stab culture gave two types of colonies, "V" and "B." The organisms in V were short and plump; those from B long and polymorphic—and often granulated or "degenerated," even in 12 hour colonies. The early V colonies remained normal for several days, but within this time the B colonies showed a "central decay" which later became macroscopically observable in the form of "fairly large holes." This lysis also occurred in V, but appeared later and was less pronounced. Seeding from these colonies gave colonies duplicating the original types. When picked to broth the V growth was homogeneous; the B growth spontaneously agglutinative.

After some daily streaks on agar each of these culture types gave rise to a new form, "M" and "Bu," respectively. On cultivation in broth V always split off M, while in agar stab it split off B. M and Bu, on the other hand, never gave any indication of returning to V and B, respectively. If V disappeared it was said to be impossible to obtain it again. The serologic reactions with these cultures are presented on an earlier page of this paper; also fermentation tests. The autolysis of the B type suggested the bacteriophage reaction and was therefore further studied. A 24 hour broth culture of B was filtered and a filtrate obtained which had a lytic effect most pronounced for culture M but also effective for a Shiga culture. "A few drops added to 10 cc. of bouillon simultaneously with the sowing of the bacteria in question completely prevented growth." A smaller concentration only delayed growth. No increase in the lytic agent was observed. These filtrates also exerted a restraining influence for growth on agar surfaces, one drop of a 1:200 dilution being able to accomplish this. After a few days scattered colonies often appeared on the lysed area, and these organisms were insensible to the same filtrate. Neither V, M, nor Bu formed such a lytic substance.

An interesting question arises as to the possible relation between the lytic substance observed by Ørskov and Larsen and the "alkaline-producing aërophilic ferment" studied in 1925 by Chiari and Loeffler.[98]

Here we observe a bacterial product of alkaline nature with which B. coli and other bacteria could be "infected" and made to perpetuate the phenomenon as revealed by whitish areas on Endo plates. Unfortunately we have no data regarding its effect on the morphologic, cultural or biochemical characteristics of the sensitive strain. Finally it may be added that de Kruif [120] found no stimulus to dissociation was given by growing his type D culture of B. lepisepticum in filtrates of type D culture. Enderlein [160] believed that metabolic products were able to incite the organisms of a culture to enter what he terms the "second reproductive stage," or the "Mochlolyse." From my understanding of this author, I believe that this is equivalent to what we have regarded as the transition from the S to the O or intermediate type. Enderlein's depiction of his various types rests purely on a morphological basis and thus renders some of his transitions difficult to follow. His culture types require correlation with certain biochemical and serologic characteristics before their relation to what we have termed the S, O and R forms can be entirely clear.

With further reference to the effect of metabolic products in producing dissociative reactions, there may be mentioned the recent report of L. Rosenthal [514] on B. anthracis. This investigator was able to change sporeforming races of this species to non-sporeforming races as a result of growing the cultures in the filtrates of previous cultures. Cultures were employed which gave abundant spores in about 24 hours under normal conditions. Such cultures were planted in the sterile filtrates of older cultures (either sporogenic or asporogenic) that had grown for about two weeks before filtration. In this filtrate the culture was grown for five days, then transplanted to a tube containing fresh filtrate. After three such passages the culture was plated and the colonies of the non-sporeforming bacilli selected. These, Rosenthal notes, can be detected after a little experience by the macroscopical features; but he fails to state what these differentiating characteristics are. If, however, his results conformed with those of Preisz,[393] Eisenberg [149, 151] and others, we may assume that the nonsporogenous colonies were of the translucent or bluish variety. The asporogenic nature of these cultures was demonstrated not only microscopically but also by the heating test. According to the view of the author, this new character was permanently acquired. In this case we manifestly have an aspect of the dissociative reaction in the bacillus of anthrax.

One of the most interesting cases which may appropriately be considered under the present heading involves the phenomenon of

"Entrainement" described by Et. Burnet [86] in 1925. As reported on a previous page (section 9) this investigator had already shown [85] that there commonly exist two serologic types of the organism of Malta fever, the "normal" and the "para" (Melitensis II) forms; and I have suggested already that the para type possesses many of the characteristics of the R form of culture, especially as described by Bassett-Smith.[39] Or at least it resembles a type of culture which shows the beginning of a transformation toward the R form. Burnet had already pointed out that one of the chief differential characters between these two forms was thermo-agglutination; that the para type gave this reaction while the normal melitensis did not. Indeed, the para types always seem to floc more readily than the normal. Heating was said to render the para form more agglutinable while it made the normal form less agglutinable.

The phenomenon of "Entrainement" appears to involve the more or less permanent modification of certain characteristics (biochemical, antigenic and serologic) of a culture through direct association with other cultures; or through submission (in the early stages of growth) of a culture to the filtered products of another culture of the same or a different species. In Burnet's elaboration of this phenomenon the first point of interest lies in the following circumstance. When normal cultures of melitensis were cultivated in association with the para form, the normal cultures became thermo-agglutinable—a characteristic of the para type. The experimental conditions necessary for performing this test of the results of the microbic association were realized in various ways. For instance, a collodion sac containing the normal melitensis culture was suspended in a tube containing a broth culture of the para form. After three to five days the normal culture was found to have acquired the property of thermo-agglutination. Similar results were obtained when Chamberland candles (porcelain) were used in place of the collodion sacs.

The first results were obtained by the use of living para culture, but Burnet was able to demonstrate that it was not necessary that the para culture should be living in order to produce these results, for cultures killed by heating at 65 C. accomplished the same effect. After three or four days the normal melitensis was found to manifest thermo-agglutination. It is of special interest that Burnet states that it is necessary in all cases that there should occur a "development" or growth of the normal culture in order that the modification may be accomplished. A few minutes contact was not found to suffice. Here perhaps we may recall that it is always necessary for cultures to grow before they can dissociate; and that it is equally necessary for them to grow before they can undergo transmissible autolysis.

Another point of interest as described by Burnet is the following. The phenomenon of "Entrainement" may not only be "specific" (paramelitensis vs. normal melitensis), but may also be "non-specific" (B. anthracis or B. pestis vs. normal melitensis). One suspends a normal culture of melitensis, already partly developed (in a collodion sac), in the filtrate of a culture (or in the broth culture itself) of B. anthracis or B. pestis, and allows the development of the melitensis culture to proceed for some days. At the end of this time the culture is tested and is found to give a positive thermo-agglutination reaction, while it did not give this reaction at the beginning of the test. The normal culture had thus been transformed into a culture possessing one of the distinguishing features of the para type.

Regarding the permanence of the newly acquired and distinctly "heritable" character, Burnet stated that, when the thermo-agglutinating characteristic had once appeared in a culture it could be transmitted indefinitely through subcultures. The modified melitensis ("entraînés") cultures were still thermo-agglutinable twenty generations away from the original, normal stock. The stimulus was always found to operate from the para form to the normal culture; never in the reverse direction.

Although "Entrainement" was found to concern in all cases the phenomenon of thermo-agglutination, it was not related exclusively to such physical properties of the culture, but to the biochemical and antigenic characters as well. With thermo-agglutinability were found to be correlated other characteristics, such as specificity in sero-agglutination, feeble production of agglutinins in rabbits, agglutination in normal sera and a tendency toward spontaneous agglutination. The melitensis culture modified ("entrainé") by association with either the para form or with cultures of other bacterial species, such as B. anthracis, B. pestis, V. cholerae or B. typhosus (also possessing the property of thermo-agglutination), was found to have gained also these other characteristics. With reference to the serologic tests, it was observed that the modified melitensis culture no longer agglutinated in rabbit serum immune to the normal melitensis culture. At the same time, such a modified culture was reported not to have been made more agglutinable with rabbit serums immune to the para culture type itself. In the light of other dissociative reactions, this result is somewhat peculiar.

Antigenic modifications of normal melitensis were also secured through association with cultures (alive, or killed by heating), or with culture filtrates of other bacterial species (B. anthracis, B. pestis, etc.). The modifications produced in this way, however, were less constant and less pronounced than those produced by association with the paramelitensis culture. When "Entraînement" was effected with killed cultures of other species, it often required several passages in series of the normal melitensis culture (in sacs or in candles) through the heterologous culture, or culture filtrate, before an appreciable antigenic modification in the normal type was obtained. Here we may bear in mind that, in the common dissociative reaction, the transformation from the S to the R form of culture is not attained at once but ordinarily by slow stages.

In his discussion of these results, and of the nature of the phenomenon which he describes, Burnet undertakes to relate them to other reported cultural transformations cited in the literature, but does not approach our present views regarding microbic dissociation, and indeed does not see his way clear to any satisfactory explanation. It seems to me, however, that just as the transformation of the normal melitensis culture type to the para form involves in all probability the essence of the dissociative reaction, so also in all his cited cases of modification through culture association, whether with the same species (para type) or with other species (anthrax, pestis, cholera, typhosus), we are confronted only with a dissociation of the normal, or S type, of melitensis into the R state, or into a form approaching the R. We have seen in this section that microbic dissociation is a phenomenon that can be "forced" by many different agents, chemical, physical and biological. That certain metabolic products given off from R type cultures are able to repress and to modify the growth of S culture of the same species has been pointed out by several investigators. Through the work of Burnet it now seems assured that another highly important "reagent" capable of instigating the dissociative reaction may be present in the growth products or dissociation products of heterologous bacterial species. The banal influence of pyocyaneus and anthrax filtrates on alien cultures is now well known through the work of Emmerich and Löw [158] (pyocyaneus) and of Malfitano [315] (anthrax); but in these cases it has never been demonstrated that the pyocyaneus and anthrax filtrates produced what we may now term cultural dissociations, although this seems probable. It is thus to the credit of Burnet's highly interesting study that we are now furnished with clear evidence of the importance of alien culture filtrates in initiating, in sensitive cultures, many of the reactions characterizing the phenomenon of microbic dissociation.

Although Burnet no doubt has presented one of the clearest instances of dissociation resulting from the use of heterologous cultures and culture filtrates, there exist in the literature other instances in which the same mechanism may be in operation, producing effects which are apparently unusual and difficult to explain. For example, Schiller [432] in 1914 studied the influence of cultures of B. acidophilus on the growth and viability of various streptococci. He observed that, if any streptococcus culture was added to a culture of B. acidophilus, the streptococci were killed in 18 hours at 37 C.; and apparently by lysis. After a week the streptococci were reduced to an amorphous mass. Schiller could not obtain similar lytic action in any other way. The fact that the results were not attributable to the acid elaborated by acidophilus, as one might expect, was shown by the circumstance that the streptococci would grow in acidophilus filtrates. But they would not grow in a filtrate of a broth culture of acidophilus in which streptococci had previously undergone autolysis. Thus: to an 18 hour broth culture of acidophilus was added a mass of streptococcus. This was destroyed in 36 hours. From this culture was then obtained a filtrate which proved just as germicidal to the streptococcus as the mixed culture of acidophilus and streptococcus. Schiller thought that the streptococcus might provoke the secretion of a defensive bactericidal substance on the part of B. acidophilus. That the phenomenon was not due to disintegration products of streptococcus itself was clearly shown by growing the latter in heated and aged streptococcus filtrates.

Also Castellani [91] in 1925 described some curious results observed in bacterial cultures grown in association. His work appeared to demonstrate that when two bacterial species grow in association, the resulting culture form comes to present biochemical characteristics which are possessed by neither of the original cultures when grown separately. B. typhosus, for example, grown under ordinary conditions, gives acid but no gas from maltose. The Morgan bacillus gives on maltose neither acid nor gas. But, when these cultures were grown in association in a peptone-maltose medium, the combination produced both acid and gas. If, however, the Morgan bacillus was added 24 hours after B. typhosus no gas was observed. In another case B. dysenteriae Flexner and the Morgan bacillus were grown in association. The former gives acid but not gas in mannitol. The latter gives neither. But, when the two grew in association, the combined culture yielded both acid and gas from the same medium. Phenomena of this sort are admittedly difficult to explain. Without the necessity, however, of calling to our assistance the "hybridization theory" of Almquist,[6] and in view of the modifying effect, biochemically and serologically, of the dissociative process (in which we know that fermentative reactions may undergo radical changes), it might perhaps be shown that the new culture form presenting the new feature of gas production was merely a dissociate (R or other) from one of the principle organisms employed in the tests. The exact nature of these phenomena is still obscure, but opens a field of study possessing much interest. The same may be said of Gratia's work showing the antagonistic and even lysogenic action of certain molds on heterologous species of bacteria.

Another interesting instance of culture modification through association is one recently reported by Jeanne Lommel.[511] Tubes of broth were inoculated with B. typhosus, B. paratyphosus, or B. dysenteriae Shiga. After the growth had attained about 500 million bacteria per cc. the tubes were inoculated with B. coli. From this point on, the method employed does not receive a clear presentation, but the statement is made that the B. coli used became modified to the extent of being able to ferment sucrose, which the original culture did not attack. The coli strain was therefore transformed into B. coli communior. Similar results were obtained by continued growth in broth containing phenol; also in broth containing formol. Since this sort of microbic association must occur in the intestinal tract whenever an infection with B. typhosus, B. paratyphosus or B. dysenteriae Shiga occurs, the author concludes that B. communior is merely a "simple biochemical deviation" from normal B. coli. In view of the results presented by Et. Burnet,[85, 86] it would be of interest to ascertain whether the transformation observed by Lommel might not also be effected by filtrates of the cultures employed, as well as by the whole cultures.

Taking these results in conjunction with those of Esther Stearn,[452] we can now conclude that B. communior can be produced from B. coli in three ways: through the action of gentian violet; through the action of phenol or formol; and through the action of microbic associations.

In concluding our consideration of the effects of microbic associations a recent contribution by Holman and Meekison [504] should be mentioned. These writers have brought together certain of the earlier observed facts relating to the effects of microbic association, together with some observations of their own. They point out anew that two bacterial species living in association may bring about biochemical changes which neither is able to produce alone. To this phenomenon the writers give the name, "synergism." Under this heading they list among others the observations of Castellani mentioned above. Prominent among the personal instances cited by them is the production of gas by the association of B. coli and S. fecalis in a medium containing saccharose. In this connection they state that the gas production was checked by increased acidity but that the addition of alkali would start the process again.

In attempting to explain these reactions Holman and Meekison, although making allowance for "changes in the metabolism of the bacteria" (implying, as I understand it, a change in the inherent physiologic reaction), seem more inclined to regard the mechanism as one involving a "co-operative" action on certain constituents of the medium. They state: "One of the pair of bacteria must be capable of splitting the test substance and forming acid. The other must be able to form gas from monosaccharides." In other words, one of the pair must be able to execute some sort of a preparing action for the work of the other.

In many of the numerous symbioses reported as existing between microorganisms of diverse nature, and also in some of the instances cited by Holman and Meekison, there seems little doubt that such an influence takes place. Moreover, that it is manifested not only by the production of gas from a certain sugar by the associated pair when neither of them alone produces gas, and only one of them produces acid, but also by the production of acid from a certain sugar by the associated pair when neither produces acid alone (Jeanne Lommel [510, 511]). The sort of reaction described by Holman and Meekison may occur in the majority of such associations; and for this phenomenon "synergism" is certainly an appropriate name. On the other hand, our present knowledge of dissociative phenomena may prompt the inquiry whether all such apparently synergistic phenomena possess the same significance. May not another reaction appertaining more closely to microbic dissociation be concerned in some of these instances? Reasons for doubting the unity of the phenomenon in question as described by the above authors may be found in the following circumstances. We know that B. coli is capable of transformation into a saccharose-fermenting (communior) type. We know, moreover, that the agencies by which this modification may be accomplished are diverse. It may be accomplished by continued growth in a medium containing malachite green (Jeanne Lommel [511]), by long submission to the influence of gentian violet (Esther Stearn [452]), by the continued influence of phenol or of formol (Jeanne Lommel [510]).

The last mentioned investigator showed further that the modifications produced by phenol and formol were less permanent than those determined by the influence of microbic association of B. coli with B. typhosus, B. paratyphosus A and B, and B. dysenteriae (Shiga) after twenty or more passages. In Lommel's case it is true that the study did not involve fermentation of saccharose with gas production; but the essential fact is that a new fermentation power was manifested by the associated pair, or by the single organism (B. coli) under the influence of an unusual environment.

It thus appears that a transformation of B. coli with respect to its fermentation abilities can be determined, not only through microbic association, but also through the presence of various antiseptic substances in the medium in which it is grown through a series of generations. In the latter cases there is manifestly no opportunity for "cooperative" bacterial action. Here we must assume either that the added reagent (dye, phenol, formol, etc.) serves to generate in the medium a new fermentable substance, fermentable by the unmodified culture; or that the organism concerned becomes modified insofar as it gains the capacity to ferment the original, unmodified substance. In the instances reported by Stearn and by Lommel, and in some of those cited by Holman and Meekison, it seems to me more likely that it is the latter reaction that really obtains. And added evidence favoring this view lies in the circumstance that such transformed cultures are likely to vary, not in one but in several respects, from the original type—undergoing a loss of some characters and a gain of others. If such transformations can result from the addition of certain chemical substances to the medium, it is not too much to expect that they may also occur under the stimulus of microbic associations.

In concluding this aspect of the matter I may say that the reactions included under the head of Holman's and Meekison's "synergism" are doubtless highly intricate, and that it may be desirable, pending the acquisition of further data, to hold open the possibility that phenomena of this apparent sort may embrace a variety of causes. Modification of the culture type through microbic dissociation (with consequent alterations in inherent fermentative capacity) determined by microbic associations may occur more frequently than Holman's and Meekison's interpretation of the reactions would permit us to be believe. Such reactions, involving dissociations, could therefore not properly be regarded as synergistic in the sense in which this term has been introduced. It seems to me that the question of synergism thus resolves itself into the true and the false. The true synergism would be that in which, as Holman and Meekison have clearly indicated, there exists a "cooperative" reaction but without necessary modification in either of the associated bacterial types. The false synergism would be that in which the outstanding results are referable to a modification of one or both of the associated pair through the functioning of the dissociative mechanism. The two sorts of reaction are manifestly quite different, and clearly to distinguish between them will be the necessary task of future workers in this interesting field of investigation.

The Influence of Certain Colony Types on the Culture Substratum. —The consideration of secondary or daughter colonies presented in earlier pages of this work serves to demonstrate that there may reside within the substrata of apparently homogeneous culture certain localized centers where a small group of organisms of a quite different type and potentiality from that of the mother culture have become differentiated and are carrying on their activities. If they remain few in number, and are not differentiated by color differences, as in the case of B. prodigiosus and other chromogens, they may escape unobserved. If they can successfully oppose the surrounding culture and increase to a sufficient extent, they produce a visible colony lying imbedded in the larger culture mass, which may be represented either by the mother colony or by an extended growth on agar surface. They may even appear as easily differentiated colonies of an agglutinative type in a liquid medium. What effect these secondary colonies exert on the primary growth might be expected to depend on their biochemical constitution, and on the kind and degree of the opposing biologic factors concerned in the association. And the same might be said with reference to the effect of the primary colony (or primary culture growth) on the secondary. It does not require a great stretch of imagination to suppose that, depending on the balance of those biologic factors underlying the lysogenic function, either the secondaries might destroy the primary culture, or that the primary culture might destroy the secondary, although this is less probable. As a matter of fact, among the great number of instances in which secondary colonies have been observed, and in a considerable number of bacterial species, neither of the reactions mentioned above is known commonly to occur. It can be stated with assurance, only that when colonies of the R type arise in a primary growth comprising the S culture, the R secondaries possess much greater vitality and still survive long after the S culture has perished. This was indicated strikingly by the studies of Atkin [22,21] on the gonococcus and the meningococcus. When, however, S type or O type cultures arise as secondary colonies in R substratum, we are not so sure of the sequel, because the exact conditions underlying such instances have not been clearly recognized or studied. Thus, although little can be said definitely on these matters, there are a few cases in which it is strongly suggested that certain repressive, and perhaps lysogenic, influences of the sort mentioned above actually exist. For this reason our consideration of the incitants to microbic dissociation would not be complete without reference to certain observations dealing with the influence of some slightly known types of colony, resulting from dissociation, upon the other S members of the bacterial society. This aspect is obviously closely connected with the matter just discussed, and is, I believe, likely to play an important rôle in further study of transmissible autolysis.

The problem in its present state is difficult to present because it often is impossible to draw sharp lines between the rôle of metabolic products and more specific inter-colonial influences. The essence of the problem was first made clear by the studies on B. anthracis by Preisz,[393] Pesch, Katzu and others, dealing with the lytic spots arising in cultures on agar. De Kruif,[120] however, was one of the first to call attention to a related phenomenon in cultures in which the chief dissociates were clearly recognized. When a culture of Bact. lepisepticum containing a mixture of S and R, but with a preponderance of the latter, was streaked parallel on an agar plate, both forms developed their characteristic colonies. But the interesting point is this: The first two or three streaks on the plate showed massed R colonies, but no S, although S colonies were numerous in the subsequent streaks. If S colonies did appear in close contact with R colonies the former were irregular and distorted, their margins being cut into by the R forms. Such pictures give the impression that the latter exert a strong repressive influence on the development of the S. I have seen the same phenomenon repeatedly in mixtures of R and S of many bacterial species, including intestinal forms, B. diphtheriae, B. proteus and B. subtilis. Mr. Nungester in our laboratories has made a similar observation, to be reported later, on dissociating cultures of B. anthracis. Pesch[387] stated that the greater development of the "blue" or transparent (R) form of this organism could "repress" the development of the smooth, opaque form (S). Mellon also has mentioned the antagonistic action of certain cell types upon the growth of other forms of the organism in the same culture. The matter has also been briefly referred to by Ischii[262] for B. typhosus. Gildermeister,[190] moreover, described for B. coli a special type of colony which he termed the "insonstant form" or "Flattenform" which changed rapidly in subculture and which possessed lysogenic tendencies. D'Herelle [248] believed this culture was "contaminated" with the bacteriophage.

Arkwright,[17] furthermore, has added evidence of the repressive rôle of certain colony types in Shiga cultures. He mentions certain "small colonies which appear to have an especially inhibitory or lytic action on larger colonies present on the same agar surface. The small colonies lie in concave notches of the larger ones." Such variants, he pointed out, like sensitive variants, may arise either from normal cultures or from cultures submitted to the action of the lytic principle. "The behavior of these variants indicates the action in the cultures of an excess of bacteriolytic function in some or in all of the bacteria present. . . ." It may be added that Mr. Nungester in this laboratory has demonstrated in his study of anthrax dissociation a third type of anthrax colony, sometimes very numerous on plates, and characterized by self-lysis. In further course it becomes transformed into the R type culture as well as continuing to perpetuate the lytic type. The lysogenic function of these colonies has not yet been studied.

Perhaps somewhat related to the antagonistic influence of certain colony types upon the mass of S type culture are the observations of Gratia [213] on the antagonistic influence of one culture of B. coli upon another coli culture. Strain V grew normally and when implanted in broth gave normal clouding after three hours. When one drop of the filtrate of V was added to a broth tube freshly implanted with another coli strain (ϕ), no growth appeared for seven hours and then a spontaneously agglutinative growth appeared. When a drop of the filtrate V was allowed to run over an agar slant recently inoculated with strain ϕ no growth appeared on the drop area. Finally the clear area yielded a few colonies which, when seeded to broth, gave an agglutinative growth. The V principle was not generated by strain ϕ, and that generated by V showed no possibility of serial transmission. Whatever may have been the cause of the reaction in strain ϕ culture, Gratia's results indicate a transformation into an R type of growth; in other words the bringing into effect of a marked dissociation under the influence of V or its filtrates. It is regrettable that so interesting a phenomenon was apparently not further studied.

Conclusion.—In concluding this section on the incitants to dissociation I wish to add that I have used the terms "to incite" and "to enforce" dissociation as if the action of the incitant were direct in determining the generation of the transitional or the R type culture. It is doubtful, however, if any such direct action occurs. It seems much more probable that the critical act of dissociation, involving what it may, is accomplished by the bacteria for themselves; and that the "inciting substance" or "inciting condition" merely sets into operation a physiologic mechanism by whose activity or by whose products the ultimate results are accomplished. Moreover, it seems that when this mechanism is once set into operation it tends to maintain its action until a certain portion of the culture is exhausted and certain transformations accomplished. My meaning in this will perhaps become more clear as we advance to later sections of this paper.

The problem of the action of incitants is undoubtedly capable of solution; but that the means and the results may be found different for each unrelated group of cultures we can scarcely doubt. We can even now observe, however, that one common factor unites all the agencies that we have thus far surveyed—namely, a condition unfavorable for growth of the S type culture. As a consequence it passes into a transitional form which is highly unstable; and from whose destruction in greater or less degree there arises a new type, the R. When the nature of this reaction is discovered it may be anticipated that the path will be revealed to the solution of another phase of the problem of dissociation —namely, the phenomenon of the bacteriophage.

Of the data that have been presented, probably the most significant relate to the action of immune serum on the virulent S type of culture. We have seen that the common effect is to produce a culture of modified antigenic nature, of diminished virulence and of increased phagocytability; moreover, that there is some evidence to suggest that this modification occurs in vivo as well as in vitro. These results are therefore naturally concerned with a very fundamental aspect of immunity—namely, the rôle of the immune serum and its actual mechanism in protection. Out of the vast amount of study which this last problem has received during the past forty years, a lamentably small nugget of truth has been derived. We have been forced to abandon many of the older conceptions regarding the bactericidal influence of immune serum so eagerly supported by numerous older writers; but little has been supplied of a direct nature to take their place. The accent given to various theories of immunity has changed with the season, although in more recent time there has appeared a definite tendency to turn more and more to the influence of the bacteriotropic antibodies in their correlation with phagocytosis. But the bacteriotropic antibodies are not ordinarily germicidal, although they are sometimes bacteriolytic. They are usually conceded to be agents which effect some sort of a modification of the virulent culture preparatory to phagocytosis, but not necessarily destroying directly the virulent organisms. What is the nature of this "preparing action" on virulent bacteria? I believe there already exists sufficient evidence to intimate that it is *incipient microbic dissociation in vivo;* and that this reaction is the underlying mechanism of all opsonic action. This possibility was first suggested by Griffith [215] in 1923.

If this view of the matter should eventually prove to be true, it is possible that the aim of much therapeutic endeavor might undergo significant modification. At the present time we often seek to destroy microorganisms which have established an infection in the blood or tissues and we have become accustomed to measure the expected efficiency of the destructive agents by ascertaining their germicidal power in vitro; we look for the killing power of the substance and trust that the effect in the body may be on the same order. It might be the outcome of such a new conception as I have outlined above, that we should strive not so much to destroy directly the microbes infecting the blood or tissues, as to stimulate dissociation in vivo; and so permit the organisms to work out their own destruction while aided by the phagocytes.

Although we know little at present regarding the influence of various substances, organic or inorganic, in effecting microbic dissociation in vitro, and even less regarding inciting the reaction in vivo, it might eventually appear that many of the substances that are at present employed because they are germicidal or bacteriostatic in vitro will be found to have less significance as stimuli to dissociation than some others that may have no or slight germicidal power. One reason for considering such a possibility is that, while the average homologous immune serum cannot be said to possess appreciable germicidal power, we know of no incitant to dissociative reaction, either in the tube or in the body, which can compare with homologous immune serum in energy, speed of action or selectivity when it is employed in sufficient concentration and for a sufficiently prolonged period.

These conceptions of the possible part played by microbic dissociation in vivo as a protective reaction, and of the homologous immune serum as a dissociation-provoking agent, are as yet far from possessing adequate support in concise experimental evidence. But it may be said with some justification, I believe, that the view is sufficiently well grounded to warrant its consideration as a modifying factor in many aspects of experimental immunization and serum therapy.

On the other hand, there is one circumstance that may cause us to be less optimistic regarding the possibility of creating therapeutic dissociations in the body except in a few infections in which we can recognize the S type culture as most virulent. Experience leads us to believe that several pathogenic organisms whose S type carries the virulence are taken into the body in this form, and in such case may eventually be dissociated into the nonvirulent and easily phagocytable R form under the influence of the blood, immune bodies or body fluids. But suppose, as might easily be the case, we are dealing with pathogenic organisms in which the virulent stage in the cyclogeny of the species does not lie within the S type culture, which in this event may be nonvirulent. It is then conceivable that such organisms, taken into the body in a harmless state, might, under the influence of the blood or body fluids, be dissociated into a virulent form. From this point of view the socalled filtrable virus might represent an ultramicroscopic and virulent dissociate of an organism which is common to us in the guise of a microscopic and nonvirulent form. It seems more than possible that these views may have a relation to several types of streptococcal and of spirochaetal infection. This conception is admittedly highly speculative, but it has some basis in fact. Furthermore, I believe the knowledge already gained of many remarkable phenomena occurring under the cloak of microbic dissociation should make us alert to eventualities heretofore unsuspected in the field of pathogenic bacteriology; and particularly with reference to the filtrable forms of pathogenic bacteria.

12. THE DEGREE OF PERMANENCE IN THE CHARACTERISTICS OF THE R DISSOCIATES AND THE NATURE OF THE R → S REVERSION

With reference to problems in epidemiology and carrier infection microbic dissociation assumes considerable significance; and in this relation one of the most important aspects lies in the question of the permanence of the characteristics of the R types and other less clearly recognized culture forms. Moreover, the question of permanence bears closely upon our current views of bacterial mutation, since the justification with which we might regard the newly-formed dissociates as true mutants, would depend upon their ability to hold permanently to their newly acquired form, and to become the center of fresh variational activity—a point considered in another section of this paper.

Evidence for Permanence of R Type.—Regarding the permanence of the R types there exists much conflicting evidence. Firtsch,[178] who in 1888 was one of the first to present evidence of the phenomenon that we now term dissociation, spoke strongly in favor of the permanence of his types ("variants") II and III of the Vibrio proteus. Dyar,[142] who in 1895 presented the details of a clearcut case of dissociation (in this case involving B. lactis-erythrogenes), stated that, of 125 subcultures made from isolated colonies of the "wrinkled mutant" (R), all but three remained true to the wrinkled type; and these three were of the normal, "soft" form (S). At the same time Dyar stated that a few of the wrinkled colonies showed "soft borders." The wholly wrinkled type, however, bred true for many generations, and in the hands of a less careful observer than Dyar might easily have been reported as involving a permanent mutation. Dyar apparently recognized nothing unusual or significant in his findings and casually remarked that the same sort of phenomenon could probably be observed in most bacterial species. After more than thirty years, we are just beginning to appreciate the truth of this view.

Loeffler [307] also reported that his four B. coli mutants remained true to type. Baerthlein [30] recorded permanence for some bacterial species but not for others. The permanence of what was probably the R type of the diphtheria bacillus has been upheld by Corbett and Phillips,[110] Slawyk and Manicatide,[443] Zupnik,[498] Bernhardt [49] and many others; and most recently by Crowell [114] working with a single cell strain. Von Lingelsheim [304] reported the permanence of his "Q-form" (possibly an "intermediate") of B. typhosus for at least five years and Feiler [167] recorded the permanence of his immune serum R strain of the same organism on agar, though not in broth. R types produced by Feiler by other means (starvation, phenol) were not so well "fixed." Bernhardt stated that no reversion of his "extreme variant" of typhoid occurred in 15 months. Arkwright [16] mentioned the permanence of the R form of various members of the colon-typhoid-dysentery group in some cases but not in all. Schütze [436] stated with reference to B. paratyphosus B that "it is impossible apparently to convert a 'rough' into a 'smooth'". Preisz,[394,395] Eisenberg,[149] Bail and Flaumenhaft,[34] Wagner [476] and others found the nonvirulent type of B. anthracis (which as we have seen seems to correspond to the S form) remained constant after long isolation and after many transfers. Weil and Felix,[483] Braun and Schaeffer [66] and others bespeak a high degree of permanence for the O form of B. proteus X19 and X2. P. B. White [487] attempted by many methods to cause a reversion in his R strains of B. cholerae suis but without success and concluded:. "On the whole we are of the opinion that the rough variation is probably an irreversible change." In the same organism Orcutt observed a considerable degree of permanence. De Kruif [110] found no evidence of the return to the S form of his R type of Bact. lepisepticum, except in a few cases of animal inoculation which, however, (because of the frequency with which normal rabbits carry the S organism) do not constitute altogether favorable evidence. All cultivation methods were reported unsuccessful in causing reversion in this case. Reimann,[405] as also Amoss,[8] have favored the view of the permanence of their R types of the pneumococcus, which Amoss regarded as "a genuine mutant." Mary Cowan [111] also considered her R type streptococci (both hemolytic and greening) as fixed stages. I may add that I have maintained for four years R strains of B. pyocyaneus without there having occurred any tendency to revert to the normal pyocyanogenic (S) form. An R type of the Friedländer bacillus (noncapsulated) in my hands has maintained its chief characteristics for more than two years.

Hans Wreschner [492] in 1921 presented evidence dealing with the permanence of the noncapsulated (R) type of M. tetragenus and its failure to revert to the capsulated S type. He found that in blood mediums such as serum broth the reversion occurred regularly in most cases; but a strain of R was ultimately found which no longer possessed the ability to revert in any medium. Further data on Wreschner's interesting case are presented in connection with capsules (section 5) and with virulence (section 8). Eisenberg [153] in 1914 also recorded the permanence of the R type of M. tetragenus but did not study the question fully.

To the views mentioned above relating to the degree of permanence of the R type cultures arising from active microbic dissociation, a statement should be added regarding the permanence of the presumably analogous R type cultures arising under the influence of the bacteriophage; for, as we shall see in a later section, the phenomenon of transmissible autolysis, accompanied by its generation of secondary, resistant cultures, in reality may be regarded as a form of dissociation of the original sensitive culture. D'Herelle does not bring out this relation, and for the reason that, throughout all of his publications, he is apparently oblivious to the existence among bacteria of the phenomenon of microbic dissociation which receives no mention, either under this caption or under any other, in any of his works. For him, a resistant culture is always one that has been made so through previous contact with the bacteriophage. He [248] notes, however, that these secondary cultures are often quite different, morphologically, biochemically and serologically, from the original culture. They are all indeed "mutants," immune strains, produced by the bacteriophage! Among them some may revert easily and quickly to the former state while others he regards as permanent in their newly acquired characters. We shall see later, in the section dealing with the relation of dissociation to the mutation theories regarding bacteria, that Bordet [61] also has raised the bacteriophage to a high eminence as an agent producing permanent mutations, and thus controlling the destiny of the species. But, as we shall also note later, both of these views are alike in their falsity. True bacterial mutations are, as yet, unknown, for we have been using false criteria for their attempted recognition.

Evidence for "Reversion" of R Type.—Regarding the reversion of the R form to S, fewer instances are on record. Bordet and Sleeswyk,[64] it is true, observed the variant of their B. pertussis return to the original form as soon as it was again placed on blood medium (from plain agar); but Bordet and Sleeswyk did not work with pure line strains. Penfold [386] found his daughter-colony mutants (the typhoid isodulcite fermenters) reverted easily unless grown for a considerable time on the same medium that gave the mutation; and, even under these conditions, partial reversion was often observed. Baerthlein [30] who gave the subject of reversion considerable study reported the return of the R type of his B. paratyphosus B after five and one-half months of uninterrupted cultivation in broth. Wreschner [492] obtained the reversion of one R form of M. tetragenus, but not of another R form of the same strain, by cultivation in normal serum broth.

In the case of the pneumococcus the reports regarding reversion are variable. Notwithstanding the permanence of the modified type obtained by Stryker [454] by the employment of homologous immune serum, so long as the culture was maintained on plain mediums, she found that inoculation into mice determined a reversion after one or more passages. The number of passages required to accomplish this seemed to depend, however, on the length of time that the modified culture had been under the influence of the homologous immune serum. If the original virulent culture had received only 6 to 12 passages, one passage through a mouse might reestablish a certain degree of virulence. If, however, the culture had received 50 to 100 passages through immune serum broth, three or more passages through mice were required to bring about a reversion to the virulent form. Here we manifestly have different degrees of "fixity" of the variant which are quite in accord with the results of much work of a similar nature on other microorganisms. Unlike Griffith, Reimann and Amoss, to be mentioned presently, Stryker did not ever observe absolutely irreversible pneumococcus variants. One possible reason for this difference is a circumstance pointed out seven years later by Griffith; namely, that the constancy of the R variant in its newly acquired characteristics is determined, among other things, by the strength of the immune serum employed. Griffith particularly was careful to impress upon his rabbits, by means of many injections, the highest possible grade of immunization. That Stryker took unusual precautions in this respect does not appear from her experimental records.

Although both Reimann and Amoss reported the stability of their pneumococcus R types, Griffith was more conservative and his results are of special interest. He [215] pointed out that the permanence of the R form depended in considerable measure upon the manner of its production, agreeing in this matter with the view of Braun and Schaeffer for proteus and with Feiler for B. typhosus. We may use his own words: "A rough colony culture which has been produced from a virulent strain by a single passage through immune serum, may retain its rough character for many generations in plain blood-broth and . . . may subsequently revert to the smooth type." Under these conditions virulence was recovered. He stated that this result obtains especially with the type 2 pneumococcus. But again he stated: "The R type characters, after three passages through immune serum broth, have remained constant for more than 13 subsequent generations in plain blood broth. The height of the serum titer seems to have an influence on the permanence of the change." One other interesting point is added. After long cultivation in plain blood broth certain of the R colony strains became in appearance almost indistinguishable from the S strain, but remained attenuated. This observation, at present unique in dissociation records, may give a hint as to the origin of nonvirulent pneumococcus S types. So far as the features of the special pneumococcus antigen involved in virulence are concerned, Griffith believed that the most important aspect of the $S \rightarrow R$ transformation was the loss of ability to produce the specific soluble substances. Evidence for this was also presented.

The study by Feiler [167] on reversion of B. typhosus is of special interest. He obtained modified typhoid cultures by means of phenol agar, "starvation agar" (see Braun and Schaeffer on proteus), and immune serum. In the first case the dissociation apparently progressed only to the intermediate O forms, while in the last the true R form was undoubtedly obtained. When Feiler attempted to obtain a reversion by returning the modified cultures to normal culture mediums, this was accomplished easily with the starvation strain, with greater difficulty in the phenol strain, and with still greater difficulty in the case of the immune serum strain. The last, after 107 passages in inactive, immune serum broth, returned to normal only after 18 passages through broth without serum at 37 C. At 22 C. twenty-six passages were required. When a reversion of the immune serum strain was attempted on agar, 25 passages failed to produce the normal type; the modification thus appeared permanent under these growth conditions. From these results we can see that growth on a solid medium may favor, not only the stability of the culture against the dissociation from S to R, but also stability against reversion from R to S. It is also of interest to note that, just as Soule [450] observed that the $S \rightarrow R$ transformation in B. subtilis occurred more rapidly in large volumes of medium (300 cc.), Feiler found the $R \rightarrow S$ transformation of B. typhosus was hastened by transfers through comparatively large volumes (50 cc.). Passage of the R typhoid through guinea-pigs showed no influence in effecting reversion.

One of the more recent instances of reversion of type R culture is found in the work of Jordan [273] who studied the problem in B. paratyphosus B. For this study he used two representative cultures which at the outset showed respectively 10 S to 1 R, and 5 S to 1 R, as indicated by examination of colonies after plating. From these cultures single cell isolations were made and pure cultures of S and R obtained. Frequent transfers were then made through veal infusion broth, P_H 7.4. One of the R strains revealed some S colonies after 16 days (32 transfers); the other after 25 days (50 transfers). The virulence of the derived S type (after it had been regained from the R) was still intact, while the accompanying R form was still nonpathogenic.

Most recently Soule [450] has accomplished the reversion of an R type of B. subtilis which had remained constant on solid media for many generations. The reversion was effected by the addition of 5% of active, normal rabbit serum to the beef infusion broth used for culture. After eight hours in such serum broth the degree of dissociation amounted to about 10% of the cells, as indicated by plating; while a control tube without serum, placed under the same conditions for growth, remained 100% R type cells. Stronger concentrations of normal serum were not found to produce a more marked reaction. When, however, a 10% homologous, R immune serum broth was employed, a reversion of about 40% of the R cells was quickly obtained. This interesting result has been considered more fully in section 11. It is a matter of some interest that something in serum is able to enforce the $R \rightarrow S$ transformation, not only in pathogenic species like the penumococcus and lepisepticum, but also in so prominently a saprogenic form as B. subtilis.

Conflicting Evidence.—A consideration of the nature of the incitants to dissociation, and the possibility of reversion, would not be complete without the presentation of certain evidence which in some measure tends to confuse the fairly definite results already established. The observations I refer to are found in the work of de Kruif [120] and Morishima.[352] The point involved in both is the following: not only was dissociation produced, but after a time, although the modified culture was maintained constantly in the same sort of medium that enforced the dissociation, a reversal to the normal form seemed to occur. This seems fairly clear in de Kruif's work; in Morishima's observations another explanation is perhaps possible. The instances may be described as follows:

De Kruif studied the influence on the $D \rightarrow G$ ($S \rightarrow R$) transformation of varying amounts of peptone, 0.2 to 2%, and made readings of the degree of dissociation, in terms of D and G colonies on agar, after periods of 24, 48, 72, 96, 120, 144 and 192 hours. Considering the results as a whole it appeared that the lower concentrations of peptone (0.2 and 1.0%), as well as the plain broth control and the undiluted rabbit serum control, gave a small degree of dissociation as compared with the high peptone concentrations (10 and 20%), as recorded after a period of 72 to 192 hours. In some cases the transformation involved nearly 100% of the organisms after 96 hours. But the important point is that, after 192 hours, there sometimes occurred a change in proportions by which the readings gave 50 D to 50 G—in other words indicating a regeneration of the D type. The same results are indicated in another experiment in which it was shown that, starting with a pure type D culture in plain broth, the type G organisms reached a maximum after 173 to 197 hours, but that the relative number was lower after 228 hours and still lower after 18 days. The same was true of a culture of D in 5% serum broth. In still another test, in 2% peptone solution, the count at 15 hours showed all D type, at 96 hours the ratio of 40 D to 60 G and at 146 hours all D again. In other words, in all of these tests there seemed to occur, after the initial $D \rightarrow G$ transformation, either a renewed overgrowth of D or a higher mortality among the G forms, which would be unusual. De Kruif presents no data that permit us to decide between these alternatives.

The second investigation mentioned above conducted by Morishima [352] differed from that of de Kruif in that, instead of growing his normal culture (B. typhosus) continuously in the same tube of serum broth, he transferred from tube to tube daily over a considerable time; moreover Morishima employed typhoid immune horse serum while de Kruif used normal rabbit serum. Since the further details of Morishima's tests have been presented in section 10, it is now sufficient to recall that, although the immune serum at first caused a rapid change of the normal culture to the nonagglutinating type (in homologous immune serum), continued transfers through the same medium caused, or permitted, a return after a varying time to the normally agglutinating culture. This seems to be equivalent to the D → G → D transformation observed by de Kruif occurring in the same tube of medium after varying intervals. We note that Feiler on the contrary was able to carry his strains of nonagglutinating typhoid bacteria (also produced under the influence of immune serum) through 174 consecutive immune serum broth passages without any tendency to reversion. Indeed, when this culture was returned to normal medium (serum free) the condition persisted for 18 to 25 generations. These results of Morishima are difficult to explain and seem to uncover a new question for which, at present, we have no answer. There is the possibility, however, that the transformation first observed under the influence of the immune serum was not the S → R transformation, but only the S → O, due perhaps to a weak immune serum. In this case the continued cultivation under the same conditions might have served to complete the dissociation to the R type. Unfortunately, Morishima did not present cultural or other data which would make an opinion on this point possible.

The Nature of "Reversion."—In concluding this section dealing with the so-called permanence of the R type, one further aspect of the problem demands attention. In our consideration thus far of the retransformation from R to S, I have employed the term "reversion," since this has been in common use in the literature and seems to cover the sort of change superficially observed in the cultures. I believe, however, that we shall do well to consider very carefully another possibility: Is the reverse transformation from R to S in reality a reversion or is it a further *progression?* In dealing with this question, which I have not found treated in the literature, although it is suggested in the work of Mellon [331] on B. coli mutabile (with reference to cell forms and their fermentation reactions), I wish first to present my own view of the nature of the changes. We may then undertake to ascertain what evidence supports this view.

Experimental data already presented with special reference to the serologic and cultural (colonial) aspects of dissociation indicate, for several bacterial species at least, three distinct antigenic components, S, O and R. Of these there can be little doubt in the cases mentioned; whether there are still other components must remain a question. These antigenic components are found united into various, but fairly orderly, combinations so as to produce culture forms (antigenic configurations) which have also become recognized as S, O or R cultures, named (according to one scheme) from the particular antigen that is most prominent. It does not need to be said that these antigenic configurations are found to be correlated with certain culture types of fairly distinct morphologic characteristics. Although some cultures exist which apparently contain only a single antigen (O or R), they more commonly contain two; and it is possible that all three may be present in a single culture. Accepting therefore the fact that the serologic culture type is determined by the nature and balance of the antigenic components which make up the configuration, we may make use of capital letters to designate the antigen which is dominant in the configuration, and of small letters to indicate the antigen that is commonly secondary. Making use, then, of the experimental data supplied by Weil and Felix, Arkwright and Goyle, White, Goyle independently, and Baltenu, as well as of many incidental observations made by others, we may set up the antigenic constitution of the three chief culture types as follows:

S type (normal smooth) becomes...... So
O type (smooth variant) becomes..... Or (or O) (or sO)
R type (rough variant) becomes...... Rs (or R) (or oR)

In other words, the S type culture always contains some O antigen, although the S is dominant; the O type culture is likely to contain some R antigen, although the O is dominant; and the R type culture may contain some S antigen, although the R is dominant. With respect to the antigenic configuration of the R type culture, White[487] (p. 80) expressed the view, based on absorption tests, that rough cultures often show a sufficient remnant of O antigen "to absorb all or nearly all the smooth O agglutinin from a smooth serum." In addition, Goyle has pointed out that both O and R culture types may appear which are apparently pure O and pure R, respectively. It might happen that the "pure R" antigen is correlated with those culture forms which show the often described, marked constancy in their adherence to the R features. The normal S type culture, on the other hand, seems never to consist of pure S antigen; it always contains O. That it sometimes may also contain R antigen (especially in old cultures) seems probable from results of cultural and serological tests, and especially in those cultures where the O form rapidly gives place to the R, as in the dissociation of the pneumococcus and Bact. lepisepticum. The points thus far presented are strongly supported by experimental evidence from B. typhosus, B. enteritidis, B. dysenteriae, B. cholerae suis and (to a slighter extent)

The Trend of the Dissociative Reaction.—The question now arises regarding the trend of the dissociative reaction, starting from the "normal" type S culture. As has been shown by considerable evidence, although the generation of the R type culture often seems highly abrupt, it is in many cases a gradual process, and the "extreme variant, R" is not ordinarily attained until a considerable time after the first rough characteristics may be noticeable in the colony or culture. Moreover, we have seen that in many instances an intermediate or transitional form of culture intervenes between the S and the R forms, although in certain cases this intermediate seems to be absent. Even here, however, we have noted that there is reason to believe it still exists, but for some reason that is still beyond us fails to make its appearance on the usual culture mediums. This is what we have designated the O type, although it would be better to say, one of the O types (intermediates), since they comprise a somewhat variable group, and it is only certain members of this group that are the most unstable. This aspect of the matter has already been presented in section 6 and it is unnecessary to repeat it here. Although the transitional form is apparently lacking in some instances, in others, as in B. proteus for example, the transforming organism seems to halt easily at the intermediate stage and to demand a greater stimulus to force it over to the R form. We might perhaps anticipate that different species would vary in the readiness with which they become transformed into the "extreme variant." Considering all the data at hand, the direct transformation from S to R, however abrupt it may appear, is highly improbable.

The establishing of an intermediate type of culture, characterized by a specific antigenic configuration, has therefore the great importance of determining a *direction of transformation.* The actual process of change is, therefore, not $S \rightarrow R$, but $S \rightarrow O \rightarrow R$. The first phase of this transition ($S \rightarrow O$), as we are led to believe from the work of Arkwright and Goyle, White, Goyle and Balteanu involves a loss of the heat labile antigen (S), and an increase in the heat stable antigen (O), which we know is always present to some extent in normal S cultures. At the same time the configuration may perhaps gain a little R. The configuration therefore changes from So (normal culture) to O or Or (intermediates). It would seem probable that sO also exists at a certain stage in transition, but this has not been reported. In any case, the S antigen has been largely lost by the time the intermediate O culture type has been reached. Among the many and various cell forms that comprise the O type culture, just which one, or ones, are responsible for the marked antigenic differentiation cannot be stated, but establishes an interesting problem.

But many observations already cited show that the intermediate type O culture, when compared with the S and R cultures, is more unstable. It can either revert to the S culture or progress to the R, depending no doubt upon whether it has lost all of its S antigen and upon the nature, intensity and duration of the stimulus that forces the reaction. Indeed, these O type cultures seem to be "opportunists" insofar as, within the limits of their antigenic structure, they are quick to seize upon any advantage offered by changing environment or circumstance. If the conditions provocative of dissociation are maintained or intensified, the transformation to the R type culture sets in, the R antigen being acquired by degrees. The type O culture thus changes from O, or Or (according to White's data), to oR or R.

Whether the culture now remains in the R state or begins to generate the S antigen, as seems to be the case with many partially stabilized R cultures, again seems to depend on the environmental conditions. We have already examined some instances in which great stability has obtained; others in which the retransformation to S culture has occurred within a short time. In any case, whatever remnant of O antigen may have remained in the R configuration is dropped and some S antigen is gained. The configuration accordingly changes from oR or R to Rs, which culturally at least may still be a typical "rough." If the "reversion" is destined to complete itself, the culture passes through rS, not apparently to S (pure), but to So, representing again the normal S culture type. And thus the antigenic cycle is completed. Whether there may exist a transitional stage between R and S in the R → S transformation, such as is found between S and R in the S → R transformation, cannot yet be stated. It has never been described, but its existence seems possible. Even if found, however, it is doubtful if a new antigenic group would be added to the three types already indicated. These three, independently or in combination, provide ample basis to afford a serologic differentiation of the three chief culture types already considered. On the other hand, it seems probable that, under varying conditions of cultivation, there may exist still other antigenic configurations incorporating the three essential antigens, or any two of them, in varying degrees, and which would be represented as other arcs about the central antigenic circle. It is clear, however, that for some unknown reason the chief culture types become stabilized near to the three antigenic points. These also mark a stabilization with respect to the colonial and cultural characteristics of the three chief types, as has already been shown.

The Antigenic Cycle.—The preceding exposition illustrates my meaning when I have said that it seems most likely that the R → S transformation takes place, not through true reversion, but through further progression; and thus effects what I have termed an antigenic cycle, rather than a "back-and-forth" variation. Any final conclusion that such a cycle exists requires more and better evidence than is now at hand in the literature bearing on dissociation; but many circumstances point strongly to such a cycle of change involving the main culture types and their varying antigenic constitution. Some of these points may be briefly referred to:

While cultural examinations, carefully made, have usually demonstrated the O type intervening between the S and R types in the S → R transformation, I know of no evidence to indicate the intervention of this type, or of any other type, in the R → S transformation. (But see Soule,[450] P type.)

At the very beginning of dissociation, in nearly all cultures that have been examined carefully, the first sign of antigenic change involves a loss of the S antigen and an increase of O antigen. The second stage involves the loss of O antigen and the gain of R. The third stage involves the loss of R antigen and the gain of S.

If the R → S retransformation were in the sense of a true reversion, then the R antigen should first lose R and gain O; and next, lose O and gain S. There is no clear evidence that either of these changes takes place. The first sign of transformation in the R culture is not the gain of O, but the gain of S. Many R cultures (not the "extreme") contain S antigen, in addition to R; just as the normal S culture contains O in addition to S.

In old broth cultures consisting exclusively of R, and from which the transitional O has quite disappeared (although it was present earlier), the S type culture may sometimes be seen to regenerate in limited degree (Bernhardt), but still without the reappearance of the transitional form.

The conclusions to which I am thus led from the foregoing observations will doubtless remind the reader who is familiar with the valuable works of Mellon, Enderlein and others, of the cycles of development which they have postulated for various bacterial species, based largely upon the changes in the morphology and reproductive function of the individual cells. While "life cycles" of bacteria will doubtless eventually be worked out on this basis, I have for some time been of the opinion that, if a cyclical development is established, it will be first on a larger pattern than that cut out by the majority of recent writers, whose work (with the exception of Mellon's) has not consulted in appreciable measure the fundamental antigenic and serologic aspects of the transitions; and which has also perhaps made too slight use of the striking cultural aspects which are found to underlie the various culture types, the significance of which we are just beginning to appreciate. In nearly all the published work dealing with the alleged life cycles of bacteria I have found it a difficult if not impossible task to ascertain what the definitely "cyclical" aspects actually were, although the fact of transitions has often been clear enough. And I believe that these cyclical phases will become fully clear only when we grasp the problem with reference to the varying antigenic and serologic features of the chief components—the general nature of which, I believe, is now fairly well established. When such fundamental data are fully in hand it will be both an interesting and a necessary task to learn what the cell changes actually are that support this diversified antigenic structure; also what relation these chief components bear to the filtrable forms of bacteria which, despite much current protest from the still surviving monomorphists, may now be recognized with certainty in many bacterial species.

Conclusion.—In concluding this consideration of the socalled reversibility of the R type I quite agree with Baerthlein [30] that the possibility of its demonstration will depend on the method employed. We know that all cultures capable of manifesting dissociation are not made to show the reaction by the same condition or environment, although there may always be a common factor in such conditions, in that they are prejudicial to what we term "normal growth." It is legitimate to believe that the conditions which enforce dissociation are not the same as those which enforce or permit reversion; and many instances bearing on this point have been presented in the foregoing pages. I believe, therefore, that we shall scarcely be justified in the view that such reversion is not to be expected in all cases until we have exhausted every means of modifying the environment of the organism concerned. It does not seem quite in accord with general biologic principles to assume that variations of hereditary significance (that is, true mutations) would be formed as easily or as commonly as we observe to be the case in the production of cultures of the O and R types; or that microorganisms in general are addicted to discarding permanently their ancient hereditary characters with such apparent nonchalance as can be observed in the frequent examples of the dissociative phenomenon occurring in every

13. DISSOCIATION DETERMINED BY THE LYTIC PRINCIPLE (BACTERIOPHAGE) AND SUGGESTIONS FOR A MODE OF REFERENCE TO THE SECONDARY DISSOCIATES

Whatever may be the nature of the bacteriophage of d'Herelle, or of its lytic influence upon the bacterial substratum, the fact has become clear through numerous observations on many bacterial species that, when employed in weak dilutions, or sometimes in strong, it acts more or less like an antiseptic substance (phenol), like some dyes (gentian violet, malachite green), or like immune serum, insofar as it produces, or intensifies, culture changes which are of the general nature of dissociations: O types and R types of culture of various grades and potentialities are generated from the "normal" culture of the substratum. There is to be observed, however, between the action of the bacteriophage and other agents of dissociation, a marked difference in the speed of the reaction. Without relating them in any way to the dissociative process, or to any phenomenon allied with it, both Bordet [61] and d'Herelle [247, 248] have recorded the origin of "mutating" culture types under the lytic stimulus; and Bordet has gone so far as to set up the lytic principle as an agent which, through its power of effecting "mutations," acts as a "director of evolutional progress"; and in this way "controls the destiny of the species." It may be that the thing which we call the bacteriophage is so bound up with the "destiny of the species," but not, I suspect, in the manner in which Bordet believes. We have already considered the nature and effects of active microbic dissociation as it appears spontaneously, or as it is forced under the pressure of unfavorable environment. But no treatment of the problems of microbic instability would be complete if limited to this aspect of the matter. In the present section we shall therefore turn our attention to the dissociation of microbic species as determined by the influence of the bacteriophage; and we shall see that many features of transmissible bacterial autolysis, as the bacteriophagic phenomenon has been termed by Bordet, resemble closely those of microbic dissociation—not merely in the nature of the results effected, but also in the mechanism of the reaction so far as we are at present able to understand it.

Cultures Secondary to Lysis.—The subject may be made clearer if we first review the simple facts established by the studies of d'Herelle and of Bordet. When the bacteriophage in appropriate concentration is added to a young broth culture of the sensitive organism lysis takes place within three or four to 24 hours and most of the bacteria are apparently destroyed. But some clearly survive. After some days these may give evidence of new growth and produce a fresh clouding in the medium. This represents the secondary or resistant culture. It is usually resistant to the same concentration of the same bacteriophage but not necessarily to a stronger concentration of the same (Bordet [61]), nor to another "strain" of the bacteriophage (Zdansky [495]). In other words, the resistance to the lytic agent is not absolute but relative, just as we have seen to be the case in the resistance of bacteria to antiseptic substances. As we shall note later, these secondary cultures arising after lysis differ in many respects from the original sensitive form (S).

Similar results in the production of secondary cultures may be obtained on solid culture mediums. If the concentration of the lytic filtrate is such as to carry only a few "lytic units" per loop of filtrate, and such a loop is streaked over the surface of an agar slant which has just been seeded with several loops of a young broth culture of the homologous sensitive organism, after incubation there will appear on the surface of the agar a continuous growth of the substratum culture, but it will be spotted with lytic areas, usually round and measuring from a fraction of a mm. to 15 mm. in diameter. Here at first there appears to be no growth. After a few days, however, there will often be seen on these patches a sprinkling of small colonies which grow only slowly. Most of these colonies also represent the secondary or resistant culture. It is resistant to the same concentration of the same lytic agent but not necessarily to a stronger concentration, nor perhaps to some other "strain" of lytic agent.

Among the colonies that arise on such bare areas on slants, or appear on plates streaked with mixtures of broth culture of the sensitive organism and lytic principle, there are likely to be several types. Some appear identical with the original sensitive strain and are round, regular and soft. These are infrequent. Others of different build are resistant to the lytic influence and "carry" the lytic agent into subsequent transfers. Still others are resistant to lysis but do not "carry" the agent in subculture. Whether these cultures are lysogenic, that is, will perpetuate the lytic action, can be ascertained by inoculating their filtrates into broth cultures of the homologous S organism and noting whether lysis or inhibition takes place. If the culture does carry the lytic agent it may be freed from it by appropriate methods of cultivation (Bruynoghe [77]); or by dilution methods (Bordet [61]). This results in a resistant but nonlysogenic culture. It appears further that the resistant character itself can be made to disappear by long cultivation on solid mediums (d'Herelle,[246] Bruynoghe). The lysogenic colonies are sometimes revealed by their irregular shape and ragged outline (Gratia [211]). When the resistant colonies are seeded to agar slants they give a slow, modified growth, always compact, sometimes mucoid and often hard and tenacious when old. Transferred to broth they always give the agglutinative form of growth like the R type from active dissociation.

Microscopically the cells of the resistant cultures differ widely from the original type. Although sometimes (especially in the mucoid colonies) giving filamentous and beaded forms, they most commonly present fore-shortened rods which frequently approximate the cocco-bacillus or even coccus type. This is true of B. coli, B. typhosus, B. dysenteriae and many others (d'Herelle [245]). Such cultures, which also possess modified biochemical reactions, may be remarkably persistant in their new form and sometimes, according to d'Herelle,[248] permanent.

What has been stated thus far relates to the action of the lytic principle on the "normal" sensitive culture. But we have seen that, through dissociation, the S and R types may be isolated and grown separately. What then is the relation of the lytic principle to these two chief dissociative types?

One might conclude from d'Herelle's [248] summary dismissal of Arkwright's [17] statement regarding the relation of S and R to bacteriophagic action, that no significant relation exists. But this would be very far from the truth as is made clear by certain studies of Gratia [211] and of Fejgin,[168] as well as by many still unpublished observations of my own. If Arkwright, as d'Herelle somewhat pointedly states, "has no idea of what the phenomenon of bacteriophagy really is," it is apparent that d'Herelle is equally lacking in understanding what constitutes one of the most obvious cultural distinctions in microbic dissociations, a subject which has received no consideration in any of his published works. The results of Gratia's [211] tests, the proper significance of which does not appear to be brought out in d'Herelle's analysis of them, agree to a considerable degree with my own findings with apparently analogous cultures and may be reported to make this aspect of the matter clear.

So far as I am aware Gratia was the first to report on the action of the bacteriophage on the well established S and R types separately. In this case they were the dissociates of a single-cell strain of a sensitive coli culture. The lysis of the S form gave the resistant SR, while the lysis of the R gave the resistant RR. Compared with the original S culture the type R culture was markedly resistant to the bacteriophage and gave many more secondary colonies than the former. No difference in resistance between the SR and RR secondary cultures is reported. In my own study of this point I have found that when S and R cultures (Shiga) are spread on agar plates, the addition of either anti-S or anti-R lytic filtrate makes a much slighter impression on the R, while lysis and inhibition of the type S culture is marked. Fejgin [168] also has found certain R types of the Shiga dysentery bacillus resistant to lysis by the bacteriophage.

It would be a misapprehension, however, to understand that the average type R culture from active dissociation is not to some degree susceptible to bacteriophage action, and this point will be considered further in the latter part of the present section.

It was perhaps the chief merit of Gratia's contribution, however, to reveal the numerous departures from normal type that can be produced by properly graded doses of bacteriophage acting on sensitive culture. Dilutions 10^{-1} to 10^{-5} of filtrate lytic for B. coli were used against a 12 hour culture in broth and the mixtures plated. The results showed not only that the number of colonies obtained varied inversely with the amount of the filtrate, but also that the amount of filtrate determined within limits the sort of colonies, and subsequent culture. In general there were three chief types of colony: mucoid (rare), non-mucoid and regular, nonmucoid and irregular. The mucoid colonies were nonlysogenic and on further plating gave rise to opaque, translucent and mixed colonies. The nonmucoid, regular colonies occurred only in high concentrations of lytic principle and bred true. The nonmucoid, irregular colonies were determined by lytic principle in moderate concentration and were lysogenic, as this term is later defined.

It may also be observed at this point that Blanc [54] reported the transformation of normal B. coli culture into a form characterized by mucoid colonies resembling the Friedländer bacillus as a result of bringing filtrates of sewage (which we know contains abundant lytic principle) into action against normal coli culture. It is not necessary to go to this trouble, however, since sewage already contains large numbers of this form of coli which are easily demonstrable on Endo plates. Similar transformations of B. coli into a mucoid aerogenes-like organism under the influence of the lytic principle have been reported by Bordet and Ciuca; [62] and by Gory [203] (using tapwater presumably containing lytic agent). D'Herelle [248] also reports having observed similar instances and assumes that all these cases have to do with "mutations" determined by the lytic principle.

It is important to note in passing that these various cultural modifications produced by the lytic agent on type S culture are not (at least with one exception) wholly peculiar to bacteriophage action. The same trend of transmutation, often revealing forms that are fairly resistant to lytic action, appears in many quite normal cultures when they are aged, or when they are brought face to face with various kinds of unfavorable environmental influence. In this respect, therefore, the action of the bacteriophage is not so unique as d'Herelle maintains. Perhaps the most striking difference that obtains between the R series of active dissociation and the R series of transmissible autolysis is that, in the latter, the resistance to lytic influence is greater; and that some of the secondary colonies produced under the influence of bacteriophagic lysis are themselves lysogenic. Even the last exception, however, we may eventually find to be gratuitous, since it already begins to appear that certain type R or other colonies of the active dissociation series exert a marked repressive (and perhaps lysogenic) influence on the growth of the normal S culture.

In his theory of transmissible autolysis Bordet was quick to seize upon the above mentioned observations of Gratia, and to confirm them, as supporting his conception of gradations in susceptibility of different organisms in the same culture to the lytic action, a point which forms the keystone to his entire theory. Fortunately the fact cannot be questioned. From many other observations there can be little doubt that the extent to which a culture proves resistant to lytic action, as also the extent to which it is capable of elaborating additional principle, depends not only on the strength of the lytic filtrate applied, but also—and perhaps primarily—on the state of the culture, the degree to which dissociation has progressed at the moment when the lytic test is performed. This, in turn, will depend upon the ability of the culture to undergo dissociation and may also be a function of time.

In connection with the forms of culture resistant to lysis by the bacteriophage, their origin and mode of generation is of interest; and especially d'Herelle's conception of these phenomena. D'Herelle's view is indicated by the following statement taken from his latest book [248] (p. 189). "The secondary cultures, then, are the result of the adaptation undergone by the bacterium which acquires an immunity to its parasite." He points out that the nitrate and acetate of lead, also nitrate and sulfate of silver, added to the culture favor the reaction. Aside from such stimuli, secondary cultures "have their origin in the operation of the phenomenon of natural selection whereby some bacilli show a greater aptitude than others to the acquisition of resistance to the bacteriophage" (p. 191). Although Gratia [211] had suggested on the basis of his own experiments that resistance of cultures to the bacteriophage might not involve the gain of resistance by the bacteria (after put into contact with the bacteriophage) so much as a selection of those organisms possessing a natural resistance (previous to contact with the bacteriophage), d'Herelle cannot accept this view, since for him the bacteriophage always acts on a "normal," homogeneous culture. For him selection operates, not on bacteria naturally endowed with greater resistance, but "through a selection of those susceptible bacteria which are more apt at acquiring resistance" (p. 191). In examining d'Herelle's position on this point it becomes clear that his objection to Gratia's view and all similar views of a previously-existing resistance to lytic action on the part of certain cells of the culture, is based on a lack of knowledge, or appreciation, of the process of dissociation of a normal culture into a culture form (R) more resistant to the action of the bacteriophage. The experiments of Gratia and my own are in agreement in indicating that the cells which are resistant to the lytic agent of medium strength are identical with those cells which comprise the R culture type, often mixed with the S forms. Their resistance to the bacteriophage is predetermined in the cyclogeny of the species; it is not produced de novo by the bacteriophage, although certainly in the course of bacteriophage action the resistance may become intensified. This is, I believe, the true explanation of the circumstance that, when an anti-S type bacteriophage is brought into action against an S type culture and an R type culture, the number of secondaries will be much greater in the case of the latter.

With further reference to d'Herelle's failure to comprehend the relation of microbic dissociation and of the R type culture to the phenomenon of transmissible autolysis, one other matter of considerable significance may be introduced at this point. The problem at issue is revealed by the following paragraph appearing in d'Herelle's latest book (p. 236):

"Strains contaminated by a bacteriophage may be recognized in that they give mutant colonies, while all the colonies of an ultrapure strain are identical. The bacteria of contaminated strains are but slightly, or not at all, agglutinable by a specific antiserum. The bacteria are resistant to the action of bacteriophages which attack homologous strains. Every time that one or the other of these characters is encountered there is reason to suspect contamination by a bacteriophage."

Nothing could be further from the actual truth than the information conveyed by this paragraph. Its composition was manifestly influenced throughout by the old, monomorphic conception of the nature of bacteria, which more than once serves to produce entanglements in d'Herelle's exposition of the bacteriophage and its mode of action. We may profitably analyze the paragraph mentioned by sentences. "Strains contaminated by a bacteriophage may be recognized in that they give mutant colonies while all the colonies of an ultrapure strain are identical." This statement is contrary to recognized facts because, as we have amply observed in the data on colony variation presented in earlier pages of this work, colony variability is one of the most fundamental and constant characteristics of any "normal," pureline, bacterial species; and, in the majority of cases certainly, there is no evidence that the bacteriophagic phenomenon, at least in the d'Herelle sense, is attendant. Evidence for this does not need to be repeated here. We may take the next sentence: "The bacteria of contaminated strains are but slightly or not at all agglutinable by a specific antiserum." This statement is true; but the circumstance is neither differential nor distinctive. While it is a fact that cultures resistant to the bacteriophage are only slightly agglutinable, or nonagglutinable in specific antiserum, it is by no means cultures alone which have proved resistant to lytic action that possess this characteristic; for it is also possessed by the R dissociates of all bacterial species that have been studied from this point of view. Neither the occurrence of spontaneous agglutinability nor failure to agglutinate with specific serums can thus be taken as a criterion of "contamination" with a bacteriophage in the d'Herelle sense.

Regarding the third sentence—merely because a culture is resistant to bacteriophages which attack homologous strains is in no way an indication that such a resistant strain is "contaminated" with lytic principle; and for the reason that any "normal" S type culture may dissociate into the R form and become more or less resistant to lytic action, even while homologous cultures, which have not undergone dissociation, are still sensitive to the lytic action. Finally, with reference to the last sentence: "Every time that one or the other of these characters is encountered, there is reason to suspect contamination by a bacteriophage." If this statement were true one would need to suspect the presence of bacteriophage in every sensitive (S type) culture capable of dissociation into the R state (which as we have seen is always endowed, among other characters, with exactly those features which d'Herelle states denote the contaminating presence of the bacteriophage). As a matter of fact,

I suspect that the lytic principle is present, potentially at least, in every such culture; but certainly not in a form that is within the scope of d'Herelle's meaning when he employs the term, bacteriophage. In every bacterial species thus far studied from this viewpoint there are stages highly susceptible to bacteriophage action, but other cyclogenic states in which the culture is much more resistant. Furthermore, when adequate tests have been made, it will undoubtedly be ascertained that there exist still other stages in the cyclogeny of a species in which the culture is actually refractory. These circumstances constitute added evidence of the close parallel, at least with reference to one important feature, between microbic dissociation and transmissible autolysis, a subject to be considered in greater detail in the following section.

Filtrable Forms of Bacteria Secondary to Lytic Action.—We have now observed the influence of the bacteriophage in producing modifications in normal, sensitive culture and in normal cells, as also in "pure" S and R strains. We may now turn to its influence in causing the generation of filtrable forms of bacteria. In the sections dealing with active dissociation in normal cultures it has been noted that such cultures may enter a stage of development in which they are filtrable through Berkefeld or Chamberland candles that are able to hold back all microscopic forms. In most of these instances no signs of dissociation have been noted at the time. It is also clear, however, that accompanying that form of dissociation stimulated by the bacteriophage, there often occurs a quick development of filtrable bodies. In 1922 d'Herelle [246] reported filtrable forms of the Shiga bacillus occurring in the filtrates of lysed cultures. Filtrable stages of B. coli, B. typhosus, B. dysenteriae and M. aureus under the influence of weak (homologous) lytic filtrates have been reported by Hauduroy [234, 235] and by d'Herelle and Hauduroy; [249] also for B. coli by Tomaselli.[463] Some of these filtrable forms were reported to yield a faint, opalescent growth in broth; and less frequently a delicate growth on solid mediums (Hauduroy). It seems also that they may propagate in the invisible state. In still other instances they may revert to the original, but still resistant, cell type; or to a modified cell type (coccoid). In such cases the opalescent growth gradually gives place to a definite turbidity. One point of interest in these cases is that, whatever the morphologic type of organism submitted to the lytic action at the beginning—rods or cocci—all show the same disintegrative trend toward granule formation; and in this end-state all meet on a common morphologic footing (d'Herelle and Hauduroy). In other words, the organisms concerned seem to have entered into a sort of morphologic cosmopolitanism, comparable with the "serological cosmopolitanism" of Schütze, relating in this instance not to the SR, but to the R types from microbic dissociation. I believe that we may eventually find in this "cosmopolitanism" or convergence of the rough types, whether morphologic or serologic, and whether occurring in connection with simple dissociation or with transmissible autolysis, a deeper significance than now attached; and one bearing upon a number of at present inscrutable problems recently introduced into bacteriological literature. What relation, if any, these and similar observations on bacterial convergence may have to the well established bacteriophagic heterogeneity of action in inhibition and lysis, remains to be ascertained. In this connection I have already shown,[227] however, how closely the bacteriophagic relations follow the serologic in the curious relationships existing between B. typhosus and the avian paratyphoids, Bact. pullorum and B. gallinarum. This phase of the subject is considered further in the following section.

But, to return to the subject of the filtrable forms of bacteria, it may be noted that Bronislawa Fejgin[171] has obtained a filtrable stage of the typhoid bacillus capable of producing an experimental disease in guinea-pigs. The virus was obtained through the action of a weak lytic principle on normal typhoid culture. From her results Fejgin concluded that "the invisible virus of typhoid is only the bacillus of Eberth lysed by the bacteriophage of d'Herelle." In a later communication Fejgin[171a] reported that, from the brains of guinea-pigs infected with typhoid virus, she could obtain by cultural means minute coccobacilli which differed from the normal culture biochemically and serologically, and which she believed represented the formation of a microscopically visible secondary culture in vivo.

By the use of a lytic principle acting on proteus X19 Fejgin[169] was also able to produce a secondary culture resembling in many respects the "natural" OX19 strain of Weil and Felix, as well as the O forms produced artificially by Braun and Schaeffer by resort to phenol and to "starvation agar." Thus, in the case of B. proteus, we observe that essentially the same results have been secured (with respect to obtaining dissociation) by five distinct methods: spontaneous dissociation; reaction to phenol; reaction to starvation; heating, and reaction to the lytic principle. Here again, therefore, the bacteriophage is not unique in its ability to produce "mutations." The essential phenomenon involved is microbic dissociation, but the fundamental reaction on the part of the bacteria may be incited in various ways.

In concluding these references to the filtrable forms of bacteria, generated under the influence of the lytic filtrates, one further allusion is required. Insofar as these filtrable particles not only survive the action of the lytic principle, but also propagate in a favorable medium (Hauduroy [235]) they might be regarded as a sort of secondary, resistant type (d'Herelle and Hauduroy [249]). If, as these workers also assume as a result of their more recent experiments, the lytic principle is bound up with some of these filtrable forms, as well as with intact organisms of the resistant type, producing a symbiosis (d'Herelle and Hauduroy), such filtrable forms would also be regarded, in the Bordet [60] sense, as lysogenic. Only a beginning of their study has as yet been made.

The Antigenic Relation Between R and SR Cultures.—The characteristics of the secondary cultures arising after lysis of the S type may also be considered in relation to serologic reactions. Few studies on this aspect have been conducted but those that exist afford interesting data. They may be considered especially with reference to the agglutinative and protective reactions.

It has already been indicated that when an R type culture is injected into rabbits the serum is likely to contain agglutinins for the R type, but not (or not strongly) for the S. The R form is different antigenically. The question naturally arises regarding the antigenic relation between the S type and those cultures that arise as secondary to the action of the bacteriophage. This point has been partially studied by McKinley.[324] Rabbits were immunized against the original culture (coli S); and against coli resistant to lytic principle (SR). Each serum agglutinated its own antigen but not the heterologous antigen. The complementary point, which is quite important, but which was not within the scope of McKinley's tests, is whether there exists an antigenic difference between cultures of the R type and cultures arising as secondaries to lytic action (SR). Also whether there exist reciprocal antigenic relations between various lytic secondaries, as Arkwright found there existed close antigenic relationships between the various R types of dysentery Shiga; or whether there are such antigenic differences as Felix [172] found to be the case in the "Zwischenformen" of B. proteus X19. Of the fact that both microbic dissociation and transmissible autolysis involve the generation of antigenic "mutants" there can be no doubt; but are the transformations along the same line of antigenic modification? This point is of much importance in any detailed comparison of the dissociative and autolytic reactions (section 14). No data bearing on this subject are yet available.

The Nature of the $S \rightarrow SR$ Transformation.—Regarding the nature and significance of the change from the normal to the resistant type under lytic stimulus, d'Herelle,[246] in the earlier stage of his investigations on the bacteriophage, was inclined to regard these changes as due to the effects of natural selection, by virtue of which only those bacteria most resistant to the bacteriophage were able to survive; and they survived, according to d'Herelle, because they were able to undergo adaptation: "Secondary cultures, then, have their origin in the operation of the phenomenon of natural selection whereby some bacilli show greater aptitude than others to the acquisition of resistance to the bacteriophage." In this reference it will be clearly seen that there is no allusion to actual transmutation in the ordinary sense. The result was due merely to a survival of some cells among fluctuating variations that existed preformed in the original culture.

But this point of view was apparently unsatisfactory to d'Herelle, for, in his second book,[247] he lays stress on the fact that some of the cultural modifications observed in bacteria under the influence of lytic principle actually possess "mutational" significance. Indeed, in his second book d'Herelle gives us to understand that "these mutations are always produced under the influence of the bacteriophage," a view warmly seconded by Bordet.[61] In 1926 d'Herelle's [248] belief in the mutation-producing power of the lytic agent is undiminished for he states in his last book, "It is indeed probable, as various investigators have suggested, that all of the fixed mutations occurring among bacterial species are produced through the action of the bacteriophage." This naturally suggests that no "mutations" can arise in bacteria, purely from intrinsic causes, or from extrinsic causes other than the bacteriophage; and that this is actually d'Herelle's view may be inferred from his further statement, "a pure bacterial strain is not subject to transformations." In view of our present knowledge of the facts regarding microbic dissociation, even if the explanation is not in hand, we can scarcely accept the truth of either of these statements. Not only pure bacterial strains, but even pure line strains, are certainly subject to transformations of a very far-reaching and remarkable sort. Moreover, although we cannot fail to recognize in the lytic principle an important agent in the production of so-called bacterial "mutations," we cannot safely regard it as the only factor; nor even assume that, in its essence, it is a fundamentally different factor from some others already at work in producing the phenomenon which we have termed "active microbic dissociation." To this subject we shall return in a following section.

Terms of Reference to the Secondary Dissociates.—The important point which the above findings bring out, however, is not that cultures dissociate into the S and R types, and that it is possible to obtain a "strain" of bacteriophage for each of these dissociates, but that, when such dissociates in turn undergo transmissible autolysis, each gives its own type of secondary and that these in reality appear to form a continuous series of transformations from the original type to the new "mutant." The transitional types punctuating this series are determined by the nature of the original culture and the strength of the lytic stimulus. Just as we may term the cultures arising from active dissociation the primary dissociates, I propose for these new cultures (arising as secondaries from bacteriophagic or passive dissociation) the secondary dissociates. To what extent these two forms of the culture resemble each other, and to what extent their generation may be dependent upon similar forces, will be considered later. The situation, however, is one that demands for purposes of clear exposition an expansion of the terms of reference already proposed. The following suggestions can perhaps be followed most clearly by the employment of concrete examples. The scheme presented has an experimental background in the studies of Bordet, d'Herelle, Gratia and Zdansky, together with some observations of my own on the behavior of the R types of culture under progressive lytic stimulus. I therefore suggest the following terms and, to make the subject clearer, I reintroduce the terms already proposed for the primary dissociates occurring in the primary reaction.

TERMS REFERRING TO THE PRIMARY DISSOCIATES

Type X. Refers to an original culture from a natural source. Likely to resemble the following type S but with somewhat different serologic reactions.

Type S. Usually a "normal" culture such as the average laboratory stock in the case of saprophytic bacteria. Contains besides antigen S, some of the antigen O of the "intermediate culture type"; if an old culture, it also contains some R. The normal, virulent type in acute infections. Exception in case of B. anthracis.

Type O. Representative of the intermediate or transitional group of cultures. Sometimes contains besides antigen O, remnants of S antigen and often some R antigen. Usually unstable culturally, and can transform either to S or to R. Probably the active culture element in the "suicide cultures" and in the S cultures in the lytic state ("Ly" of this list). Its natural end product in further dissociation is the R.

Type R. A dissociate from S through the intermediate O. Found particularly in old laboratory cultures of saprophytic bacteria and sometimes comprising the chief element in laboratory stocks of pathogenic organisms not maintained on fresh blood or tissue mediums. Contains R and often some

S antigen. A natural product of the influence of homologous immune serum. Especially stable and often reported permanent, but commonly transforms to the normal S.

Type Ly. Superficially an S type culture in the active state of dissociation (acute lytic tendency); actually an S type culture transforming rapidly to the intermediate O, with the rapid dissolution of this culture form, leaving as its product the R. In all but suicide cultures the lytic state can be perpetuated easily by transfer. Plating yields S, R and (sometimes, but not always) Ly colony forms. The Ly colonies are probably identical with what has been described as the "third colony intermediate." May also be slightly lysogenic.

TERMS REFERRING TO THE SECONDARY DISSOCIATES

Type SR. The secondary and somewhat resistant culture arising from the action of a lytic filtrate (bacteriophage) on its homologous substratum of the S type. Fairly stable. Plating gives mainly SR together with some S forms. Common in chronic and carrier infections in which lytic principle is also present. Antigenic configuration unknown.

Type RR. The secondary and resistant culture arising from the reaction of the lytic principle on R type culture. Equally or perhaps more stable than SR. Plating gives mainly RR forms. Usually a laboratory product. Antigenic configuration unknown.

Type Lg (lysogenic). An SR or RR culture resistant to lysis but which "carries" the lytic agent. Filtrates may initiate serial lysis in the homologous S substratum. Probably stable for many generations except on glycerin agar or sugar media. Plating gives SR or RR and Lg forms, but probably no S. Antigenic configuration unknown.

Types $SR,^2$ $SR,^3$ $RR,^2$ $RR,^3$ etc. SR and RR cultures of increasingly greater resistance to the lytic principle. Correspondingly increased stability. Usually laboratory products.

Types SR^n, RR^n. SR or RR cultures possessing maximum resistance to the influence of the lytic principle. An hypothetical type representing the most resistant, and pari passu, most stable, form of R, possessing "absolute" resistance to the lytic principle.

Resistance to the Lytic Principle is Relative, not Absolute.—No special reference need be made to these various culture types except perhaps in the case of SR^2 and SR^n. As is now well known, many cases of cystitis and pyelitis or pyelonephritis are due to infections with B. coli. From such cases the causative agent can be isolated and in a certain number is accompanied by a lytic agent present in the patient. Although one may suppose that the original infecting agent was an S type coli, an organism ot this sort is seldom isolated from the urine; it is sometimes an R type, but when lytic principle is also present, it is an SR or an RR culture. The same may be true of some chronic typhoid infections, particularly urine-carriers. Two such carriers which I have examined showed SR or RR types, another showed only the R type. In such cases (presence of SR or RR) we are therefore confronted with the incongruity of a strong but subacute infection existing in the face of a lytic principle sufficiently powerful to destroy laboratory strains of typhoid culture, but quite powerless to cope with the case strain.

Even under these circumstances, however, as Bordet [61] has shown by in vitro tests, and as Zdansky [495] has shown in clinical tests, we do not need to conclude at once that the case strain of B. coli or B. typhosus maintaining the chronic infection is not susceptible to lysis. It is merely required to discover or to produce another lytic agent possessing greater capacity to attack the SR or (probably less often) the RR culture. Such effective strains may usually be found in mixed city sewage as pointed out by Zdansky and confirmed by myself. Experiments involving this point have more recently been performed by Bordet [61] under well controlled laboratory conditions and using, not different lytic agents, but different strengths of the same lytic agent. Bruynoghe [77] has also demonstrated the wide range of resistance among the cultures secondary to lytic action.

But the circumstance that an SR or RR culture, resistant to one lytic principle, can be influenced by another or by a stronger one indicates to us that the resistance of the SR and RR types is not absolute or final. It is merely a relative resistance; for when tests are performed in vitro, we may find further resistant strains arising from these already somewhat resistant substrata after the application of the second or the stronger lytic filtrate. Such cultures are, in a way, "super-resistant" and accordingly may be designated SR^2 or RR^2 depending on their origin from S or R. Whether such cultures are still susceptible to another lytic agent cannot at present be stated. It seems probable, however, that the time arrives when no discoverable lytic agent has the power to effect further dissociation in the super-resistant culture; and this hypothetical "ultra-resistant" form might be termed the SR^n or RR^n. Its actual existence is merely postulated, for experiments leading to its demonstration have never, to my knowledge, been performed.

Regardless of the existence of such R^n types, however, present evidence is sufficient to demonstrate that the so-called "mutational" progress as "dictated" by the lytic principle is cumulative, just as we have found that the development of the R types of normal active dissociation is cumulative. High resistance to transmissible autolysis is seldom if ever gained in a single "stepping-up," except perhaps under the influence of a lytic agent of extreme energy; and under these circumstances the result is more likely to involve the complete disappearance of the bacterial culture than to make possible an adaptation.

Conclusion.—Many of the matters presented in this section are of much interest and significance in the formulation of any theory of transmissible autolysis; and the omission of several of them in d'Herelle's various expositions of his theory of the nature of the bacteriophage constitutes a failure to establish defense at one of its most vulnerable points. Although d'Herelle, at the time of his earlier studies,[246, 247] was able to recognize differences in the relative numbers of secondary colonies that arose from lysed-out cultures on solid mediums, and noted that old cultures, particularly broth cultures, gave more secondaries than young cultures, he apparently did not recognize the phenomenon of microbic dissociation, appreciate its significance in the production of "mutations" nor understand its influence in determining, in part at least, the varying degrees of sensitiveness or resistance to lytic action. If, however, we "read between the lines" of many of his experimental reports, we can observe the influence of the R types in determining the results; and experimental evidence for this has been presented particularly by Bordet[61] and Gratia.[211] D'Herelle's chief mischance in connection with this phase of his subject lies in the circumstance that he has attempted to understand and to depict the causes and trends of "pathological" variation and autolysis before he had adequately surveyed the nature and range of the "normal," cyclogenic variations involved in dissociative behavior. It may be that the keystone to the grand structure that d'Herelle has reared about the bacteriophage lies in his statement, already quoted, "a pure bacterial strain is not subject to transformations." If this stone crumbles, the edifice can no longer stand, at least without far-reaching reconstructions. We have here touched on perhaps one of the most important aspects of the dissociation problem, and one to which we shall return in a subsequent section.

14. COMPARISON OF THE MECHANISM AND RESULTS OF ACTIVE MICROBIC DISSOCIATION WITH THOSE OF TRANSMISSIBLE BACTERIAL AUTOLYSIS (D'HERELLE PHENOMENON)

Although it is not primarily within the scope of this review to consider the nature of the bacteriophage or the mechanism of transmissible autolysis, the circumstance that both of these phenomena appear to play so important a rôle in microbic instability, and that both involve cultural reactions and other modes of expression that are, at least superficially, so much alike, makes it impossible to omit a special reference to certain points of similarity as revealed in more recent literature.

The only writers who, to my knowledge, have referred to any distinct points of similarity between the phenomena of what we have come to term microbic dissociation and the phenomenon of transmissible autolysis are Eastwood [144] and Hoder.[253] Eastwood's conclusions of his brief review may be put in his own words. After noting the confusing mass of data to be explained, he states: "It would simplify matters greatly if one could recognize some general principle underlying this general confusion and could attribute the different types of variants to different phases in the operation of this principle."

And again: "In the search for a common factor, it is observed that these changes only occur with living and actually growing bacteria and that, therefore, they are probably due to some influence which operates at the nascent stage of growth. Hence the common factor can only be explained in terms of vital processes; but about these no precise information is available."

In attempting to formulate a general explanatory statement, Eastwood succeeded in pointing out merely that bacterial protoplasm has two functions—catalytic and synthetic; and that any disturbance of the balance (which is very minutely adjusted) will tend to the production of variants. He did not advance toward the fundamental facts of dissociation.

Hoder apparently noted clearly the similarity between "bacterial mutations" and the secondary cultures arising from lysis by the bacteriophage. He concluded that the same sort of influence or injury which brings about the one also produces the other. He accepted this circumstance as favoring Bail's [33] "splitter" or "chromosomal" theory of transmissible autolysis but did not recognize the phenomenon of dissociation as underlying the "mutations" observed.

It thus appears from the literature that practically no far-reaching attempts have been made to investigate the relation that actually exists between microbic dissociation and transmissible autolysis. Even those that have indicated a partial relation with mutation phenomena have not penetrated beneath the superficial aspects of the reaction. For this reason we may now attempt to ascertain for ourselves to what extent there may exist analogies between these seemingly quite distinct phenomena.

Relation of Active and Passive Dissociation with Reference to Incidence.—A survey of the bacterial species that have been observed to manifest noteworthy dissociative reactions and a similar survey of the species for which a lytic principle has been obtained, reveals a striking fact. Although there are some species showing active dissociation that have not yet been found susceptible to bacteriophage action, I know of no species clearly susceptible to the bacteriophage that has not also revealed, and in a particularly striking manner, the phenomenon of active dissociation. Moreover, a further study of the literature makes it apparent that those bacterial species most commonly and most easily used for an elaboration of a specific lytic principle are, without exception, the ones that most easily manifest microbic dissociation. Indeed, our present conception of this phenomenon has been built up to a large extent on the strength of data derived from B. coli, B. typhosus and B. dysenteriae Shiga; and, as is well known, it is precisely these organisms that have served as the ground work for the huge structure that d'Herelle has raised in his study of the bacteriophage and its behavior, as well as for the greater part of all subsequent study on this phenomenon by others. Practically all of Bordet's studies have been based on that classical "mutating" form, B. coli.

In further reference to the above statement I would mention especially the following bacterial groups:

Colon-typhoid-dysentery.
Paratyphoid-enteritidis.
Avian paratyphoids: (Bact. pullorum, B. gallinarum).
Capsulated: (Friedländer's pneumobacillus).
Pasteurella:[1] (B. pestis, Bact. lepisepticum, B. cavisepticus, B. bovisepticus).
Suppurative: (Micrococcus albus, M. aureus, M. citreus).
Proteus:[2] (B. proteus).
Moro-Tissier:[3] (B. acidophilus).
Diphtheria:[4] (B. diphtheriae).
Asiatic cholera;[5] (Vibrio comma).

[1] In the Pasteurella group dissociation has been observed by Gotschlich,[204] Dudschenko[187] (B. pestis); by Bernhardt[49] (B. avisepticus); by Bernhardt[49] and de Kruif[119] (Bact. lepisepticum). A lytic principle for B. pestis has been obtained in 1923 by Villason[474] and in 1925 by d'Herelle;[246] for B. bovisepticus by d'Herelle.[240] The B. pestis caviae for which Bronfenbrenner[78] has isolated a lytic principle is not a member of the Pasteurella group.

[2] It is certain that Weil and Felix,[483] Baerthlein,[27, 30] Braun and Schaeffer[66] and others were dealing with an actual dissociation of B. proteus X19. I have been able to dissociate a laboratory strain of proteus. A lytic principle has been obtained by d'Herelle[243] and by Fejgin.[169] I was unsuccessful in an attempt to obtain from sewage a lytic principle for this culture.

[3] In this group a lytic agent has been reported by Sierakowski and Zajdel.[442] In two attempts I have not been able to confirm this, but I have obtained the dissociation of a culture of this organism isolated from a carious tooth.

[4] The dissociation of B. diphtheriae has undoubtedly been observed many times, most recently by Crowell.[114] I have observed the dissociation of the "Park 8" strain. A lytic principle has been reported by d'Herelle;[248] also by Fejgin[170] and by Blair.[51]

[5] The dissociation of this organism was probably observed by Baerthlein,[30] and more recently by G. Petrovanu[388] in 1924. The latter also succeeded in obtaining a lytic agent against the S, but not against the R type culture. A lytic principle was also reported by d'Herelle,[246] Jötten,[275] Meissner[330a] and by Flu.[181]

Streptococcus:[6] (Streptococcus hemolyticus, Streptococcus viridans, Streptococcus fecalis).

Lactic acid:[7] (Streptococcus lacticus).

Spore-forming aerobes: (B. anthracis,[8] B. subtilis[9]).

Thermophilic:[10] (T-60, Illinois Collection).

With reference to the foregoing tabulation two points may be considered before proceeding further. They deal with the situation in the anthrax bacillus and in B. pyocyaneus. For neither of these organisms does d'Herelle thus far (1926) admit the discovery of a bacteriophage, as seems to be implied by their omission in his summarizing table (p. 269),[248] although a lytic principle for the former species had been mentioned by Pico[390] in 1922, and a lytic agent for the latter species had been described by Canzik[69] and several others, including myself,[224] in 1923. Pico's lytic agent perhaps accomplishes little more in the way of lysis of the anthrax culture than the organisms can often accomplish for themselves, as seems to be indicated by the work of Preisz[393] and the more recent studies of Pesch and Katzu. Similarly, the lytic agent in pyocyaneus manifestly effects no more in the way of erosive reactions (autolysis) than these organisms can accomplish spontaneously when surrounded by suitable growth conditions. I have attempted several times to obtain from sewage (from which I have seldom failed to isolate a bacteriophage for many other species) a lytic agent for B. anthracis and for B. pyocyaneus. With the former I have in no case succeeded; while with respect to the latter I have readily secured

[6] A bacteriophage active against streptococcus was reported by Piorkowski[391] in 1922; also by both MacKinley[328] and Clark.[101] In the last two cases, however, the data as yet published have not served to make clear the actual facts involved. Most recently, Dutton[141] has reported a streptococcus bacteriophage. From the data presented, however, the lytic action seems to be of a low order and does not correspond with the principles for Shiga, coli, etc. Evidence is present that the dissociation phenomenon accompanied the reaction. With the assistance of Eugenia Dabney and E. M. Brill, I[228] have obtained from sewage a lytic principle of the typical sort active against Streptococcus fecalis, and Faith Hadley[406] has separated the S and R forms of this organism in colony and in culture. The relation of Streptococcus fecalis to the more typical streptococci (hemolytic and greening), however, is not perhaps clearly established (Hadley and Dabney[228]). Beckerisch and Hauduroy[44] have reported a lytic agent for the "enterococcus," which many investigators regard as identical with Streptococcus fecalis.

[7] I have observed the dissociation of Streptococcus lacticus, which was also probably observed by Joseph Lister[306] as early as 1873, and Eugenia Dabney, a student in my laboratory, has succeeded (1926) in isolating a lytic principle for this organism (Hadley and Dabney[228]).

[8] Many older records indicate the dissociation of the anthrax bacillus. The phenomenon has, however, been clearly verified in our laboratory by W. Nungester,[409] with complete recognition of the two fundamental culture and colony types and probably several others besides the S and R forms. A lytic principle has been reported by Pico[390] in 1922, although this circumstance is not recognized by d'Herelle[248] (1926).

[9] M. Soule[430] was the first clearly to demonstrate the dissociation of the hay bacillus (subtilis) and to show the characteristic S and R colony and culture types. A bacteriophage for B. subtilis has been reported by d'Herelle.[246]

[10] Dr. Stewart Koser,[285] while working in my laboratory during the summer of 1926, succeeded in isolating from sewage a bacteriophage active against a true thermophilic microorganism (growing well at temperatures of 60 C. or above), and effective in producing typical lysis and inhibition in broth or on agar slants at a temperature of 59 C. within the usual time limits. He also succeeded in obtaining the dissociation of this culture. The exact bacterial species concerned is not known, but passes under the symbol of "T-60" of the University of Illinois collections.

a filtrable principle which produces in fresh, sensitive cultures much the same type of erosive action that is commonly observed to occur spontaneously in sensitive laboratory strains. In view of these circumstances, it seems to me strongly suggested that the erosive phenomena occurring in B. anthracis and in B. pyocyaneus cultures, but not in other bacterial cultures (except in the case of Monilia perhaps, as recorded by Sonnenschien [448]), so far as I am aware, represent a reaction analogous to bacteriophage lysis; and that this is the only form in which bacteriophage can manifest itself in these bacterial species. I have repeatedly insisted [224, 226] that the phenomenon of transmissible bacterial autolysis must be conceived as sufficiently broad in its power of manifestation to include these "aberrant" cases, which d'Herelle [248] has unfortunately relegated to the field of "bacterioclysis," a "bacterial disease," without foundation in fact or theory, but highly useful to the proponents of the virus theory of the bacteriophage as a convenient repository for illegitimate and otherwise troublesome phenomena.

Whatever other common factor may support a similarity between active microbic dissociation and transmissible autolysis, as exemplified particularly in the cases mentioned above, there thus exists one such factor in their mutual relation to the phenomenon of microbic instability, and the general trend of the dissociative reaction. In the operation of both there is significantly concerned the disappearance of a relatively unstable type of organism (S), together with the birth of new, different, and usually more resistant types (R), usually characterized by foreshortened coccobacilli or coccus forms. In both, moreover, the unstable form disappears by a process of lysis or of transformation. In active dissociation this lytic action may be so slow as to be almost inappreciable, or it may be fairly rapid. It may extend through the "lifetime" of a single culture and continue in that succeeding (lytic type), or it may culminate within a few days or weeks. It involves the complete or partial disappearance of the "normal" or S type of culture, either over broad areas or in isolated spots on solid cultures, and usually leaves behind a residue whose nature and appearance varies with the intensity of the dissociative process and which contains the modified, resistant forms. In transmissible autolysis (passive dissociation) on the other hand, the disappearance by lysis is sudden; it may cover a day or two or it may culminate in a few hours. Such an autolysis likewise involves the actual or apparent destruction—at least the disappearance—of organisms of the S type, and leaves behind a population of modified, often filtrable and usually resistant, secondary forms which, for a time at least, may undergo no further change. These similarities are sufficiently striking to anyone who makes a detailed comparative study of the fundamental reactions of the two phenomena.

Relation Between the R Types From Active and Passive Dissociation.—But the similarity may be traced further to the relations between the respective secondary or resistant cultures, already referred to as the primary (R), and secondary (SR, RR) dissociates, a point which has been called attention to by some interesting considerations of Eastwood. The most significant points of comparison involve the following: increased resistance to certain unfavorable conditions of environment and to the lytic principle, together with greater viability and increased susceptibility to phagocytosis (Bordet and Sleeswyk,[64] de Kruif,[121] Griffith,[215] Reimann,[405] Amoss [8] and others); agglutinative or sedimentary form of growth in broth together with instability in the usual concentrations of salt solution; modified antigenic and serologic qualities; and diminished virulence while sometimes still maintaining antigenic protective power. We have already seen that the R type of culture usually shows these characteristics in contrast to the S form which is sensitive, grows homogeneously in broth, retains its original antigenic power and serologic attributes and, if pathogenic, is usually more virulent. We have also seen that somewhat similar changes occur in the transition from the S to the SR type arising after lysis, and probably also in the double transformation from S to RR. In these cases, however, whether we focus attention on dissociation or on autolysis, we observe that the original culture was the same at the start. We also see that the products of change bear close resemblance to each other in many important respects. Indeed, the chief differences between the two processes lie in the following points: the speed of the reaction, and the degree of its transmissibility in vitro.

One exception that might be taken to the foregoing statements regarding the analogies between the R and the SR types relates to their virulence. In this respect d'Herelle [248] has pointed out that cultures resistant to lytic action are more virulent and less phagocytable than their respective normal sensitive progenitors. On the other hand, it has been amply demonstrated in the preceding pages that the R cultures from normal dissociation are both less virulent and more phagocytable than the original S cultures—at least in many pathogenic species. In order to support his view d'Herelle mentions several instances. Bordet and Ciuca [63] for example reported that a resistant coli culture was less phagocytable and more virulent for laboratory animals than the normal culture. Similar observations were made by Gratia [211] for B. coli. Davison [116] noted a similar phenomenon in B. dysenteriae Shiga, and d'Herelle [248] for B. pestis. In the last instance 0.000.2 cc. of the SR culture killed guinea-pigs in 46 to 50 hours while this result was obtained only by the injection of 0.1 cc. of the normal culture (S). D'Herelle stated, however, that among the secondary colonies there occurred great variation with respect to virulence; and that associated with colonies of increased virulence, he found others completely avirulent. These differences, according to d'Herelle, were due to various "mutations" produced by the bacteriophage. "These mutations do not involve all the characters of the bacterium, but only a certain number of them, varying from a single bacterium to another, even in a single culture. With one bacterium the character of 'agglutinability' will be modified, with another, the character of 'virulence,' and with a third both of these will be changed at the same time. All that may be predicted is that, usually, there is a reduction in agglutinability and an increase in virulence" (p. 219).[248] Thus it appears that, among the organisms resistant to lytic action, d'Herelle admits that nonvirulent forms do appear; and the question at issue therefore resolves itself largely into a matter of the predominance of these or of the virulent form. The answer will need to await further study. It may be noted, however, that Fejgin [168] has presented evidence opposed to d'Herelle's conception. She was able to demonstrate that her Shiga secondary cultures that were resistant to lysis not only lacked virulence, but also showed no power of toxin production. Clearer evidence opposed to d'Herelle's view is found in certain experiments performed by Bronfenbrenner and Korb,[78] having this point particularly in mind. These investigators had already obtained a bacteriophage for B. pestis caviae (not one of the Pasteurella group) and ascertained that lytic principle therapy did not influence the course of experimental mouse typhoid produced with this organism. In studying further the reason for these negative results they had occasion to compare the virulence of the resistant, secondary strain with that of the original culture. To state the results briefly, the cultures that were resistant to lysis were also quite lacking in virulence. Neither through animal inoculation, nor by passage in vitro, did these resistant strains easily recover their susceptibility or their virulence. In this respect, however, there appeared to be two groups of resistants. In broth culture one group gave an agglutinative form of growth; and these organisms reverted to susceptibility and virulence after five to seven daily transfers. The second group in broth gave homogeneous clouding; and these failed to become susceptible (or virulent) even after 120 daily passages. Both groups were stable after 25 passages on agar. In analyzing these results the authors pointed out (also in opposition to the view of d'Herelle) that the production of resistant strains is the result of a selection of variants already existing in the parent culture. This conception was also voiced by Gratia; and, it cannot be doubted, represents the situation that actually exists with relation to the generation of the SR forms of culture. Similar results indicating the nonvirulence of the lytic-resistant cultures of B. pestis caviae were published subsequently by Bronfenbrenner, Muckenfuss and Korb.[500]

In summarizing these somewhat conflicting statements regarding the virulence of these cultures that are resistant to the bacteriophagic influence, it must be said that the issue remains open. Although such an increased virulence of the resistant types as d'Herelle pictures in his latest book is the outcome naturally expected in accordance with d'Herelle's theory of the bacteriophage, its truth is not supported by sufficient evidence to permit its acceptance at the present time. Indeed, I believe it might be stated without exaggeration that present evidence favors the opposite view; namely, that the cultures resistant to lytic action are commonly less virulent than the normal culture type, and in this respect correspond with the majority of the R forms arising through the course of normal active dissociation. Moreover, I have little doubt that, when the problem has been studied with sufficient care, and in a sufficiently large number of of bacterial species, it will be found that the partly stabilized secondaries to bacteriophage action possess some residual virulence, as also is the case with the partly stabilized secondaries from active dissociation; but that the fully stabilized secondaries arising under the lytic stimulus are commonly nonvirulent, corresponding perfectly in this respect with the socalled "extreme" R types from microbic dissociation, so far as these have been studied up to the present time. To the above it may be added that, with reference to lytic principle prophylaxis and therapeutics, this situation is the most to be desired. If the power of the bacteriophage is not sufficient to destroy all of the virulent S forms, it is of advantage that the remaining organisms should be modified into a nonvirulent, rather than into a more virulent, culture type. We should, however, perhaps hold ourselves open to the view that the relation between virulence and the form of culture resistant to lytic action (SR) may eventually be found to vary with the bacterial species

Similarity with Reference to Serologic Reactions.—Among the changes revealed by a bacterial culture in the state of "active or latent resistance" to the bacteriophage, d'Herelle [244] has mentioned as most common the loss of agglutinability with specific antiserums. Also in 1926 he states: "The loss of agglutinability seems to be related to the degree of acquired resistance, for the refractory state is accompanied by a complete inagglutinability and there is only a diminution if the resistance is partial" (p. 218).[248] D'Herelle cites as instances of this fact observations on B. coli, the dysentery bacteria, B. typhosus, B. gallinarum and B. pestis. Gratia [211] has given a similar instance for B. coli strains possessing different degrees of resistance to the bacteriophage. D'Herelle [248] has further pointed out that typhoid bacteria when first isolated from the patient are not only inagglutinable but also resistant to the lytic principle; also that, after agglutinability has been restored by some passages on artificial culture mediums, the organisms become susceptible to bacteriophage action. He (p. 254)[248] has also mentioned obtaining from cultures of B. paratyphosus B mucoid colonies which were resistant to lytic action. These were undoubtedly the mucoid transitional (O) often described in the literature.

From the instances mentioned above, and many similar experimental data, one may certainly conclude that the bacteriophagic reaction is such as to generate cultures which possess marked inagglutinability, and at the same time greater or less resistence to the bacteriophage. But we have also observed in earlier pages of this work that these are exactly the characteristics of cultures of the O or R type which have been derived as dissociates from normal, sensitive culture; and that similar R type strains can be obtained at any time as the direct result of either "spontaneous" or of "forced" dissociation. In other words, the bacteriophage, at least as interpreted by d'Herelle, possesses no exclusive influence in the creation of such culture types. Their appearance, in the culture tube or in the body of the animal, has no other significance than to indicate to us that microbic dissociation has occurred. The force that determines these reactions is intrinsic in the cyclogeny of the species concerned; and their expression is often facilitated, but not exclusively determined, by that agent which we term the bacteriophage

One point of similarity between microbic dissociation and transmissible autolysis relating to serologic aspects of the R types of culture not yet considered is the following. In the section on the relation of microbic dissociation to serologic reactions, attention has been called to the phenomenon of bacterial convergence and to what Schütze has termed the "serological cosmopolitanism" of the R types. It is now important to note that the reactions which bring about these results are by no means limited to normal, spontaneous dissociation, but are observed equally in the dissociations produced by the lytic principle. The following case will suffice to illustrate the point. So-called mutations of B. paratyphosus A under the influence of the bacteriophage have been described by Bachmann and de la Barrera.[26] They mention considerable variation in colony morphology, and some of these forms were undoubtedly the resistant type. Cultures made from some of these colonies revealed marked agglutination in antityphoid serum. With further transfers of the mutant line this was increased to 6,000 while there was no increase of agglutinability in the original paratyphoid A serum (1600). In addition, an antiserum was prepared against the fourth subculture generation of the "mutant." This serum agglutinated the homologous (mutant) culture at 3,200, B. typhosus at 800 and the original paratyphoid at 800. A serum prepared against the sixth subculture generation of the "mutant" agglutinated this culture, also B. typhosus, but quite failed to agglutinate the original paratyphoid culture. These results were upheld by absorption tests. Commenting on these findings, d'Herelle[248] states, "These facts can only mean that there had been a transformation of the antigenic properties brought about through the action of the bacteriophage." D'Herelle also calls attention to Baerthlein's somewhat similar case of alleged transformation of B. paratyphosus B into B. typhosus and wonders whether the bacteriophage may not have been present in this "mutation." With reference to B. coli and its noteworthy "mutations" he mentions the fact that symbiosis with the bacteriophage in the intestinal tract is the normal state of existence for B. coli, and suggests it is for this reason that coli is a "mutant species." From these several observations, unless one wishes to assume that the bacteriophage is present in all dissociative reactions, and is the determining factor in the "mutations" that accompany them, one must believe that, for some reason, the changes which a culture undergoes in the course of dissociation occurring "spontaneously" are duplicated in nearly every point by the changes which a culture undergoes when submitted to the influence of the lytic principle. This result can leave us only with the thought that the two phenomena are closely related in effect; and, in all probability, in respect to cause.

Similarity with Reference to the Generation of Filtrable Forms of Bacteria.—The comparison of the secondary cultures arising from these two reactions may be carried a step further to deal with an aspect of the subject that is of considerable present interest and significance; namely, the granular, and sometimes filtrable, stage of the dissociates or related forms.

As in the case of the Much granules in the tubercle bacillus, the Neisser granules in B. diphtheriae and B. malleus, the Babes-Ernst corpuscles in the typhoid bacillus, the granular bodies observed in spirochetes (Wolbach [489]), and others of an apparently similar sort, most granules of bacteria and most "globoid" or "granular stages" (so-called) of bacteria, have been regarded with suspicion so far as representing viable forms, or playing a rôle in the further life-history of the culture, is concerned. The same attitude toward "involution forms," as maintained by Loeffler, and by Fischer [176] many years ago, remained quite unquestioned up to, and well beyond, the work of Hort [254] a decade later. The severely critical attitude toward such forms, which has descended to bacteriologists along with the mantle of Koch, has doubtless been justified by the circumstance that, although they may be found in quite young cultures, they have often been observed in old or otherwise degenerating cultures, or in organisms submitted to the prejudicial action of normal or immune sera. If any significance at all has been attached to such bodies (particularly small granules) it has usually been in conformity with the teachings of Günther [218] and of Loeffler; namely, that they were "Absterbenerscheinungen," concerned with degeneration changes, and not otherwise of significance. A few investigators (Leishman, Balfour, Hindle and Fantham) have held—for spirochetes—different views, as also did Fontès [182] for the granular bodies which he observed in cultures of the tubercle bacillus in 1910. Of their frequent association with apparent degenerative processes in the cell there can be little doubt, and some of them probably possess no other significance; but it is also becoming clear that some of these "degenerative" changes may, themselves, be of significance as evidence of dissociation in the life of the culture. Moreover, some of these granulation and related changes may occur even in the first hours of growth of quite fresh culture. Indeed, it appears that more than one phenomenon may be hidden beneath Loeffler's "Absterbenerscheinungen." But let us turn to certain facts observable in cultures of several bacterial species.

In active dissociation the reaction is often accompanied or preceded by the generation not only of long filamentous but also of multitudes of minute coccus forms and still smaller, granular, "globoid" or gonidia-like bodies, apparently resembling some of those first described as gonidia by Conn [103] for Crenothrix. These have been pictured frequently, and especially by Fontès,[182] Almquist,[3] Hort,[254, 257, 258] Löhnis,[310] Mellon [337] and Enderlein.[160] I have observed them often in dissociating cultures of B. coli, B. typhosus, Bact. pneumoniae, B. lepisepticum, B. malleus, B. diphtheriae and in some unidentified organisms from air and water. Faith Hadley [400] has observed them in Streptococcus fecalis. Their mode of origin has been followed with special care by Almquist,[3, 4] and by Enderlein. In cultures of B. malleus and B. diphtheriae these bodies may be detected in normal cultures in small numbers but they are often the preponderant form at certain stages in dissociating cultures. They vary in size from the limit of vision up to a size merging with the distinct coccus and coccoid forms. Unless they are present in large numbers, or are seen in the course of formation, they are often admittedly difficult to differentiate from granular bodies present in known degraded cultures. As I have observed in B. malleus and B. diphtheriae, the transfer to broth from an agar slant culture in the state of active dissociation, and containing many filamentous forms, together with granular bodies and cocci, may result in a few hours' time in the almost complete disappearance of the bacillary and filamentous types and the survival (and apparent active multiplication) of the gonidia-like bodies only. This does not occur in normal cultures similarly treated. These cultures may be stabilized in the coccoid state and apparently remain so for a considerable time. Culture reactions of this sort have been described by Hort,[258] Mellon,[326, 327] Enderlein and others. When transfers of the granular or small coccus stage are made to agar slants growth in the first generations may be absent or it may take the form of an almost invisible growth-film. Microscopic examination of such cultures usually reveals granules or "globoid bodies" mixed with small cocci and occasionally rod forms (Breinl,[70] Fejgin,[169] for B. proteus). I have filtered through Berkefeld N candles such cultures coming from the dissociation of B. diphtheriae, B. malleus and B. coli, and have observed that, after a few days, the filtrates sometimes revealed a faint opalescent growth which showed itself microscopically in fresh preparations as a mixture of minute granules, cocci and small bacillary forms. I have not succeeded as a rule in obtaining such growths in Berkefeld W-filtrates. The demonstration of filtrable forms of bacteria has, however, not been uncommon in recent times, as we have noted in the case of the tubercle bacillus, the bacillus of Johne's disease, B. proteus and other forms; and these results have done much to validate older observations on the "filtrability of bacteria." Some of these include the filtration of the organism of Schweinepest by Lourens,[313] of the relapsing fever spirochete by Novy and Knapp [374] in 1906, by Nicolle and Blanc,[369] Breinl and Kinghorn,[72] and by Wolbach [489] at a later date; the filtration of microgonidial forms of the tubercle bacillus by Fontès in 1910, the filtration of forms of the typhoid bacillus by Almquist [3] in 1911, the filtration of the meningococcus by Hort [259] in 1915; of Azotobacter by Löhnis; [310] of a diphtheroid culture by Mellon; [325] of Streptothrix by Mellon; [325a] of his "spore-forming coccus" by Pryor; [398] of Streptococcus strains from encephalitis by Evans,[161, 162] confirming earlier reports of the same phenomenon by Rosenow.[412] The work of Fontès, Hort,* Almquist, Mellon and Enderlein is such as to relate the filtrable forms of bacteria to certain developmental stages (microgonidia, gonites) commonly observed in the process of microbic dissociation. In other cases we are not aware of the existence of such a relation from the nature of the data reported. In the tubercle bacillus distinct cultural dissociative changes have not yet been clearly reported, although cell transformations which usually accompany dissociation have been pointed out by Fontès,[182] Enderlein,[160] Karwacki,[507, 508] and perhaps by Vaudremer.[472]

The preceding records of filtrable forms of common bacteria, arising either with or without evidence of dissociation, are perhaps of sufficient significance in themselves; but it is their counterpart in the products of bacteriophagic action that I desire particularly to make clear. And here we shall see that there exists an interesting parallel.

First Hauduroy [234, 233] and later d'Herelle and Hauduroy [249] pointed out the common existence of filtrable forms of B. coli, B. typhosus, B. dysenteriae, Vibrio comma, Micrococcus aureus and Micrococcus albus arising under the influence of a weak lytic principle working on its homologous substratum. Fejgin [171] has also shown the same for B. typhosus. Accompanying the filtrable bodies were observed minute

* Hort's conception of the nature of the meningococcus and its relation to infection is of special interest. In 1915 he [259] had performed filtration tests which led him to believe that the clinical form of the organism was filtrable. In 1917 he [256] again reviewed the situation and came to the following conclusions, largely supported by single cell observations.

The socalled "giant meningococcus" is not a bacterium but an ascal stage in the life cycle of an organism allied to the ascomycetes. The ascus may vary in size from 0.2 to 5.0 μ. The meningococcus of Weichselbaum is not a bacterium but an ascospore derived from the giant meningococcus by a process of endosporulation. He believes that the filtrable form previously described is a stage in the life cycle of this organism, and suspects that the reason why the true nature of the meningococcus has not been recognized earlier is because of the exclusive use of solid culture mediums for identification purposes; and also because the organisms have not been carefully studied in the living and unstained condition. In respect to the possible establishing of the meningococcus as one of the Ascomycetes, however, it is perhaps permissible to believe that this organism might incorporate into its life cycle some of the reproductive methods of the Ascomycetes without the necessity of becoming a member of this group. However we may come to regard Hort's conclusions as a whole, the data which he presents to support the view of endosporulation are, to say the least, very striking and entirely in line with similar observations on B. typhosus.

granular bodies and small coccus or coccoid forms. D'Herelle and Hauduroy reported that the granular bodies arising in the filtrates might or might not grow on ordinary mediums. If they did grow the culture might perpetuate the granular form, show a reversion to the original form or grow in a new form such as coccus or coccoid. Such results were not uniform: they occurred only in a small proportion of the tubes. They called special attention to the circumstance that, whatever the morphologic type of the organism first submitted to the action of the lytic filtrate—spirillum, rod or coccus—the end was the same, a granular or coccus type. It will be borne in mind that Bail [33] has seen in some of these filtrable stages a significance ("splitter" theory) uniting them with the actual agent of bacteriolytic action. According to his view these particles propagate only when provided with nutrient material from the living organisms. Otto and Winkler,[380] moreover, have regarded the bacteriophage as constituted of minute fragments of bacterial substance possessing the properties of enzymes. Also, Bail, Otto and Munter [379] and many others have reported obtaining the bacteriophage from filtrates of old normal cultures.

The recent work of Enderlein [160] has given a view of much interest regarding the significance of small coccus stages and granular or "globoid" bodies in cultures of many common bacteria, including Vibrio comma, B. diphtheriae, B. tuberculosis, B. anthracis, and also the spirochetes; and particularly with reference to the so-called "death" of the old culture mass. According to Enderlein, there has existed a great misconception as to what really constitutes the death of a culture. Ordinarily, when we attempt to subculture from an old agar slant, for example, to another slanted medium, and observe that no growth results, we conclude that the old culture is dead. Not necessarily so, according to Enderlein; and the reason is as follows: In the old culture the organisms have undergone a transformation. If a rod form, the normal bacilli or the filaments generate coccus or coccoid bodies (gonidia) which are still viable and easily cultivable on fresh medium. Later, however, as the culture ages (and particularly if placed in a window in diffuse sunlight), the gonidia give rise to smaller, granular elements (gonites) which cannot be cultivated on a solid medium but which, when planted in broth, become transformed further into the sex cells (spermites and oites). In a similar manner, neither of these sexual forms is independently cultivable; but after fertilization has taken place (in a liquid medium), as it does about five and one-half to seven hours after seeding the gonit culture, the newly fecundated oites are cultivable either in broth or on agar and proceed to generate the normal cell type (mychit), thus completing the developmental cycle or cyclode. The terminology employed by Enderlein in delineating the detailed features of his comparative bacterial cytology is extremely elaborate. The fact remains, however, that whatever we may term these granular bodies and the elements from which they arise, they have been observed repeatedly in many bacterial species and have been traced to the gonidia as a point of origin (Almquist,[3] 1911, for B. typhosus). Enderlein's conception of the actual mechanics of this phenomenon, isolated phases of which have been reported by numerous workers, to say the least, offers the most logical hypothesis that has been made available to explain the alternation of coccus and bacillary phases of growth in the same culture; and for the alternation between filtrable and nonfiltrable stages in the life of the culture, as observed by d'Herelle and Hauduroy and many others. Thus, to conclude the reference to the work of Enderlein, the most common of the "filtrable forms" of bacteria are not "fragments" endowed with "regenerative" power, but definite stages in the cyclogeny of the species, and for the most part the microgonidia (Cohn, 1870) and the gonites (Enderlein, 1925); or in some cases the spermites, according to Enderlein's interpretation.

There is thus seen to be a marked similarity between some of the ultimate products of active and passive dissociation with reference to these granular, secondary types and their capacity for, and manner of, regeneration; and the same is true of the filtrable forms of bacteria. Although, as d'Herelle states, filtrable forms certainly do arise under the influence of the lytic agent, manifestly the lytic agent is not a necessary factor in the process. In this connection it is interesting to note that in earlier days one of d'Herelle's chief objections to regarding the Twort phenomenon as an example of actual bacteriophagic action was that some of the organisms, rather than being wholly destroyed in the lysing, transparent colony, were transformed into minute cocci, still observable in the lysate. Later he points out a similar transformation (in broth) of M. albus and M. aureus as evidence of the modifying influence of specific bacteriophage, together with the ultimate production of "symbiotic types" (filtrable cell fragments parasitized by the bacteriophage). Here we have the remarkable picture of two filtrable organisms living together in a state of symbiosis!

In all these studies dealing with granular and filtrable forms of bacteria we are naturally treading on difficult and uncertain ground. From the results thus far attained, however, I believe we may be reasonably certain of two things: first, that aside from the lifeless colloidal fragments which undoubtedly constitute the chief background in many lysate examinations, there actually exist living, ultramicroscopic and filtrable forms of bacteria; second, that the generation of these bacterial "fragments" or invisible corpuscles stands in some relation to processes involved in microbic dissociation, either active or passive. Although themselves often filtrable, and capable of generating experimental infections, as now shown by many investigators, the relation of these bodies to the class of recognized filtrable viruses, as suggested in the interesting contribution of Ch. Nicolle [368] in 1925, must be left an open question. In any case, however, we cannot close our eyes to the possibility that some of the bacterial forms which we obtain for cultivation and study may, as Hort has expressed the matter, and as Enderlein has attempted to demonstrate, merely be the "vegetative side issue" of a life story, the nature of which, for the most part, is still beyond our knowledge.

Comparison with Reference to the Medium.—With further reference to the similarities between microbic dissociation and transmissible autolysis, there may be noted a relation to certain conditions in the medium which permit or favor the reaction. It is commonly accepted that transmissible autolysis does not occur in a medium which does not permit growth. The same appears to be true of active dissociation, although it does occur, indeed is sometimes favored, on mediums much impoverished in nutrient qualities (Braun and Schaeffer,[66] Feiler,[167] de Kruif [118, 119]). While dissociations may proceed rapidly in plain broth or peptone water (de Kruif for Bact. lepisepticum), Griffith has stated, with reference to pneumococcus dissociation, that the organisms must grow before the S to R transformaton can occur. Soule [450] also has ascertained quite recently in his study of dissociation in B. subtilis that, when the S type is suspended in distilled water or physiologic salt solution and plated after varying intervals, no dissociation is found to have occurred. While the microbes live, but do not multiply, they do not dissociate. Control broth tubes similarly inoculated revealed growth and from 20 to 30% dissociation in a few days time. Quite in line with these results, I have found that the S forms of B. coli, Bact. pneumoniae, and B. pyocyaneus, when suspended in salt solution, adhere to type as long as they live. Although additional tests of this matter in other dissociating species are needed, all present evidence points to the view that neither transmissible autolysis nor microbic dissociation can occur in a medium which does not permit growth of the bacteria concerned. This circumstance might lead us toward the conclusion that both dissociation and transmissible autolysis are functions of growth, as is also intimated by much other evidence.

Again, it is to be noted that a similar reaction of the medium is important for both. The reaction most favorable for autolysis was established by Gratia [209] as about P_H 8.5 for B. coli. D'Herelle,[248] however, has stated most recently that P_H 7.8 is the most favorable reaction for the action of the bacteriophage. The same reaction was found most desirable by Griffith [215] for the dissociation of the pneumococcus; also by Atkin [21, 22] for dissociation of the meningococcus and the gonococcus. De Kruif [119] found an alkaline reaction most favorable for the reaction in Bact. lepisepticum. For his study of dissociation in B. typhosus Hort [258] used an alkaline agar; and the same was true in the case of Eisenberg for various species. Preisz [393] obtained his dissociating anthrax cultures in a slightly alkaline medium. I have been able to obtain "spontaneous dissociation" (accompanied by macroscopic areas of lysis) in more than a dozen cultures belonging to various groups, merely by growing on rich beef-infusion-agar containing proteose peptone and adjusted to a reaction of P_H 7.8. In most cases the dissociation begins after two to four days on the free borders of the growth, thus confirming Hort's [258] observation that in his typhoid cultures the "mutants" were always found most numerous on the free edges. In my experience, I have not been able to obtain dissociations accompanied by macroscopic areas of lysis on mediums of low P_H value, and Atkin showed that P_H 7.4 or 7.5 did not commonly permit the appearance of papillae on his colonies of meningococcus and gonococcus. B. pyocyaneus, as I have ascertained, may be maintained permanently in the lytic and lysogenic state by monthly transfers on mediums of 7.6 or above. If, however, the reaction drops to 7.0 or below, the lytic phenomena rapidly disappear or are represented merely by metallic flecks on the surface of the culture. Although, as d'Herelle [248] has pointed out, the lytic reaction caused by the bacteriophage may be adapted to conditions of acidity, there is universal agreement that alkaline conditions favor in a marked way both microbic dissociation and transmissible autolysis.

Comparison with Reference to Proliferative Growth.—Another observation in which the similarity between the two reactions is shown relates to the phenomenon of proliferative growth. This is manifested as a rule preceding the critical reaction in transmissible autolysis, and is usually observed in such acute dissociations as reveal themselves by the formation of macroscopic areas of lysis on solid mediums. The highest point of excess growth terminating the proliferation, and just preceding lysis, has been termed by Bordet the "critical stage"; and in a former paper I have termed it the "lytic threshold" (in B. pyocyaneus and other cultures). Attention to the same point in the dissociation of B. anthracis has been called by both Preisz and Wagner. In both instances one may note a degree of increased growth-energy which in many instances is so great as to justify the term, proliferation. It may be borne in mind, however, that in such slow-going dissociations as occur on "starvation agar" or phenol agar (Braun and Schaeffer,[66] Feiler [167]), or in organisms undergoing adaptation to dyes (Revis,[404] Stearn [452]) such exalted growth may not occur. In transmissible autolysis the period of proliferative growth is terminated by the sudden disappearance of the majority of the S organisms present, while in active dissociation it is terminated by a much slower and progressive disintegration and transformation of the S type, some of which may, for a variable time, continue to support a changing bacterial population in flux toward the O or the R form of culture. It seems clear that, in dissociation at least, the proliferative growth marks the development of the intermediate or O culture type.

Comparison with Reference to Lytic and Lysogenic Cultures.—Next, in pursuing our consideration of the similarities existing between active microbic dissociation and transmissible autolysis, it is necessary to mention a matter of considerable theoretic importance relating to the rôle played by a certain type of cell (and sometimes, colony) in the culture in the perpetuation (and perhaps in the initiation) of, not only the simple dissociative, but perhaps also the frankly bacteriophagic, reaction. On an earlier page I have called attention to certain observations of Preisz, Katzu and also Pesch on lysing colonies of the anthrax bacillus; also to the observation of Pesch that certain translucent colonies exert a "repressive influence" on the growth of the whitish, more opaque S type. I have also mentioned the observations of Firtsch and of Bernhardt on the curious reactions of the "third colony intermediate." There also has been pointed out the observation of de Kruif on the repressive action of the R colonies of Bact. lepisepticum on the growth of the S colonies. Also the observations of Ørskov and Larsen on the inhibiting action of filtrates of broth cultures of their curious self-lysing variant (paradysentery) upon the growth of normal culture. Also some observations of Arkwright and of my own on the same point. I have in addition called attention to the peculiar colonies which I have seen in cultures of B. diphtheriae and staphylococcus, and which I have termed "invisible colonies" because of their ability to grow in the midst of the culture mass in an almost invisible form; and in such a manner as to produce the semblance of lytic areas, together with the subsequent characteristic appearance of secondary colonies. Indeed, anyone working much with dissociating cultures on solid culture mediums cannot fail to be impressed by the inhibiting, and sometimes apparently lytic, effect produced by certain kinds of colonies in the culture mass. Our conclusions from these observations were that, through active dissociation there may arise certain cells that are instrumental in reinitiating, and perhaps in perpetuating, the dissociative reaction. Indeed, Otto, Munter and Winkler, according to d'Herelle,[248] have reported raising the level of such an antagonistic reaction to the grade of an actual transmissible autolysis merely by passing normal Shiga cultures through heated filtrates of homologous broth cultures. I believe that studies in this field are among the most important that can be directed toward the solution of both acute microbic dissociation and transmissible autolysis.

Turning the line of our inquiry now more directly to the problem of transmissible autolysis, we have also seen that the primary rôle of the bacteriophage, apart from its more obvious lytic and inhibitive influence, may be regarded as that of an agent effecting a dissociation of the sensitive culture into various components, among which the O and R types come into special prominence. We recall that, for Bordet, the bacteriophage is a factor in the production of "mutations," and that for Arkwright [17] "it is a substance which disturbs the balance of the growing bacteria in the direction of autolysis." There certainly exists little if any direct evidence that the bacteriophage, itself, is the thing that accomplishes lysis. Among the variants produced by the action of the bacteriophage in a sensitive culture some may still be sensitive (Bordet,[61] Arkwright [17] and others), some may be resistant without "carrying" the lytic agent, while still others are resistant and do "carry" the lytic agent (d'Herelle,[248] Bordet,[60] Gratia,[211] Lisbonne and Carrére [305]). The last are called lysogenic cultures and are able, through contact or through their filtrates, to precipitate lytic or growth-inhibitive reactions in normal, sensitive culture. Furthermore, in so doing, they cause the generation of more organisms like themselves, as well as more sensitives (perhaps) and more nonlysogenic resistants. D'Herelle and Hauduroy [249] have made special provision for these lysogenic forms, either as organized cells of normal SR type, or as filtrable forms, "existing in symbiosis with the bacteriophage." While this conception of symbiosis between a bacteriophagic virus and filtrable forms of resistant bacteria requires a stretch of imagination for which many bacteriologists may not be qualified, I believe this last point mentioned above is an admission, on the part of d'Herelle, which represents the greatest, recent, single advance toward our understanding of the nature of the bacteriophage and of the probable mechanism of lytic action; and to this point I shall subsequently return.

In summarizing the most essential points of the last two or three paragraphs I would lay stress on the following: In ordinary, active dissociation certain cells may arise which exert a repressive and inhibitory or even lytic influence on the normal, sensitive components of the culture; in transmissible autolysis also, certain cells are found which are themselves resistant to lysis but which, at the same time, are capable of generating lytic reactions in sensitive culture; these lysogenic cells may exist in the common, organized form or in a filtrable form, and sometimes may not show obvious growth on solid or in liquid mediums; it may be quite gratuitous to link into symbiotic union with these cells or cell "fragments" a hypothetical bacteriophagic ultravirus in order to explain their lysogenic potentialities.

Regarding the similarities between dissociation and transmissible autolysis one final point may be considered; namely, the existence of what has been termed "border-line types" of reaction. In most cases it is not difficult, through the examination of a culture, or of its "built-up" filtrates, to reach a conclusion as to its status—whether more closely related to active dissociation or to transmissible autolysis, as these terms have been defined. Such clear differentiation, however, may sometimes prove difficult in such cases as represented by colony lysis in B. anthracis (Preisz,[393] Pesch, Katzu), in B. pyocyaneus (Canzik, myself and others) and in Monilia (Sonnenschien). There are, furthermore, other instances in which a decision is even more difficult; namely, in the "suicide cultures." In all of these, we perhaps observe "border line cases." Among them are some which lean to the side of transmissible autolysis by reason of the fact that they give rise to filtrates which have a sort of lytic action that is easily transmissible in series. Others, however, approximate active dissociation by reason of a more tardy reaction, together with the impossibility of isolating filtrates which possess the outstanding features of the typical bacteriophage. It is somewhere in this series, perhaps, that Twort's case might be placed;

also the case of pyocyaneus lysis. By several, including myself, these have been regarded as examples of actual bacteriophagic action although d'Herelle [248] has excluded both from his category of classical lysis. In his latest contribution he still maintains that there are two "diseased states" in bacteria, indicated by the phenomenon of bacterioclysis, and the phenomenon of the bacteriophage (bacteriolysis). The first, involving merely bacterial "fragmentation," is typified by Twort's [468] case; only the latter, involving complete dissolution, is related to the bacteriophage. Therefore the autolysis of B. pyocyaneus would also be an instance of bacterioclysis according to d'Herelle. The improbability of this view I have already discussed.

In this connection, however, I wish to digress sufficiently to add that further observations which I have made on pyocyaneus have only increased my conviction that the mode of lysis which I described for this organism in 1924, while differing in certain respects (mainly speed of reaction and degree of transmissibility) from the classical autolysis of d'Herelle, belongs to the same category of phenomena. But in pyocyaneus the lysis is both active and transmissible; it is of the borderline type. The fact that it is spontaneous under certain cultural conditions has been impressed upon me especially by the circumstance that, of seven typical pyocyaneus cultures received from other laboratories (where lytic manifestations had never been observed) six became lytic within a short time after transfer to the rather alkaline beef-infusion-agar commonly used in my laboratory for all dissociation work. From several of these cultures a lytic filtrate has been obtained which in recent tests has sometimes accelerated the autolysis of the lysogenic culture itself, and in addition has precipitated a fresh, sensitive culture into similar lytic manifestations. And most recently I have been able to show that there exists in sewage-polluted river water an agent which is able to induce in other sensitive pyocyaneus cultures, as well as in many other bacterial species, the same sort of dissociative change. I therefore regard the case of autolysis in pyocyaneus as one which stands near the borderline of reaction, and one which is interpretable under the heading either of transmissible autolysis or active dissociation. Of two points there can be no doubt: that it is often spontaneous; and that it is weakly transmissible. It is for this reason that I have spoken of this organism as being both lytic and lysogenic.

To the foregoing considerations I would also add one further point regarding the relation of the agent of transmissible autolysis to the culture undergoing dissolution, an aspect of the matter which has been mentioned on a previous page. In all cases thus far reported in the literature on the bacteriophage (d'Herelle [248]) the lytic principle has been found to operate only at a temperature which is close to, or below, the maximum growth temperature of the common bacterial species. Thus, in certain especially heat-resistant strains of B. coli d'Herelle has reported lytic action at 45 C. The question therefore arises, what would be the situation with respect to a lytic principle operative on a strictly thermophilic organism growing at 60 C. or perhaps higher, but only slightly at 37 C., and not appreciably at lower temperatures? Dr. Stewart Koser,[285] working quite recently in my laboratory, has studied this problem. He has obtained from sewage contaminated water a lytic agent highly active on a still unknown organism of the thermophilic group ("T-60" of the University of Illinois collections) which grows luxuriantly at 60 C. or higher, but less vigorously at 37 C. The lytic principle obtained operates well at 37 C. but much more effectively at 56 to 58 C., and possibly higher, since this aspect of the study has only been begun by Dr. Koser. It is transmissible indefinitely in series and gives small lytic areas at 37 C., but much larger areas (3 to 6 mm.) at 59 C. Broth cultures are promptly lysed overnight at 56 to 58 C., and perhaps to a lesser degree at 37 C.

The complete study of this interesting and, I believe, unique example of bacteriophage activity will be published by Dr. Koser at a later date. For present consideration it is however important to note that, in this case, the thermal requirements of the lytic agent are distinctly correlated with the thermal requirements of the organism of the substratum. Moreover, in the production of this bacteriophage there was no occasion for the slow "adaptation" of a "low-temperature" bacteriophage present in sewage to the thermophilic culture. The bacteriophage revealed itself at once and directly. Shall we conclude from these facts that there exists in sewage a special sort of bacteriophage in the form of an ultravirus, especially qualified to deal with its thermophilic host? Or shall we accept these data as suggesting the view that each strain of bacteriophage stands in some sort of a close biologic relationship to the culture of the substratum, and is perhaps a specialized form of this culture, partaking equally of certain of its biochemical and biophysical growth-characteristics? As Koser [285] has well pointed out, if we continue to assume that the bacteriophage is of the nature of a filtrable virus, foreign to the cell, then we are forced to postulate the existence of a thermophilic ultravirus.

Summary of Section.—In summarizing this section it may be said that there exists a considerable body of scattered evidence revealing a distinct similarity between active microbic dissociation and transmissible bacterial autolysis with respect to the nature of the substratum, and the mechanism and results of the reaction. In detail these similarities appertain to the following points.

In both phenomena the reaction involves the disappearance by transformation or by lysis of all, or of a part, of the "normal" organisms. Those that remain are likely to be of modified type (R or SR).

Transmissible autolysis has been found to occur only in those bacterial species which have also been observed to manifest outstanding symptoms of microbic dissociation.

Both the reactions start at the same end of the dissociative scale—that is, with cultures of the sensitive or S type. The secondary or resistant forms of both, moreover, are much more stable, but a "reversion" is possible in both.

In both phenomena the degree and the speed of the reaction, as well as the physiologic state of the product, depend on the degree of concentration or the intensity of action of the inciting stimulus—whether lytic filtrate, antiseptic, immune serum, or some unfavorable condition of environment. In both, "intermediate" cultures are observed.

There exists a striking similarity between the intermediate and the end products in the two reactions, respectively. Among the intermediates are mucoid forms, while the R and SR forms appear as cocco-bacilli in the majority of instances.

The R and SR types differ antigenically and biochemically from the mother culture; and these departures in both, so far as studied, seem to be of the same order, even with respect to bacterial convergence.

The R (from active microbic dissociation) and the SR (from transmissible autolysis) types of pathogenic cultures agree further in that they are commonly nonmotile; also often nonvirulent, or less virulent than the original S form.

The influence of filtrates of dissociated cultures on the S type is sometimes such as to parallel, within limits, some of the effects of the lytic principle on its homologous substratum; and this relates to inhibition of growth and a low grade of transmissibility.

In many instances the conditions of the medium which favor dissociation are the same as those which favor transmissibile autolysis; and this is seen particularly in the reaction, which, for best results, should be alkaline (about P_H 7.8).

Neither microbic dissociation nor transmissible autolysis appear to occur in a medium in which the organisms concerned cannot grow.

In both phenomena it is observed that a heightened growth activity or "proliferation" in cultures is present just preceding the acquirement of the lytic threshold or "critical stage."

In both cases the culture changes involve the generation of "filtrable forms" of bacteria which may or may not grow in artificial culture mediums, and which may sometimes regenerate either the original form or a modified form of culture.

Conclusion: A New Theory of Transmissible Autolysis.—The final conclusions and inferences that may be drawn from the observations reported are mainly reserved for the concluding section of this work.

I wish to say at this point, however, that I believe that the apparently intimate relation existing between many aspects of microbic dissociation and transmissible autolysis justifies the formulation of a new hypothesis of the bacteriophagic phenomenon (especially with reference to the mode of formation of lytic areas) quite different from that of either d'Herelle or Bordet, but somewhat more in harmony with the view of the latter. D'Herelle as we know believes that the lytic area is a colony of bacteriophage (ultravirus), produced from the growth of a single "corpuscle" on its proper bacterial substratum (parasitized culture). Bordet, on the other hand, holds the opinion that the lytic area is the result of the action of a nonliving growth product (chemical substance) on a certain type of bacterial cell possessing a sort of hypersensitiveness; and from whose growth, but subsequent destruction by lysis, is set free more of the same "substance" which, in turn, effects the dissolution of neighboring, but previously more resistant, cells. The destruction of these cells en masse in localized areas produces the areas of lysis. As will be seen below, my own view, an alternate to both of those mentioned above, shares nothing in common with d'Herelle's but depends equally with Bordet's upon the existence of certain cells of unusual "sensitiveness." In my case, however, I believe that I have adduced a certain degree of evidence for the identity of these "sensitive" cells with some of those belonging to what I have termed the "third colony intermediate" occurring in microbic dissociation. Beyond this, my theory calls to its support, not the participation of "chemical substances" as the direct cause of lysis, but of certain biologic elements operative in some suspected, though still unproved, phases of bacterial reproduction, which have recently been introduced into the science of bacteriology through the cytological studies of Enderlein. My own theory of the reaction may therefore be stated roughly as follows: I make use of the lytic areas in illustration, because I believe that, if these can be explained, the explanation of related phenomena will easily follow.

Lytic areas are in reality "vanishing" secondary colonies arising, and as quickly disappearing, in the mother culture. They are formed by the same organisms as those of the culture of the substratum, but by cells *in a different cyclogenic state,* or stage of development, from others in the same substratum. Having multiplied to a slight extent on the surface of the medium, they quickly disappear, leaving an apparently bare area in which there may develop, after a time, the sprinkling of colonies which d'Herelle, as also Bordet, regards as the resistant secondaries, but which are in reality *tertiary* colonies. In other words, a lytic area represents the site of disappearance of a certain type of intermediate colony which is both lytic and lysogenic: lytic, because it has the power to undergo a transformation characterized by lysis; lysogenic, because it can impart the lytic stimulus to certain neighboring cells. But why, then, do not these lytic colonies arise more commonly in normal, sensitive culture without the instigation of the lytic filtrate (bacteriophage suspension)? This brings us to the second aspect of the theory.

Although the secondary, lytic colonies arise through the multiplication of certain sensitive cells of the substratum, a stimulus to growth is required. What is the nature of this stimulus? It is the bacteriophage—a filtrable stage in the cyclogeny of the same bacterial species (or of a closely related species); and it accomplishes this end—not directly by growth and multiplication itself (for, alone, it is nonpropagating), but indirectly by serving as some sort of stimulus—perhaps fructifying—to the rapid development of certain, young specialized cells present in the substratum, as mentioned above. These cells react first by rapid multiplication (proliferative stage), thus attaining the lytic threshold; then immediately by lysis, accompanied, however, by the generation of a fresh brood of filtrable forms. These, in turn, serve as the exciting stimulus to a similar reaction on the part of other sufficiently young, specialized cells. It is conceivable that the same mechanism might operate as well in liquid mediums as on solid.

Without attempting to force this theory too far at the present moment, certain eventualities may perhaps be anticipated. First, how can such a theory explain the clear relation always shown between the number of lytic areas produced and the amount or concentration of the bacteriophagic suspension employed? It may be assumed this is because the number of cells in the substratum susceptible of receiving the stimulus of the "bacteriophagic corpuscle" is, in young, sensitive (S) cultures always sufficiently great. The number of the filtrable "units" establishing a "contact" will thus determine the number of sensitive cells that are to receive the stimulus. Only the cells receiving this stimulus will develop in such a manner as to form the colonies endowed with the "vanishing" (i. e., lytic) characteristic. If they do not receive the stimulus, they may still develop, but perhaps along another path in the cyclogeny of the species.

Again, one may ask: Why should the reaction (i. e., the formation of lytic areas, lysis in broth, etc.) manifest itself most definitely in young cultures? It might be because only the youngest cells are susceptible of receiving, or being influenced by, the stimulus derivable from the filtrable bodies. If too old, their course of development is perhaps turned into another line.

Further, why are the R type cultures (from microbic dissociation) less susceptible to lytic action, as shown by the reduced number of lytic areas, their smaller size, and the failure of complete lysis in liquid culture mediums; and why are certain SR type cultures (arising from bacteriophage action) refractory to lysis? It seems probable that this is because the type R culture has entered a cyclogenic stage which is nearly destitute of those specialized, sensitive cells capable of receiving the stimulus from the filtrable forms; and because the SR (or, as I have previously termed it, the SR^n) type culture is entirely destitute of those same sensitive cells whose further development (in lysis) depends on the stimulus derived from the filtrable bodies; and, it may be hazarded, upon whose further development, at least in a certain direction, the generation of additional filtrable bodies themselves, depends.

Again, one may inquire, why, if the filtrable form of the organism serving as the stimulus to development of the sensitive (mother?) cells is a normal constituent of bacterial cultures, is not the lytic phenomenon manifested commonly in "normal" cultures? In answer to this I would say that there can be little doubt that the reaction, in a much reduced degree, is often occurring in such cultures; and that it has at times been raised to the level of the classical bacteriophagic phenomenon. Such instances have often been mentioned in the literature and with increasing frequency; but d'Herelle has disposed of them all as cases in which the cultures concerned were "contaminated" with the lytic agent; in other words these investigators have, according to d'Herelle, worked with "mixed cultures." Such rebuttal is incongruous but apparently effectual. Wherever lytic phenomena possessing transmissible characteristics appear spontaneously, there stands the bacteriophage; and, if not explainable otherwise, as a "contaminant."

Finally, the question may arise: If the thing we term the bacteriophage is a filtrable form of the same bacterial culture as that of the substratum, how is it possible that a bacteriophage suspension of B. dysenteriae Shiga, for example, is able to produce lytic areas in a substratum culture of B. coli, or perhaps of B. typhosus? It may be freely admitted that this at present constitutes something of a difficulty. Pending the acquisition of further data, however, we can perhaps regard the matter in the following light.

The sensitive cells of the substratum culture may be stimulated not only by filtrable bodies of the homologous bacterial strain or species, but also by filtrable bodies arising from fairly closely related heterologous species. A diversity of biologic relationship is possible, but it must not be too far distant. Evidence of bacterial antigenic relationships that might suggest such a possibility has been presented in the preceding pages, and concerns the phenomenon that we are coming to know as bacterial convergence, particularly with reference to the R types. We have seen that the R forms arising from microbic dissociation in a variety of organisms, often manifest distinct antigenic relationships, which are not so commonly, if at all, revealed by the corresponding S forms of culture. This community of antigenic structure, as indicated by serologic tests, has been referred to by Schütze as the "serological cosmopolitanism" of the R type. But we have also noted from the observations of Bachmann and de la Barrera [26] that a partially analogous phenomenon occurs with respect to certain SR cultures arising from bacteriophagic action. Considering this apparently close antigenic relationship between the resistant forms of bacteria belonging to heterologous, but somewhat related, species it might seem possible that any stimulating influence (perhaps fructifying) exerted through the filtrable bodies of the homologous culture, could be shared with analogous, filtrable forms from closely related but heterologous bacterial species. If such a circumstance as this were possible, one might anticipate that there would be some correspondence between the specific limits of antigenic relationship, as manifested by serologic tests, and the specific limits of bacteriophagic heterogeneity of action, as manifested by cross tests involving lysis and inhibition. In the data presented in the foregoing pages relating to the serologic aspects of bacterial convergence, there is certainly inherent the suggestion that the serologic relationships of the R forms of heterologous species are roughly characterized by about the same latitude of action that is also observed in the heterologous reactions creditable to the bacteriophage. To ascertain in detail the actual limits of this possible parallel is a highly important matter.

In indulging in considerations of this sort, one unfortunately feels himself drawn further and further from the beaten paths of current bacteriologic thought; and sees here and there the faint shadows of possibilities which, on more sober reflection, he is likely to believe can have no basis in fact. So it is perhaps with the last mentioned aspect of the present theory of bacteriophagic action. One has glimpses of things possible but still unrecognized in the bacterial world—perhaps hybridization of a sort among bacterial species—who can tell?

Although such a theory of bacteriophagic action as I have presented above finds some support, so far as its major premises are concerned, in many of the facts demonstrated in microbic dissociation, other aspects can be sketched only roughly; and it is scarcely feasible, in our present lack of knowledge, to pursue the conception in further detail at the present time. Whatever its inadequacies may be, I believe it is no more "speculative" than any other theory of the bacteriophage that has thus far been advanced, although d'Herelle's parasitic virus hypothesis is perhaps the more obvious and easier to believe; moreover, it is far simpler in its mechanism. The theory which I have outlined may, however, possess the advantage of affording a somewhat new viewpoint for future experimental work, in which we might regard the bacteriophage "unit" or "corpuscle," not as a foreign filtrable virus parasitizing the bacteria, nor as a chemical substance capable of regeneration by the organisms which it affects, but in the light of a functionally incomplete and ordinarily non-selfpropagating stage (perhaps the "spermit" of Enderlein) in the cyclogeny of the bacterial species. It is obvious, however, that such a conception as I have presented is dependent on the existence of a definite stage, characterized by sexual reproduction, in the cyclogeny of all bacteria manifesting the phenomenon of transmissible autolysis; and even this circumstance is not yet proved. For this reason I anticipate that the theory will have few adherents. However this may be, I may say in concluding this phase of the subject that the views expressed above serve, in part, as a basis for the opinion I have presented elsewhere in this work—that the whole problem of the bacteriophage may ultimately come to appear to us merely as a side issue of the vastly greater and more significant problem of microbic dissociation, in which it will sometime find its place.*

15. RELATION OF MICROBIC DISSOCIATION TO THE PROBLEM OF BACTERIAL MUTATION

Regarding the furtherance of problems of adequate bacterial classification it is easy to observe in the literature of the past thirty years that the most common stimulus activating bacteriologists has been the desire to perfect a scheme of classification rather than to study the causes and

* Some of the statements and conclusions presented in the latter part of the present section are so closely related to conceptions developed in section 16, dealing with the biologic significance of dissociation, that these two portions of the work should be considered together. At least, the reading of section 16 will make clearer certain aspects of the foregoing treatment of the subject of transmissible autolysis.

limits of variation among bacteria themselves. Much has been written about simple variations, impressed variations and mutations; but, with the possible exception of such works as those of Buchanan and Traux,[78] Wolff,[490] the Winslows [486] and a few others, little has been of value and the majority positively misleading from the viewpoint of genetics. Cole,[104] himself a geneticist and not a bacteriologist, was one of the first (with Wright) to point out the inadequacy of the biological viewpoint of most bacteriologists in these matters; and to draw attention to what he then regarded as the probable nature of the commonly observed changes in bacterial types, as well as to the significance of Johanessen's pure-line concept for bacteria as a group. While it can scarcely be doubted that this concept is applicable to bacteria, that it operates in the manner assumed by Cole ten years ago is unlikely, as he himself would now undoubtedly be the first to admit. It is perhaps true that, of all the variations shown by different strains, many are due to the isolation of biotypes, each possessing its own range of variation, and probably overlapping other ranges.

But Cole assumed further, as many bacteriologists have also done, that if in any pure line originating from a single cell we can detect an hereditary departure from the mode of the biotype, it must be conceded that we are dealing with a mutation in the original meaning of the term. This conception I now believe to be inadequate. Such variations (mutations) Cole believed to be rare among bacteria, at least in proportion to the number of variants properly attributable to the isolation of biotypes from cultures representing a heterogeneous mixture of pure lines. He pointed out, however, a number of instances which he admitted to the mutant group; and he was ready to admit others but for the circumstance that their isolation had not involved single cell methods of cultivation. Since the time of Cole's publication many isolations of "mutants" have been made under conditions which perhaps guarantee variation from a pure line; and which, since they often seem to breed true to the new type, would naturally be regarded as mutants in the usual sense. Indeed the belief in the frequency of bacterial mutation has grown remarkably within the past fifteen years, during which time it has been a common view that the "mutants" have arisen in the progress of "normal reproduction"—meaning, of course, simple fission. In the analyses of the nature and stability of these new accessions to the already burdened archives of bacteriology, some have been reported as returning, sooner or later, to the original type; while others, it has been insisted with equal vehemence, have remained true to the new characters (section 12). This aspect of stability, it may be noted, has done much to support and even to increase the scientific zeal of the mutation discoverers. Indeed the frequency of alleged mutations has been so reflected in the bacteriological literature of recent years that it has come to be believed by many bacteriologists as established truth.

Influence of Conception of Dissociation on Mutation Theory.—But, along with the general increase in mutation literature following 1907, there have been interjected into the science of bacteriology other views regarding the reproductive methods of bacteria which may well cause us to hesitate in accepting these new forms as actual mutations. As will be pointed out later in greater detail, these newer conceptions of the nature of bacterial reproduction and of the complexity of forms which it involves, are beginning to lead us away from older beliefs regarding the simple monomorphic nature of bacteria as a class, and to supplant them with notions of a plurimorphism underlying some sort of cyclical development in every bacterial species. With such possibilities in mind it is clear that the inadequacy of the biotype conception, as applied by Cole to the problem of bacterial variation, can become fully apparent only when we are able to recognize the actual thing that varies. And naturally the same is true of the socalled mutations whose assumed importance has been reinforced in most recent time through certain statements coming from both d'Herelle [246,248] and Bordet.[61]

So long as we believed implicitly in the usual textbook statement that "the mode of reproduction of bacteria is by binary fission," aught that departed from the regulation cultural type has been classed as an involution form, a mutant, or a contamination. Usually such departures from "normal" have been regarded as contaminations and accordingly discarded. If regarded as "involution forms" or as "Absterbenerscheinungen" in the sense of Loeffler, they have been tolerated in cultures (and for good reason) but viewed as elements possessing neither special interest nor significance. If the variation was of sufficient magnitude and apparently stable, it has been forced into the literature of the day as a "mutant." Such summary treatment of the unorthodox members of the bacterial society has undoubtedly done much to keep in a well domesticated state the cultures which chiefly comprise our present laboratory stock-in-trade. As a result of all this, modern bacteriology has become impaled on the monomorphic conception. All of our laboratory methods have been such as to breed uniformity and constancy.

Dating from the year 1910, however, when Fontès first demonstrated the filtrability of the tubercle bacillus, or from 1916, when the English bacteriologist, Hort,[254] first clearly directed attention to the possibility that bacteria possess, upon occasion, other means of reproduction than simple fission, and when Enderlein [159] first introduced his views of bacterial cyclogeny, we have begun to see that some of the apparent departures from the accustomed "types" may possess a significance other than that of contaminations or of insignificant involution forms. Without here entering into the details of what may be involved in this conception (which will be considered later in greater detail), it suffices for the present to say that much evidence exists to show that at least many of the peculiar bacterial forms, now commonly regarded as mutants, stand in some relation to the operation of this reproductive mechanism as it has been revealed, though still vaguely, through the studies of Almquist,[4] Jones,[270] Hort,[254] Löhnis,[310] Mellon,[325] Enderlein [159] and a few others. In other words, the truth is gradually being forced upon us that in bacterial reproduction there is involved a process, perhaps cyclic in nature, and responsible for at least many of the often observed eccentricities in bacterial behavior. This view has been furthered not only by observed changes in the form of bacteria, but also by observations indicating the temporary disappearance of an infecting organism, perhaps for some days, as for example the relapsing fever spirochete in the body of the louse. The organism unquestionably remains in the louse, but apparently in a form in which it is not recognized. Soon, however, it reappears in its "normal" shape. Such observations, coupled with certain observed changes in morphology in the blood of the infected individual, and especially perhaps the finding of the peculiar granular bodies as reported by Wolbach [489] and others, have served to lead several investigators (Leishman, Balfour, Hindle and Fantham) to consider the view that there might exist a "cycle" of development in the life history of the relapsing fever spirochete. The demonstration of the filtrability of this organism by Novy and Knapp [374] in 1906, confirmed later by several others, does much to support this view.

If we may therefore accept for the moment the possibility of this view (namely, that there may exist among bacteria a possibly complex reproductive mechanism, lying quite outside the range of simple fission, and in the course of whose operation new and apparently "unusual" forms of bacteria arise from the parent culture), let us examine the effect of such a view on the theory of bacterial mutation—particularly as related to active microbic dissociation.

If we come to regard the new reproductive mechanism as one of cyclic nature, as depicted so clearly in the Bakterien-Cyclogenie Prologomena of Enderlein,[160] or even if we merely take into consideration certain different and changing developmental phases in the life of the bacterium, the problem requires little further consideration; and this for the reason that we could scarcely regard as a true mutant any organism transiently typifying a single stage in the life-story of a species. The important questions to be answered are: does the assumed mutant ever return to the original form, either directly, or indirectly through an intermediate type of organism; and does the mutant become the center of new variations? Although it is perhaps too early to attempt any final answer to these questions, there can be small doubt that many of the assumed mutants manifest marked adherence to the newly-acquired form, and are so reported in the majority of instances. Their stability is by no means certain, however; and all such reports lose weight in the face of Jordan's [273, 274] recent results dealing with the reversion of the R type of B. paratyphosus B to the sensitive form, as well as by the still more recent work of Soule [450] on the dissociation of B. subtilis. This question is considered further on another page. On the second point we know nothing; but it is safe to say that we possess no evidence to indicate that the new forms have created a new center of variation.

In view of these circumstances it becomes clear that we cannot yet accept the fact that the many and various cultures reported as mutants of good standing actually deserve the name. If one wishes to regard as mutant any culture which exhibits apparent hereditary differences from the parent stock, persisting in culture for some weeks or perhaps months—then all of these new discoveries are mutants. It seems to me, however, that we are justified in applying this term only to those cultures fulfilling more strict requirements. We should perhaps regard as mutants only those hereditary variations whose mode falls outside the limits of the species; and this should be construed as meaning outside the range of types represented by the normal changes (cyclic or otherwise) characteristic of the organism in question. Until, therefore, we train ourselves to detect the limits and range of cyclogenic variation characteristic for each bacterial species, we are scarcely in a position to recognize a mutation when it appears. It is safe to say, however, that at the present moment we do not know of one unequivocal or authentic case of mutation among the bacteria.

One may object that, when an organism possessing the features of B. paratyphosus B becomes transformed into one endowed with the characteristics of B. typhosus, this must be a mutation. But that will depend on whether the transformation of the "mutant" can also be effected in the opposite direction. If that should be found possible, what should be said regarding the integrity of the "species" as now established? Only that they should be consolidated and supplied with different terms of reference. Without meaning to attribute truth to the illustration given above, I do think it probable that, when the limits of normal variation and dissociation of many more organisms are ultimately known, numerous such consolidations will be required. In the meantime we must be skeptical regarding the significance of all "good bacterial species," on whatever grounds they rest; as also regarding the many new types that are appearing, stamped with the "mutation" trademark. Before we can detect the variable we must know from what it varies.

The Bacteriophage as an Agent Producing Mutations.—Regarding the rôle of the bacteriophage in effecting dissociations (and mutations) another point is of interest. In his second book d'Herelle [247] raises the bacteriophage to a point of high biologic significance with respect to its power to determine bacterial mutations; indeed he states, "these mutations are always produced under the influence of the bacteriophage." There might be some misapprehension from this phraseology, but the matter is stated more clearly in his third book in which he remarks (p. 222), "It is indeed probable, as various investigators have suggested, that all of the fixed mutations occurring among bacterial species are produced through the action of the bacteriophage." With the publication of his third book [248] d'Herelle's belief in the mutation-provoking power of the bacteriophage had thus not been diminished in intensity, as is also indicated by the following quotations (p. 225): "In summary, then, the most important facts to be derived from all of these studies is that, exposed to the action of the bacteriophage, bacteria undergo mutations, usually unstable ones, but that these may become fixed under conditions as yet undetermined. These mutations are associated with a state of resistance acquired by the bacteria." And again: "Moreover, a deeper study of these mutations is sure to completely revise our present concept of the fixity of species."

In opposing such views as these, which can serve only to make more difficult the solution of the problem of bacterial species, I can only say once more that the evidence which I have assembled in this work is sufficient to demonstrate that, with respect to its mutation-furthering ability, the bacteriophage can accomplish no more striking modifications than those clearly recognized as occurring in all bacterial cultures when placed under adverse growth conditions such as may be determined by physical and chemical agents of great variety; and, indeed, no more than bacterial cultures often accomplish for themselves in a small way perhaps, even if grown under conditions which bacteriologists may regard as the most favorable. The important point—indeed the crucial point, which d'Herelle fails to comprehend—is that the cyclogeny of a single bacterial species embraces many strictly normal forms of culture growth, each one of which is endowed with different biochemical, serologic and antigenic characters; and, it may be added, with quite different degrees of resistance to the action of the bacteriophage. To isolate and to study these varied forms is the task of the student of microbic dissociation. In all of these natural cyclogenic changes the bacteriophagic ultravirus, as d'Herelle understands it, plays no part other than its rôle of accelerating such cyclogenic reactions as the culture is already able to accomplish for itself in a rather less energetic manner. That a revision of our conception of the nature and degree of fixity of bacterial species is demanded can scarcely be doubted by any bacteriologist with laboratory experience; but that an analysis of the problem will be furthered through the intercession of the bacteriophage, as such, and of its mythical, mutation-provoking powers, is altogether impossible. The one study that can assist in establishing the new groundwork for this hoped-for revision is the study of microbic dissociation and of the sequence of cyclogenic changes, on which the dissociation must depend; and of which (as we shall without doubt sometime come to see) the phenomenon of the bacteriophage is only a part. In these matters d'Herelle's theory, like modern bacteriology, is impaled upon the old monomorphic conception of the nature of bacteria; and accordingly his views on the nature and origin of bacterial mutations are as valueless as those of all his predecessors who have failed to grasp the true significance of the new biology of the bacteria and its relation to many of the foundation stones upon which d'Herelle's theory of the bacteriophage now rests.

In the exposition of his theory of transmissible autolysis Bordet [61] also finds a mutation-furthering power in the bacteriophage. He refers to the lytic agent as something endowed with powers of "directing the trend of bacterial evolution" and of "controlling the destiny of the species." For him it possesses a sort of "regulating" influence; and he even sees the possibility of its activity, not among bacteria alone, but in the cells, organs and tissues of higher animals as well. But, whether we are justified in attributing to the lytic agent an influence so far-reaching (for the bacteria at least), will depend on whether the "new forms of bacteria" (i. e., the "mutants") generated under its "creative" stimulus, are actually outside the circle of normal, cyclogenic variation. In earlier pages I have referred to the striking similarity between the R forms of culture arising from active microbic dissociation and the SR (or RR) arising as secondaries to transmissible autolysis; and the analogy is rendered only more perfect when we consider the filtrable forms or the "ultrabacteria" of d'Herelle. We have seen that none of these forms can be regarded as a mutant in the strict meaning of the word. Thus, although Bordet makes some provision for a sort of variability (in degree of sensitiveness) among the cells of his sensitive cultures, his position is not far different from that of d'Herelle when he considers the relation of the lytic agent to mutation. To his conclusions the same answer must therefore be given. While it may yet be shown that the influence of the lytic principle can create forms of bacteria not produced through the mechanism of active microbic dissociation, and lying outside the range of normal, cyclogenic variation in its widest sense, I see no evidence of this at the present time; and, until such evidence is forthcoming, we may well be cautious in attributing too far-reaching an influence to this still unknown agent.

Conclusion.—In concluding this subject of bacterial mutations it may be said that, although new "discoveries" of bacterial mutants are still flowing in abundance into the archives of the science, none of them bear the light of careful scrutiny; and for the reasons that I have already advanced. Firmly adhering to the old monomorphic conception of the nature of bacteria, bacteriologists as a group have not put themselves into a position to recognize a mutation even if it appeared to them. Not until we have become able to recognize the wide range of cyclogenic variation to which all bacterial species are susceptible, shall we be able to detect the permanent departure from the specific cycle. This will take time, and I suspect that it will be many years before we shall be able to speak intelligently regarding bacterial mutations. In the meantime it might be well if the term "mutation," were banished from the vocabulary of the bacteriologist.

16. THE BIOLOGICAL SIGNIFICANCE OF MICROBIC DISSOCIATION

In any natural phenomenon manifested so clearly and so commonly as the dissociative process, and along such parallel lines among various species and genera of bacteria, one is naturally driven to seek for the biologic significance of the reaction—just as the discovery of the common act of conjugation among the ciliated protozoa led to an attempt to develop an explanation. Although many of the physiologic reactions occurring in bacteria, as well as many biochemical reactions determined by bacteria, seem quite fortuitous, so far as advantage to the organisms themselves is concerned, we should not blind ourselves to the view that some of these reactions are in reality advantageous, while others cannot be regarded as without significance in the struggle for existence merely because we are not yet able to appreciate it. Regarding the varied reactions involved in the phenomenon of microbic dissociation, the first hypothesis which presents itself is, therefore, that it is an adaptive reaction, or one aspect of such a reaction. In partial support of this view there could be marshalled much evidence drawn from the data presented in the foregoing pages to demonstrate that dissociative phenomena of some sort invariably accompany the adaptation of bacteria to new or unfavorable conditions of environment, as for example to the influence of drying, heat, starvation, moisture, antiseptics, oxygen tension, immune serums, the bacteriophage, etc. The newly formed R type, which we believe is a stabilized end product of the dissociative reaction, can live under the new conditions while the old form was unable to do so. Such a view would therefore indicate that a sort of "transmutation," accompanied by the sacrifice of many old characters, is the price paid for survival.

But, on the other hand, we observe that dissociation sometimes occurs in young cultures and often on rich culture mediums affording, as we assume, the best possible conditions for growth. It may occur under environmental conditions which we regard as ideal, while it may appear to be absent on an apparently inferior medium, as I have found to be the case with B. pyocyaneus. Yet we may not be justified in assuming that the sort of food and other conditions which encourage the fastest and most luxuriant growth are always most advantageous for the bacteria, although they may be most satisfactory from the viewpoint of the bacteriologist. Life processes, to proceed normally, must proceed at an optimum rate, and it is conceivable that this rate may be made too rapid for safety as well as it may be too slow. It is possible that the van t'Hoff rule relating to the influence of heat in activating enzyme action may have a parallel in the chemical stimulation of bacterial growth by an "over-rich" medium. In the one instance, as the other, the result may be a depression or annihilation of the active agent. If we take this view, we might conceive that either poor or rich mediums could produce harmful effects to which the same sort of response on the part of the bacteria might be given. This view is supported by the work of Braun and Schaeffer on B. proteus (starvation) and my own on B. pyocyaneus (rich feeding).

Views Regarding the Nature of Dissociative Variation.—In the literature bearing consciously or unconsciously upon dissociation an explanation has never been attempted; and with few exceptions (Preisz,[393] Hoder,[253] Eastwood[144]) it has not been related either to other phenomena involving the lysis of bacterial cultures or (with the exception of Mellon) to other apparent eccentricities in bacterial behavior. If, however, we regard a marked tendency to variability, and the production of more or less permanent "mutations," as evidences of the dissociative reaction, it has in this manner been linked indirectly with variation-stimulating effects of environment. Gotschlich, for example, in 1903, held that infinite variation is always a characteristic of organisms placed in a new environment, and Kruse (1896) emphasized the fact that unfavorable environment and conditions which permit only slow multiplication (as for instance growth in old cultures) are effective in producing variations. With reference to the relation of dissociation to eccentricities often observed in bacterial behavior, however, the primary studies of Jones, Hort, Eisenberg, Mellon, Löhnis, Löhnis and Smith, Almquist and Enderlein are of marked interest and will be considered further on a later page.

In general, however, so far as the brief explanatory statements regarding dissociation are concerned, they have seldom passed beyond the view that the reactions described are "variation phenomena" or "mutation phenomena" (Neisser,[362] Baerthlein[27, 28]), perhaps stimulated by unfavorable environment. From time to time the older views of Loeffler and other members of his school regarding degeneration phenomena and "Absterbenserscheinungen" have been revived and the variants have accordingly been regarded as merely of teratological significance. Limiting our consideration, therefore, to those more recent studies in which the S and R types have been clearly recognized, we observe various views expressed. De Kruif comes forward definitely to support the mutation view, and bases his opinion on the conception of Dobell[130] who referred to a mutation as "a permanent change—however small it may be—which takes place in a bacterium and is then transmitted to subsequent generations." Amoss also, with reference to his pneumococcus R type, stated that "apparently there has been a genuine mutation." Reimann, also dealing with the pneumococcus R, stated more conservatively that this form is a variant from the S culture type. Griffith was inclined to attribute the dissociative reaction in the pneumococcus to "degenerative changes" in the culture. Krumwiede and his collaborators attributed the generation of R forms of B. paratyphosus and other organisms to "degradation phenomena," and Julianelle [505, 506] has made use of the same manner of interpretation in reference to the R form of Friedländer's bacillus. Regarding the origin and nature of the R form in B. dysenteriae and related organisms Arkwright gave consideration to several possibilities: that they may be "inseparable contaminations;" that S and R forms may preexist in all cultures as elementary forms of the same species; that the S and R forms are merely "modifications" due to the influence of environment; that they may be due to "variation within the limit of the species"; and that they may be actual mutants (in the accepted meaning of the term). Among these possibilities Arkwright came to no definite conclusion but merely stated that "R forms undoubtedly readily arise under artificial surroundings." The circumstance that modifications seemed to arise from so many strains, and so frequently, made him hesitate to evoke the mutation hypothesis. P. B. White in his excellent study of dissociation in Salmonella gave us still another point of view—namely that "roughness" constitutes a deficiency disease of bacteria. He stated: "roughness appears in the light of a disease in which certain antigenic factors may be lost and others altered. . . ." And again: " . . . it is above all probable that roughness is largely conditioned by some nutritive deficiency."

So far as may be ascertained from his major contributions, d'Herelle [248] has not grasped either the fact nor the significance of microbic dissociation, although he has commented on one or two instances which I believe involved this reaction. With reference to the Twort phenomenon and also to the autolytic reaction in B. pyocyaneus he introduces the term "bacterioclysis," a disease of bacteria, which he assumes to be quite different from the phenomenon of the bacteriophage. What sort of a disease is referred to, we are not informed; nor what the cause of the malady may be. But we are left to infer that the condition is at least pathologic and not a normal reaction.

In reviewing these various attempts to appraise the significance of the peculiar reactions in cultures occurring in the progress of microbic reactions in cultures occurring in the progress of microbic dissociation, dissociation, we thus see that they have been related to: (1) "Variation phenomena"; (2) "mutation in bacteria"; (3) degeneration of cultures; (4) "degradation of cells," and (5) "disease in bacteria."

Regarding these views, little need be said. They are all developed out of the false monomorphic conception of the nature of bacteria and of bacterial reproduction. To call these changes "variation phenomena" evades the question and does not further our knowledge in any way. The "mutation view" has been considered in section 15, and as I believe, eliminated as a possible explanation. To view such changes in cultures as manifesting "degeneration" or "degradation," in the usual meaning of these terms, finds little support in the facts observed; and receives, moreover, definite contradiction in the circumstance that the "degraded" forms are commonly more vigorous in growth, and possess greater longevity in the face of unfavorable environmental conditions, than the culture types not so "degraded." To look upon these reactions as due to a "disease of bacteria" ("bacterioclysis"), is perhaps interesting, and it cannot be disputed that any living cell, in whatever phylum, may be susceptible to disease abnormalities in some form; but such conclusions, in the case of the fission fungi, finds no evidence in fact and can only lead us unnecessarily far astray from the point at issue.

In view of this manifest dearth of adequate consideration of the nature and significance of microbic dissociation, and because many aspects of the complex subject of dissociation require some final correlation, I shall attempt, in the present section, to state my own concept of the significance of the reaction; and to introduce, at appropriate points in the discussion, the views of those few authors who have given the matter serious consideration. In so doing I shall attempt to conform, so far as possible, to the experimental data introduced in the earlier pages of this work; and to make it clear where speculation is necessarily called upon to bridge over gaps unspanned as yet by complete or reliable experimental evidence.

A Possible Interpretation of the Dissociative Phenomenon.—In entering the field of inquiry, it must be frankly admitted that we are advancing into an unmapped region but one, the general aspects of which, have been hinted at by a few adventurous workers who have journeyed far from the beaten trails of current bacteriological thought. It is possible nevertheless to formulate an hypothesis which, though having perhaps none too secure a support in facts yet accomplished, may serve to direct exploratory approaches into the deeper problems underlying the acknowledged fact of microbic dissociation. A tentative thesis may therefore be put forward as follows:

Microbic dissociation, as we superficially observe its mechanism and effects, involves the partial or complete elimination by autolytic and transformatory processes, of the S type of culture when changed conditions of environment become sufficiently prejudicial to a continuance of the same type of growth; and it is accompanied by the generation of new bacterial forms better qualified to perpetuate the stock in the changed environment. In reality and fundamentally, it involves: a process of nuclear reconstruction or rejuvenation effected by the conjugation of certain cells, resulting in the production of zygospores or their physiologic equivalent (and perhaps sexually differentiated cells), and effected at the sacrifice of part or all of the old culture type, which may largely disappear; also a process of division or "fragmentation" or budding of the conjugate cells, resulting in the formation of new elements, cocci or granular bodies, some of which may be ultramicroscopic and filtrable. Some of these filtrable forms are noncultivable; others may continue to propagate in this form (invisible or barely visible growth); while others possess the ability of again entering into the "normal" bacterial aggregate or into some ultimate modification of it (R type). What morphologic type becomes stabilized after its emergence from the dissociative process depends largely upon the extent, intensity or degree of the reaction and upon the selective nature of the environment. It is at this moment that the "evolutionary trend" of the species (in the Bordet sense) might be determined, if it can be said to be influenced at all by such a means.

From this point of view, dissociation as we observe it in its grosser aspects, would be regarded as a normal adaptive reaction made possible through the intervention of a special type of, reproductive mechanism in bacteria. This process is doubtless operative at times in all cultures; but, under changed or unfavorable growth conditions, becomes sufficiently intensified to present the more striking manifestations that have attracted special attention, as in the O type cultures. The R forms of bacteria are, therefore, the newly stabilized or partly stabilized types arising from the germination of special cell structures such as the zygospores, or through the production of gonidia. The S types of bacteria sometimes surviving dissociation are, in a similar manner, the remnants of the original culture which have not entered into the modified reproductive process, but have persisted in a limited vegetative reproduction of the same form. To prevent dissociation, therefore, one must eliminate those environmental conditions leading to the conjugative or sexual phases of reproduction and its associated phenomena; and to stimulate dissociation one must bring about those conditions favoring conjugative and related cell activity and its products, living or nonliving. In transmissible autolysis (passive dissociation) whatever thing first precipitates the reaction, the substance that perpetuates it in cultures is manifestly a product or a form of the bacteria themselves; and probably a result of conjugative or other related cell activity. It may be a metabolic product, but it is more likely to be a living, ultramicroscopic form of the organism itself, endowed with greater resistance and able to produce in the parent culture centers of proliferative growth. This is at once followed by lysis, which is accompanied by liberation of a fresh generation of filtrable forms.

It is to an hypothesis of this sort that I am led from a consideration of such facts as I have presented earlier in this work. Since such a conception brings into relation several phenomena which have usually been regarded as distinct and unrelated, the proof of an hypothesis of this sort is an undertaking of considerable proportions. But there already exists some evidence that supports at least certain phases of such a view; and there also exist certain theoretical considerations and deductions which, though scarcely serving as evidence, are not entirely without force.

The Problem of Growth Cycles.—Regarding the observed facts which support such a view of the existence among bacteria of some form of reproduction involving a reproductive process quite different from binary fission and underlying the phenomenon of dissociation, one naturally goes back to the primary observations of Hort [254] and Enderlein.[160] It is now more than ten years since the former investigator in England and the latter in Berlin made their first fundamental observations in this field and reported on the complicated "life-story" (to use Hort's words) of B. typhosus (Hort) and several other organisms (Enderlein). It was at this time that Enderlein introduced into bacteriological literature the term, "Cyclogenie" (cyclogeny), referring to the series of progressive and degressive changes through which an organism passed in its life-cycle, departing from, and returning to, its basic morphologic form. The facts developed by Hort in his first and later studies, sufficiently remarkable at the time, were amply confirmed by Leishman, Adami and others appointed as a special committee of the British Medical Board to review his entire work. Hort's essential conclusion was that there existed in B. typhosus a form of reproduction, hitherto unrecognized, presumably involving stages of conjugation and endosporulation. He traced the cell changes in warm-field studies and presented a large volume of clearcut photomicrographic evidence. The views of Hort were manifestly a continuation of the conception of the

"Entwicklungscyclus" introduced into bacteriology first by Fuhrmann[186] in the year 1907 and developed to a much greater extent by Enderlein in the years from 1916 to 1925. Evidence of a sort similar to that of Hort was supplied by Löhnis in his monumental volume on variation in many bacterial species; and in later years by Mellon in his long series of papers, by Almquist,[4, 5] Lieske[303] and others. Enderlein[160] in his comprehensive work (Bakterien-Cyclogenie Prologomena *) of 1925, not only seemed to confirm much of the work of these earlier workers, but also introduced for the first time certain apparently significant data bearing on a distinct mode of sexual reproduction by bacteria, lying outside the simple amphimixis obtained through resort to the production of zygospores and endosporulation. I believe that the studies of these major workers have not only been of value in themselves but have served the purpose of lending greater significance to the host of minor observations over a period of many years dealing with phenomena that are manifestly closely related; and especially with reference to conjugation and to gonidia formation, as observed by Cohn in the early days of bacteriology. To these observations should be added the phenomenon of reproduction through the agency of symplastic structures, first called attention to by Jones[270] in 1913 and mentioned again in 1920. This observation has apparently been confirmed by both Löhnis and Enderlein.

Such reports as those mentioned above and the radical inferences that have led from them have, for the most part and until quite recently, however, fallen upon the ears of an unsympathetic bacteriological world. They are sufficient, however, even if theoretic considerations were not also in its favor, to establish not only the probability, but as I believe the fact, that conjugation, or some physiologic equivalent of amphimixis, accompanied by a form of endosporulation, gonidia formation and perhaps budding, actually occurs more or less regularly in the life of the

* Enderlein's comprehensive work, correlating and interpreting many earlier observations, as well as adding many that are new, and dealing with the assumed cyclic changes in bacterial reproduction, is unquestionably the best presentation of those cell changes (comparative bacterial morphology), some of which without doubt underlie the dissociative reaction. Enderlein, however, concerns himself with the cytological aspects rather than with the colonial and cultural, biochemical and serological, features of bacterial variation; and in this respect has attacked the problem of bacterial life cycles from the under side, so-to-speak, much as Hort, Löhnis and Mellon have also done. It is impossible within the scope of this review (which is concerned with the broader and admittedly more superficial aspects of the dissociation phenomenon) to give an adequate account of the complex comparative cytology of bacteria as presented by Enderlein. Suffice it to say, that the subject under his hands becomes involved with the description of cell phenomena and developmental stages which few bacteriologists have ever seen, much less followed; and for which he coins a large vocabulary of new and unusual terms. As such studies are now progressing among bacteriologists at large, it will probably be many years before a true apporaisal of Enderlein's contribution can be made. In the meantime, however, we may regard with no little admiration his manifestly careful and sincere attempt to put some degree of order into the at present chaotic state of bacterial cytology. I believe that Enderlein has blazed a trail which, at least in the main lines of advance, other bacteriologists sooner or later are sure to follow.

majority of bacterial species. As for true sexual reproduction, involving the highly specialized sex cells (spermites and oites of Enderlein) we may perhaps be justified in holding our conclusions in reserve, although I do not regard the circumstance as improbable. The pictures presented by Hort, Eisenberg, Löhnis, Mellon, Almquist, Enderlein and others are in almost perfect agreement and can scarcely have any other interpretation than that first proposed by Fuhrmann twenty years ago, and by Hort almost a decade later. Although I have not followed the successive cell changes on the agar block as was done by Hort, I have seen in fresh and in stained preparations of B. diphtheriae many of the same elements. I have also observed repeatedly that the clearest pictures of zygospore-like bodies, containing distinct chromidia, occur in preparations taken from actively dissociating cultures; and that the best pictures of the stabilized secondaries, on the other hand, are derived from cultures in which the dissociative process has passed the height of its reaction. Moreover, I have found it distinctly observable that the prelude to dissociation in several species (B. coli, B. typhosus, Bact. pneumoniae, B. diphtheriae) yields much the same microscopic picture as that observed at the beginning of the new reproductive stages outlined above. It is namely a proliferative growth accompanied by a tendency to develop granular or coccus forms (gonidia) within the rods; also to form long, and often beaded filaments, frequently branched as detailed by Gardiner [195] in his studies on "three-point multiplication." This sign was first mentioned by Hort as preluding what he termed, for B. typhosus, the "reproductive explosion;" and was again emphasized by Mellon as a phenomenon clearly announcing the initiation of a new reproductive phase in the culture, marked chiefly by the production of conjugative cells and subsequently of zygospore-like bodies; and often terminating in a new or modified culture type.

Not only in B. diphtheriae (Park 8 strain) but also in B. malleus and in Streptococcus viridans, I have been able to recognize the large pale spheres ("giant cocci") measuring up to 6 or 7μ in diameter and containing chromidial bodies to the extent of two to twelve as often described for the diphtheria bacillus, as described for the diphtheroids by Mellon,[326, 327] for B. coli by Mellon,[330] for B. typhosus by Eisenberg [147, 154] and others; as described and pictured for the same organism by Hort,[258] and by Löhnis [310] for many species, by Jones [270] for azotobacter, as most recently recorded by Enderlein [160] for Vib. cholerae and as followed in their subsequent development particularly by Hort, Mellon and Enderlein. In Eisenberg's case these forms were produced during the apparent dissociation of the typhoid bacillus under the influence of normal, but germicidal human serum. In some of Mellon's cases ascitic fluid in the medium seemed to have an influence; while most of Hort's special forms arose on an alkaline agar.* I believe there is no other logical explanation for these frequently described "involution" forms than that they are related to a form of reproduction involving the production of gonidia, also conjugation and zygospore formation. Regarding the existence of a distinctly sexual form of reproduction ("oites" and "spermites") as pictured by Enderlein,[160] there is less certainty. Whether, as suggested by Mellon and by Hort, the zygospores are the mother cells of the filtrable forms of bacteria, for the existence of which in many species sufficient evidence may now be said to exist, or whether the filtrable bodies are the microgonidia or the "gonites" as suggested by Enderlein, Almquist[3] (for B. typhosus) and Mellon[337] (for B. fusiformis) cannot at present be stated. Nicolle,[368] however, has voiced the possibility of the origin of the filtrable viruses from bacterial cells, a conception which appears to be gaining ground through the circumstance that filtrable forms are rapidly being discovered for many bacterial species. But, whatever the actual significance of these minute bacterial forms may eventually prove to be, we may be certain of three things: they occur regularly, consistently and in great numbers in many bacterial cultures under certain conditions of growth and at a certain stage of development; they do not always long endure as such, but after a brief development often disappear, apparently passing into other developmental stages; and although they may closely resemble certain artificial structures on the slide, they are not artefacts.

Under the present circumstances, when one is somewhat at a loss to shape a difficult hypothesis, there may be some justification for adding to what has preceded certain theoretical considerations which favor the concept of the existence of at least a conjugative reproductive process among bacteria. These considerations relate to two fundamental aspects of protoplasmic behavior: the necessity of some form of rejuvenation for any continuous stream of protolasm; and the possibility of regeneration from nuclear fragments or chromidia.

* It is scarcely feasible in the present work to undertake a review of the many interesting details of culture and cell modification reported in recent studies. Many of the facts still require confirmation, but those interested in pursuing the subject further are referred to the series of papers by Mellon, in the course of publication, and to the comprehensive work of Enderlein in his Bakterien-Cyclogenie Prologomena (1925). See previous footnote.

In many respects it is a most remarkable circumstance that bacteriologists as a rule have been so long content to regard with equanimity the assumed absence of any rejuvenating mechanism in the germplasm of bacteria, while it is observed, at least as an occasional necessity, in nearly all other forms of germplasm with which biologists are familiar; and may even be regarded as a universal need of living things. And the same holds true for some form of nuclear "fragmentation" and subsequent cytoplasmic regeneration observable in nearly all forms of plant and animal life. It is true that in some yeasts an exclusively asexual process of reproduction seems to obtain permanently, but this fact is not conclusive. Respecting this point, we may perchance do well to remind ourselves of the history of the conception of sexual reproduction of the yeasts as it concerned the botanists. At that early date when sexual phenomena were first described for yeasts they were regarded as very rare occurrences—just as most bacteriologists believe at the present time with reference to the bacteria. But it eventually became recognized by the botanists that such sexual phenomena were in reality very widespread. Each succeeding year has added fresh instances and new, important data on the reproductive mechanism until, at the present time, some botanists and mycologists entertain the view that sexual reproduction, though of a very primitive type, may occur, at least at some time and under favorable conditions, among all yeast species. Such a history of a biological conception with reference to organisms showing many characteristics in common with the bacteria is not without significance; but still the truth is that, among bacteriologists, nearly all have united in denying to the bacteria as a class, presumably because of their minute size and structure, the possibility of participation in a form of reproduction and nuclear rejuvenation which has long ago been clearly proved and accepted in other simple living things; and particularly in forms closely related to the bacteria.

But, merely because bacteria are minute and apparently simple in organization, is no logical reason for attributing to them a reproductive mechanism that is correspondingly simple; and for denying to them the possibility of sharing with related forms a reproductive process that is intricate and difficult of clear observation—detectable only by those who have the patience and perseverance to search the most diligently. In a way, it seems to me that there may be less justification for our present belief in the existence of the filtrable virus as a distinct biological entity than for our acceptance of a view favoring the existence in bacteria of a reproductive mechanism making possible nuclear rejuvenation;

and of a process in the final explanation of which, the nature of the filtrable virus itself may ultimately find its solution. Bacteriologists as a group, ardent in their support of the living, self-propagating nature of the ultravirus against current biological views which bound all life by the organized cell, have themselves, until most recently, failed to greet with appreciable understanding the only genuine attempts through which it seems possible that a fuller knowledge of the ultravirus might be derived. I am convinced that many fundamental problems in bacteriology today, and among them those of microbic dissociation, transmissible autolysis and the origin of the filtrable virus, will receive their solution only when we shall have taken up the task of studying the actual mechanism of bacterial reproduction from a broader biological viewpoint than that which has characterized the majority of work on bacterial reproduction and variation up to the present time.

Relation of Microbic Dissociation to Reproductive Phenomena (Bacterial Cyclogeny).—If, therefore, we may tentatively accept the view that some sort of a cyclical reproductive process, involving conjugative reactions, exists for bacteria (perhaps in the sense indicated by Enderlein's [160] term "Cyclogenie"), the question must arise: What is the actual relation of microbic dissociation and of the S, O and R forms of culture to this reproductive system? It can only be answered at present that we do not know. In the light of many observations, however, a suggestion may be hazarded although it may be long before either its truth or falsity can be proved. This view may be stated as follows:

When bacteria grow on a medium to which they are well adapted they multiply, at least for the most part, by a simple vegetative process, binary fission; and in consequence present collectively a high degree of uniformity in appearance, as well as in most of their biochemical, antigenic and serologic reactions. This is the normal S type, carrying the double antigen, S (in preponderance) and O (in smaller measure).

When, however, the environment surrounding these organisms changes, and particularly when it changes in a manner unfavorable to the continued vegetative growth of the same culture form, the first effect of such a change is to set into operation (or at least to intensify) in the culture a new reproductive process including opportunity for nuclear reorganization, and looking forward to an adaptation to the new conditions of growth.

This process of nuclear reorganization involves primarily the suppression of the vegetative reproduction by the usual fission and the bringing to the front a form of conjugative reproduction calculated to afford a fresh basis for variability and consequent adaptation. This culture stage is characterized by the generation of filamentous forms, beaded (gonidial) and "involution" structures, and zygospores or analogous bodies produced by conjugation. This varied form of culture represents the transitional or O type. It carries as its fundamental antigen the O, sometimes pure, but often mixed either with S or R.

The O antigen, at least in certain of the transitional forms, is characterized by heat stability. The same is true of the R antigen.

From the zygospores or related bodies embraced in the O culture there arises, as a result of budding or endosporulation or in some still unknown manner, a new generation of cells or other bodies, some of which may be filtrable, but many of which are endowed with a higher degree of adaptability to new growth conditions than was the parent form.

Although these derived bodies (cells or invisible elements) may, under favorable conditions, hold to the same type of growth through several or perhaps many generations, they gradually conform to the demands of the new environment and become temporarily stabilized in a new morphologic and biochemical type the nature of which is determined, within limits, by the existant environmental conditions, as well as by the specific cyclogeny.

The gradual stabilization of the new form is usually, but not always, accompanied by a return to a vegetative mode of reproduction and by a corresponding suppression of sexual or conjugative reproductive methods which were instrumental in its generation. The new culture type, if perpetuated in the new environment, again becomes uniform in composition and, within limits, constant. It now possesses the new heat stable antigen, R; and may have generated a little of the antigen, S. It is not, however, a true mutant since it carries the potentiality for "reversion."

In considering the modified nature of the O and R types of bacteria and their apparently frequent stabilization in culture, the question will undoubtedly be asked, whether these changes are all actually related to the genetic mechanism or whether some of them may be "impressed" variations; or even forms possessing only teratological significance. It is impossible to answer this question with any degree of finality at this time. I believe however that the majority of evidence favors the former view, although actual cases of polymorphism may exist—(Mellon [341]). Of such evidence I believe the most significant is the following. While many of the O and R culture forms are known to arise under conditions which appear to enforce dissociation, such as heating, contact with antiseptics, drying, submission to immune serum, etc., it is an important fact that no variant cultural, biochemical or serologic types appear to be produced in this manner that cannot also be discovered in cultures existing under conditions of growth which we commonly regard as "normal" and physiologically the most favorable; and under conditions far removed from any "enforcing" influences, so far as we can observe. In other words, we can note no significant qualitative difference between the effects of "spontaneous" dissociation and "enforced" dissociation so far as the diversity and nature of the variants are concerned. This circumstance would seem to relieve such variants from the suspicion of being "diseased," or of having been produced as "pathologic" forms under the pressure of unfavorable environment, although we know well enough that such an environment produces these results most quickly and most markedly. It must be freely admitted, however, that there still remains a question as to the extent to which environment may modify any single stage in the cyclogenic series. Although the majority of the variants that have been described in the literature are undoubtedly produced through the agency of the genetic mechanism, "impressed" variations are as yet by no means ruled out of the field of bacterial variation, and future workers will have the task of attempting to distinguish between these possible forms of culture and the true cyclogenic types.

What is the "Normal" Bacterial Type?—From the foregoing considerations one may be inclined to ask—what is the normal bacterial type? Or even—is there such a thing as a normal bacterial type? It is true that through long years of laboratory study we have come to regard as the normal type that form of the organism which grows best for us on artificial culture mediums. But, even in this regard, there have always been grounds for confusion, since it has clearly been recognized that the kind of culture obtained on one medium may be different from the same culture grown on another medium. Even the "same" medium, in dry or moist states, may produce deepseated changes in the culture. What, then, under these varying conditions is the "normal?"

Without the need of multiplying the fallacies of many of our old-time arguments, I believe that a careful consideration of the data assembled in the preceding pages tends to establish the view that "normal culture" or "normal type," in the absolute meaning of these terms and as commonly employed, is something of a myth. The stock typhoid organisms which we maintain in the laboratory are not the same in all respects as those found in the blood or intestines of typhoid patients at the time of isolation; nor necessarily the same as those isolated from the urine or gall bladder of typhoid convalescents. What, then, is the "normal" typhoid organism and culture? The same question may legitimately be asked in the case of other pathogenic bacterial species; and also, though perhaps to a smaller extent, in the case of saprogenic forms. Shall we regard as "normal," the disease form, the convalescent form or the old laboratory form? Also, what shall we say regarding the "intermediates" and the filtrable forms? We have become accustomed, I believe, to regard as normal that form of the organism or culture which occurs most prominently in the sort of reaction or environment that interests us most; and bacteriologists for the most part have been chiefly interested in their culture tube collections. It is these organisms

—often tame, domesticated things—that have been set up as "types" and as standards of normality. It would seem to me much more accurate, however, to refrain from speaking of normality in such an absolute sense, but to regard a culture as normal relative to a given condition of environment; or from the viewpoint of a definitely conditioned reaction. Thus in typhoid we might well refer to the "normal laboratory type," to the "normal disease type" or to the normal type of growth on phenol agar, or at 42 C. or in homologous immune serum—meaning in all these instances the form of typhoid culture which experience has shown usually to be correlated with them. In this manner many years ago certain German writers ceased to speak so much of normal and abnormal cultures of B. anthracis as of the normal, "tierische" or the normal, laboratory forms.

All this must not be construed as meaning that we may not have normal or abnormal species or races of bacteria. It means only that, when we speak of normality or of abnormality, we must make our comparison, not between different stages in the cyclogeny of the strains, but between corresponding stages. In other words, we must compare S type culture with S type, R type culture with R type, and intermediate with intermediate—insofar as we are able to recognize these forms. And to succeed in recognizing them, must be our first business if we are going to attempt comparisons. I have no doubt that some of the greatest sources of confusion and error, not only in systematic but also in applied bacteriology (particularly in serology and immunology), are to be found in our well entrenched habit of attempting to make comparisons between fundamentally different things—that is to say, between cultures in different cyclogenic phases of development. I believe, moreover, that until this viewpoint is changed no successful advance can be made through the commonly used methods of classification of bacterial species or races or strains by seeking to establish "serologic types" as an end in themselves. For advancing the fundamental problems involved, such limited methods are, at the present stage of our knowledge of cyclogenic variation, of little value, except perhaps (if need be) to emphasize more quickly their own futility. Indeed, it seems to me that we have arrived at a point in the development of the science of bacteriology where much work on elaborate superstructures must be suspended until we have made more stable the foundations.

Conclusion.—In concluding and summarizing this section we may return to our original question—What is the biological significance of dissociation? In the present state of our knowledge, although there exist sufficient data from experiment and observation to formulate an hypothesis which no doubt contains some elements of truth, in its finer details it still must be left an open question. At the same time it is clear that the wide extent of its occurrence in diverse bacterial groups, the orderly and usually parallel nature of its manifestations, and the fairly constant nature of its effects upon the cultures concerned, can lead us only to the conclusion that it is not a meaningless reaction, but one possessing considerable significance in the life of the culture if not in the species. If this is true, the significance, whatever else it entails, must be one of a reproductive nature. When, moreover, we examine critically the morphologic changes in the cells occurring during and just preceding the dissociative reaction, we are forced to the view that the reproductive significance of this phenomenon must concern a generative mechanism quite different from simple binary fission in the traditional sense; and something far more complicated. And when, finally, we observe that the microscopic pictures preceding or accompanying microbic dissociation reveal an unusual wealth of those peculiar cell structures which have been recorded persistently for more than two decades in the handful of studies either suggesting or definitely dealing with new reproductive mechanisms among bacteria, we have added evidence that microbic dissociation, as an adaptive reaction, stands in close relation to a type of reproduction about which we as yet know little. More than this cannot be said at present regarding its biological significance. But, if we can accept this much as a working hypothesis, it can scarcely be doubted that the important facts, in their proper relations, will follow in the course of further persistent inquiry.

17. GENERAL CONCLUSIONS

From the varied assortment of data on experiment and observation which I have brought together in the foregoing pages of this review, I believe it becomes clear that many important problems in modern bacteriology and pathology have their roots in the phenomenon of bacterial instability, and especially in that aspect which has been termed microbic dissociation. We begin to appreciate, moreover, that this phenomenon, far from being a sign of chaos, as many have believed, is in reality merely the necessary manifestation of certain more or less orderly processes that are correlated with a physiologic and reproductive mechanism the nature and significance of which we are just beginning to observe and, perhaps to a limited extent, to understand. In the operation of this mechanism we can scarcely doubt that there are concerned significant issues dealing with bacterial relationships, with serological behavior, biochemical reactions, immunologic reactions, with pathogenicity and virulence; and, in all probability, with the problem of transmissible bacterial autolysis as well.

Regarding the relation of dissociation to the systematic aspects of the science of bacteriology it may be safely predicted that the time will soon arrive when, in discussing the nature of bacterial cultures isolated from infected tissues or maintained in laboratory stocks, it will not be sufficient merely to name the organism. The specific name of a microbe, as names now stand, may mean little; for it is becoming increasingly clear that we shall never know what a bacterial species really is until we acquaint ourselves with the outermost limits of its variability; and this must mean not merely its "normal" vegetative form, but all of its dissociated types or cyclogenic stages as well. Our exact knowledge of an organism may be slight even after we have noted its morphology, recorded its measurements, registered its reactions, ascertained its common serologic and antigenic characters and given it a name. And this is because these superficial considerations, based only upon the "normal form," give us slight understanding of its potentiality for, or mode of, variation under changed conditions of growth. We do not perhaps see the bacillus masquerading as a streptococcus, the branched filament hidden beneath the cloak of the common rod, the true meningococcus in its "giant coccoid" state, or the granule of the invisible virus latent amongst the microscopically visible elements. The truth we shall eventually come to, however, is that the free-living microorganism is potentially a kaleidoscopic thing, in which the power of responding successfully to a changing environment by alterations in body state, both morphologic and biochemical—and even by self-destruction, if need be, in order to generate another and more stable type—stands as its one most important attribute.

It is of course quite true that we can remove an organism from its customary environment, confine it for years in tubes, and enforce upon it thereby a certain kind of stability. Some such "domesticated" cultures, particularly saprophytes, may under such conditions retain perhaps indefinitely their original proclivities, while others, notably pathogenic species, may quickly lose much of their ancient heritage. In so doing, however, they become increasingly amenable to consistant laboratory findings; and have therefore become the favorite subject-matter of the systematists. Long ago, Darwin recognized this situation with respect to variation in higher forms, and consistently avoided the impasse to his study of variation determined by too continuous an existence of experimental stock in an unchanging environment.

But, outside the field of cultures possessing such enforced stability, it must be admitted that most cultures, when first secured from their natural habitat and placed upon the usual culture mediums, possess great potential variability. Each apparent species is surrounded by its small group of satellites to each of which we unwisely attempt to assign a classificatory niche. We have meningococcus and pneumococcus types which we usually number in Roman; paratyphoid and streptococcus types for which we employ the Greek; B. coli and Bact. aerogenes types for which we employ the small Arabic; B. dysenteriae types to which we assign the name of the discoverer; and still other "types" which we have difficulty in figuring out at all. But such procedures do not yield an advance toward a clearer understanding of the fundamental nature of the species concerned, nor of their actual genetic relationships. Although sometimes of practical value, they are for the most part makeshifts only; and we should not long be content to permit their endurance without demanding a more complete recognition of their meaning and significance.

What we ought to wish to know about these cultures is, for instance, how the various types of pneumococci, of streptococci, meningococci, of B. coli, of the paratyphoids, of dysentery and other bacteria, have been (we might even say "are being") formed. Are types 2 and 3 of the pneumococcus variants of type 1; and, if so, under what conditions do they arise? What genetic relationship or sequence in origin exists between B. coli a, b and c? And the same for the paratyphoids, the diphtheria bacilli and other bacteria. These are questions of practical significance to be answered; but it is difficult to see how they can be answered so long as we concentrate all our attention on methods of classification rather than on the one thing that is most essential—the problem of the nature and origin of variations. This single problem, for higher animals and plants, is the stumbling block in the path of evolution inquiry; but we are not so certain that the difficulty exists, in the same way at least, for bacteria. Although we may not learn how to create a "species," there is no class of organisms more favorable than bacteria for studying the possible influence of environment in determining the trend of hereditary variation.

One object of this discussion, however, is not only to emphasize our present inadequate conception of the nature of bacterial species and of the systematic relationships of the bacteria, but also to point out certain observed grounds for microbic instability and to suggest possible methods of reorganizing the system. It seems to me that this must consist first of abandoning, for the present at least, our vain attempts to perfect schemes of classification. This seems logical, especially in view of the fact that we do not yet know exactly what it is that we have to classify. Secondly and more important, it consists in initiating a somewhat new branch of bacteriological study in which we shall strive to recognize bacterial species relationships, not by a comparison of isolated single cultures of this or that, but by a study of all the various types or stages comprising the cyclogeny of the species in question—a difficult but necessary task. I believe it is only in this way that we shall come to understand the limits and the relative significance of cyclogenic, fortuitous and impressed variations; and so be able to recognize the bacterial species in its entirety. Bacillus diphtheriae, as assumedly typifying a definite bacterial species, should eventually come to mean to us —not merely a rod of fixed size and shape (possessing certain constant biochemical, serologic, immunologic, and staining reactions, by means of which it can conveniently be recognized and cataloged), but in reality a host of things, which we must be able to recognize individually and collectively before we can affirm that we know the "species." Bacillus anthracis, as a "bamboo rod" of certain shape, size, colony form and pathogenicity, has occupied the attention of countless bacteriologists for fifty years; but who at present would dare affirm that we know the anthrax species? Our ignorance of some of its most intrinsic characteristics is nearly as great today as it was half a century ago. In these species, and in others, important problems are awaiting solution.

Regarding the mode of approach to the species problem, as also to several others, I believe the simplest and most direct way in which the average bacteriologist can render service is to attack the subject from the viewpoint of microbic dissociation, as the outstanding features of this phenomenon have been outlined in these pages. And this refers mainly to an attempt to recognize and study cultures arising from the two chief colony types, S and R, together with such intermediates (O) as may be observed. While such a mode of approach does not necessarily bring us face to face with the biological realities underlying dissociation itself, it should serve to make available a mass of evidence which may not only prove of value in useful bacteriological procedures, but also have the merit of indicating more clearly what the exact nature of the more remote problems underlying dissociation really is. In facing such a complex question it may be best that the beginning work should not dig too deep. Here again, perhaps, we must ascertain more clearly just what we have to explain before we attempt to explain it. On the other hand, much may doubtless be gained by those few adventurous workers, such as Mellon and Enderlein, who are engaged so-to-speak in attempting to solve the problem from the bottom up. Theirs is at present the more difficult task. But, when the two groups ultimately meet, it may be predicted that we shall have an understanding, not only of the significance of microbic dissociation, but also of still largely unknown reproductive phenomena among the bacteria.

For a partial indication as to what success may be expected to arise from such endeavors one needs only to refer to such studies as those of Stearn, Gratia and Mellon on B. coli; of Mellon on the diphtheroids; of de Kruif on the Pasteurella; of Weil, Felix, and Braun and Schaeffer and others on B. proteus; of Bernhardt on B. diphtheriae; of Cowan on the streptococci; of Stryker, Griffith, Reimann and of Amoss on the pneumococcus; of Goyle on B. typhosus and B. enteritidis; of Balteanu on the cholera vibrio; of Enderlein on the cholera vibrio and the diphtheria bacillus and of Julianelle on Friedländer's bacillus. All of these studies begin to show, for the first time and in one way or another, how, in what sequence, and under what conditions of environment, cyclogenic variations have been, and can be, produced. Whether such changes in general, or any stage in particular can, by artificial means, be made permanently hereditary, may be left an open question. Undoubtedly many of them possess remarkable stability. We see that the geneticist, working with multicellular forms, has been able, by the manipulation of his unit characters and with the aid of the Mendelian principles of segregation, as also of sex-linkage, to blend these available units into new configurations; but he has not been able to synthesize a new species. It is permissible to believe, however, that bacteriologists, working with a less differentiated and more impressionable protoplasm, may yet be able to produce from known stocks, new bacterial forms possessing at least the equivalence of what have heretofore been termed species; but for which new designations would be required. At the present time our convictions are such that the conception of bacterial hybrids as introduced by Almquist [6] in 1924 appears to most bacteriologists as the height of absurdity. While it is quite true that his remarkable results, like those of Castellani,[91] may be explained on other grounds than that of an actual crossing of species, and that all possible attempts should be made so to explain them, it also may be remarked that their mirth-provoking power is today much less than it would have been a decade or more ago. It must be frankly admitted that we still have little intimate knowledge of the private life of the bacterium.

In this connection there should be mentioned another field in which I believe microbic dissociation may eventually play an important part. This is in connection with the rickettsial bodies, both pathogenic and nonpathogenic. In the study of these elements, whether intracellular or extracellular, there is now ample support for the growing suspicion that they are related in some way to the better known bacterial forms. This is perhaps indicated most clearly in the case of Rickettsia prowazeki and B. proteus X19 of Weil and Felix. The results of many histologic, serologic and immunologic studies seem at present to combine in suggesting that the typhus fever rickettsiae may be a dissociated stage of the proteus organism. With further inquiry, as the earlier histologic approaches give way to cultural studies, it seems probable that new and significant facts bearing on the dissociative reaction will be brought to light.

That the brief knowledge already available regarding the dissociative reactions among disease-producing bacteria is even now in a position to give us a more exact and critical view of certain serologic and immunologic reactions there can be no doubt. It not only makes clear many previously observed inconsistencies, but helps both to indicate and explain the limits of serviceability of many serologic reactions as used for diagnosis. In addition it presents the most tangible basis for an understanding of the confusing data that have gathered about the subject of the "double" and "single" antigens, as well as the "major" and "minor" agglutinins. In relation to still other methods of bacteriologic diagnosis (cultural, microscopic, biochemical) it affords a new and important point of view by virtue of which we are in a position more carefully to observe and more intelligently to interpret the results of bacteriologic examinations. Although its bearing upon immunological theory and practice is as yet hardly touched upon, it already shows us one possible reason for frequent failure to obtain expected results from current immunological procedures; and thereby opens the door to a perhaps helpful reconsideration of several significant problems relating to vaccine therapy in man. Regarding the rôle of pathogenic organisms in infections, a knowledge of dissociation phenomena offers, for the first time, a rational and exact basis for developing virulence by individual colony, rather than by mass culture, selection. It affords us, moreover, a closer view of the origin and the far-reaching significance for serology and immunology of the socalled specific soluble substances of bacteria. Through the now more clearly recognized mode of action of immune serum on bacterial cultures in vitro it offers a new conception of one perhaps highly significant mechanism of humoral and cellular defense in the living body against invading micro-organisms; and thus may afford an explanation, heretofore lacking, regarding the actual mode of protective action of the bacteriotropic antibodies. In view of the dissociation-furthering influence of high growth temperatures on several pathogenic bacterial species, microbic dissociation may supply the grounds for a better understanding of the rôle of hyperthermia in infectious disease; also, regarding the the cause of relapse in fevers as being related to incomplete dissociations, or to the reversion-provoking power of R type immune serum, as demonstrated by Soule for the nonpathogen, B. subtilis. It fixes in our mind, moreover, the vast, but heretofore unrecognized, importance in the treatment of disease, of the control of the cyclostage in the individual case, as already emphasized by Enderlein; and persuades us that it is toward this end, rather than toward the often impossible direct elimination of the infecting organism, that we should look.

In epidemiology our knowledge of the dissociation phenomena in relation to specific, pathogenic, bacterial species is certain to cause a revision of many of our older views, especially those relating to the "carrier state." It may be that we shall sometime learn that the progress of the epidemic is merely the gradual development of the infective agent into its virostage; and that the remission of the epidemic is due to the passage of the infecting agent through the virostage, and beyond it, into a nonvirulent form. Indeed, we may conclude by saying that there are few fields in either systematic or applied bacteriology that are not in a position to be illumined and augmented through the knowledge that has been and will be gained regarding microbic dissociation.

Regarding the joint relation of microbic dissociation and the phenomenon of the bacteriophage to microbic instability, we have seen that there are good reasons for suspecting that both of these reactions are inseparably united in the reproductive mechanism of the bacterial cell. It seems to be the "normal" functioning of this mechanism, operating in a favorable, and one might add accustomed, environment, that determines traditional microbic stability; while it is the "abnormal" or "pathological" (Bordet) functioning of the same mechanism that conditions mutation-like changes, accompanied by lysis of varying grades and possessing great significance in the life of the culture—perhaps in the life of the "species." At the same time, we may not be quite justified in regarding these dissociative transformations, (whether active or passive) as "abnormal" or "pathological" in any other sense than that they may be manifested to an unusually high degree or that they may be forced—sometimes intentionally (as in provoking transmissible autolysis), and sometimes unintentionally (as in "spontaneous" dissociation). We have seen that there is much evidence that the reactions of microorganisms are much the same in both; and that these reactions possibly represent the only means by which bacteria are able to insure the survival of their germplasm in the face of age or unfavorable environment. In this presumably adaptive reaction we may also eventually come to see that supremely important, but still largely unrecognized, events in the life of the bacterial species are occurring during that stage of the dissociative reaction lying between the disappearance of the old form and the generation of the new.

When, moreover, we pass in review the many and varied manifestations of autolysis and regeneration, whether in appearance "transmissible" or merely "dissociative," and observe the frequent connecting links between these reactions, we can scarcely doubt that we are dealing in all instances with a single biologic phenomenon which, depending on the speed of reaction and correlated features, gives us the varied lytic manifestations that we are trying to explain. May it not be probable that some organisms, well adapted to their environment, live and die without presenting, or presenting only obscurely, outstanding lytic and regenerative tendencies? Or that others of different constitution give us the moderate changes seen in the slow-going, active dissociations? Or that others, unstabilized by new environmental conditions, react by giving the rapid dissociations seen in the typical "suicide cultures?" Or that still others afford a sort of dissociation that is transmissible through the agency of filtrates (rather perhaps than from cell to cell) and characterized by great suddenness, as typified by the classic bacteriophage phenomenon of the Shiga dysentery bacillus?

If these conceptions should prove tenable, we must then come to regard transmissible autolysis merely as a normal reaction carried to a "pathological" extreme, as has been held from the beginning by Bordet and Ciuca.[62]. In such a situation we should expect to be confronted in nature by lytic phenomena of such diversity and overlapping grades that no arbitrary line could be drawn between transmissible autolysis and certain other autolytic manifestations of less acute nature; and this, I believe, is the exact situation that I have shown to obtain in the bacterial world today. I therefore regard the suggestion justifiable that active microbic dissociation and the phenomenon of the bacteriophage may represent merely two different stages in a single phase of normal reproductive and physiologic behavior which exists for a purpose presumably adaptive. Obviously, however, even to be convinced of the truth of this hypothesis, is still to be far from solving in its entirety the mystery of the bacteriophage; but it may well have the advantage of showing us in what direction we must look for the final solution.

And, finally, if for a moment we abandon ourselves to speculation, though not perhaps without some evidence in fact—it may eventually be demonstrated, not that a foreign filtrable virus gives rise to dissociation and to autolysis in the d'Herelle sense; but, on the contrary, that the fundamental physiologic reaction, of which both microbic dissociation and transmissible autolysis are only different modes of expression, gives rise to the filtrable virus.

ADDENDUM

In reviewing the final proof of this review, I am impressed with several important and quite unintentional omissions. Among these, one of unusual interest relates to the subject of "receptor analysis," developed especially at the hands of Felix[517] and Felix and Olitzki.[518] If, however, the interested reader will substitute the term, S antigen (labile), for their "large-flaking antigen" (or serums), and the term, O antigen (stable), for their "small-flaking antigen" (or serums), much of this highly important investigation on bactericidal serum action and qualitative receptor analysis may be brought into relation with certain aspects of the dissociation problem as presented particularly in sections 9 and 10 of the present work. With the possible exception of the older work on anthrax bacillus, I know of no other study dealing with the immunologic significance of cultures (or antigens) of the intermediate or O type, as opposed to the R forms. The latter were clearly not involved in the experiments of Felix and Olitzki. Even with this lack, however, their work is a singularly clear example of the vast significance of microbic dissociation for outstanding problems in serology and immunity.

REFERENCES

[1] Adami, J. G.: Abbott, M. E., and Nicholson. F. S.: On the diplococcoid form of the colon bacillus, J. Exper. Med., 1899, 4, p. 349.

[2] Ali-Krogius: Du rôle du Bacterium coli commune dans l'infection urinaire, Arch. de méd. expér. et d'anat. path., 1892, 4, p. 66.

[3] Almquist, E.: Studien über filtrierbare Formen in Typhuskulturen, Centrabl. f. Bakteriol., I, O., 1911, 60, p. 167.

[4] Almquist, E.: Wuchsformen, Fruktification und Variation der Typhusbakterie, Ztschr. f. Hyg., 1916, 83, p. 1.

[5] Almquist, E.: Variation and life cycles of pathogenic bacteria, J. Infect. Dis., 1922, 31, p. 483.

[6] Almquist, E.: Investigations on bacterial hybrids, ibid., 1924, 35, p. 341.

[7] Altmann, K., and Rauth, A.: Experimentelle Studien über Erzeugung serologisch nachweisbarer Variationen beim Bakterium coli, Ztschr. f. Immunitätsforsch. v. exper. Therap., 1910, 7, p. 629.

[8] Amoss, H. L.: The composite nature of a pure culture of a virulent pneumococcus, J. Exper. Med., 1925, 41, p. 649.

[9] Andervont, H., and Simon, C. E.: On the origin of so-called pellucid areas which develop on agar cultures of certain spore-bearing bacteria, Am. J. Hyg., 1924, 4, p. 386.

[10] Andrewes, F. W.: Studies in group agglutination: I. The salmonella group and its antigenic structure, J. Path. and Bact., 1922, 25, p. 505.

[11] Archard, C., and Renault, J.: Sur les bacilles de l'infection urinaires, Sem. méd., 1892, 12, pp. 136, 490.

→ Arkwright, J. A.: Varieties of the meningococcus with special reference to a comparison of strains from epidemic and sporadic sources, Brit. J. Hyg., 1909, 9, p. 104.

[13] Arkwright, J. A.: The serum reactions (complement fixation) of the meningococcus and the gonococcus, ibid., 1912, 11, p. 515.

[14] Arkwright, J. A.: Grouping of the strains of the meningococcus, Brit. M. J., 1915, 2, p. 885.

[15] Arkwright, J. A.: The bacteriology of cerebrospinal meningitis, Brit. M. J., 1920, 2, p. 420.

[16] Arkwright, J. A.: Variations in bacteria in relation to agglutination both by salts and by specific serum, J. Path. and Bact., 1921, 24, p. 36.

[17] Arkwright, J. A.: The source and characteristics of certain cultures sensitive to the bacteriophage, Brit. J. Exper. Path., 1924, 5, p. 23.

[18] Arkwright, J. A., and Goyle, A.: The relation of the "smooth" and "rough" forms of intestinal bacteria to the "O" and "H" forms of Weil and Felix, ibid., 1924, 5, p. 104.

[19] Arloing, F., and Dufourt, A.: Contribution a l'étude des formes filtrantes du bacille tuberculeux, Compt. rend. Soc. de biol., 1925, 93, p. 165.

[20] Ascoli, A.: Ueber den Wirkungsmechanismus des Milzbrandserums: Antiblastische Immunitat., Centralbl. f. Bakteriol., I. O., 1908, 46, p. 178.

[21] Atkin, E. E.: Some cultural characteristics exhibited by serological types of meningococci, Brit. J. Exper. Path., 1923, 4, p. 325.

[22] Atkin, E. E.: The significance of serological types of the gonococcus, ibid., 1925, 6, p. 235.

[23] Axelrad, C.: Ueber Morphologie der Kolonien pathogener Bakterien, Ztschr. f. Hyg., 1903, 44, p. 477.

[24] Babes, V.: Ueber Variabilität und Variaten des Typhusbacillus, Ztschr. f. Hyg., 1890, 9, p. 322.

[25] Babes, V.: Erklarende Bermerkungen über "näturliche" Varietäten des Typhusbacillus, Centralbl. f. Bakteriol., I. O., 1891, 10, p. 281.

[26] Bachmann, A., and de la Barrera, J.: Quelques variations serologique du bacille paratyphique A, Compt. rend. Soc. de biol., 1923, 89, p. 756.

[27] Baerthlein, K.: Ueber Mutationserscheinungen bei Bakterien., Arb. a. d. Kaiserl. Gsndhtsamte, 1912, 40, p. 433.

[28] Baerthlein, K.: Weitere Untersuchungen über Mutationserscheinungen bei Bakterien, Centralbl. f. Bakteriol., I, Ref., 1912, 54, Beilage, p. 178.

[29] Baerthlein, K.: Untersuchungen über Bact. coli mutabile, Centralbl. f. Bakteriol., I. O., 1912, 66, p. 21.

[30] Baerthlein, K.: Ueber bacterielle Variabilität inbesondere so-genannte Bakterienmutationen, ibid., 1918, 81, p. 369.

[31] Bail, O.: Untersuchungen über Cholera und Typhus Immunität., Archiv. f. Hyg., 1905, 52, p. 272-377.

[32] Bail, O.: Ueber die Korrelation zwischen Kapselbildung, Sporenbildung und Infektiosität des Milzbrandbazillus, Centralbl. f. Bakteriol., I, O., 1914, 75, p. 159; 76, pp. 38, 320.

[33] Bail, O.: Der Stand und die Ergebnisse der Bakteriophagenforschung, Deutsch. med. Wchnschr., 1925, 51, p. 13.

[34] Bail, O., and Flaumenhaft, T.: Verandergung von Bakterien in Tierkörper: XIII. Versuche mit abgeschwachten Milzbrandbazillen in Meerschweinchenkorper, Centralbl. f. Bakteriol., I. O., 1917, 79, p. 425.

[35] Bail, O., and Rubritus, H.: Veränderungen von Bakterien im Tierkörper, ibid., 1906, 43, p. 641.

[36] Bainbridge, F. A.: Some observations on B. anthracoides, J. Path. & Bact., 1903, 8, p. 117.

[37] Barber, M. A.: On heredity in certain microorganisms, Kansas Univ. Sc. Bull., 1907, 4, p. 1.

[38] Balteanu, I.: The receptor structure of Vibrio cholerae (V. comma) with observations on variations in cholera and cholera-like organisms, J. Path. & Bact., 1926, 29, p. 251.

[39] Bassett-Smith: Paramelitensis infection in man and animals, J. Trop. Med. & Hyg., 1921, 24, p. 53.

[40] Beham, L. M.: Die agglutinatorischen Eigenschaften der Kapselbacillen und die Anwendung der Serumagglutination bei den Tragern von Kapselbacillen, Centralbl. f. Bakteriol., I, O., 1912, 66, p. 110.

[41] Behring and Nissen: Ueber bakterienfeindliche Eigenschaften verschiedener Blutserum Arten, Ztschr. f. Hyg., 1900, 8, p. 412.

[42] Beijerinck, M. W.: Mutation bei Microben, Folia Microbiol., 1912, 1, p. 4.

[43] Beijerinck and Minckmann: (Quoted from Enderlein [140]).

[44] Bekerisch, A., and Hauduroy, P.: Au sujet de l'obtention de Bacteriophage par antagonisme microbienne, Compt. rend. Soc. de biol., 1922, 86, p. 881.

[45] Beneke, W.: (Review of work of R. Müller), Ztschr. f. indukt. Abstamm. u. Vererbungsl., 1909, 2, p. 215. (Cited after Dobell.[129])

[46] Benians, T. H. C.: A record of an inagglutinable form of Shiga's dysentery bacillus experimentally produced from an agglutinable culture, J. Path. & Bact., 1920, 23, p. 171.

[47] Benington, Ida: Separation of toxic and non-toxic cells from cultures of an anaerobe isolated from larvae of the green fly, U. S. P. H. A., Pub. Health Rep., 1922, 37, p. 2252.

[48] Berger, E., and Englemann, B.: (Serology of the pneumococcus), Berl. klin. Wchnschr., 1926, 5, p. 599.

[49] Bernhardt, G.: Ueber Variabilität pathogener Bakterien, Ztschr. f. Hyg., 1915, 79, p. 179.

[50] Beyer, H. G., and Reagh, S. B.: Further differentiation of flagellar and somatic agglutinins, J. M. Res., 1904, 12, p. 313.

→ Blair, J. E.: A lytic principle (bacteriophage) for Corynebacterium diphtheriae, J. Infect. Dis., 1924, 35, p. 401.

[52] Blake, F. G., and Trask, J. D.: Alternations in virulence and agglutination reactions of pneumococci induced by growth in immune serum, Proc. Conn. Branch Soc. Am. Bact., Abstracts, 1923, 7, p. 353.

[53] Blanc, J.: Transformation de Bacillus pyocyaniques en bacillus sans pigments, Essai d' interpretation, Compt. rend. Soc. de biol., 1923, 88, p. 52.

[54] Blanc, J.: (Transformation of B. coli into an organism of the mucous type), ibid., 1923, 88, p. 49.

[55] Block: Technique of diagnosis, Brit. M. J., 1897, 11, p. 1777.

[55a] Bongert, J.: Beitrage zur Biologie des Milzbrandbacillus und sein Nachweis in Kadayer der grossen Haustiere, Centralbl. f. Bakt., I, O., 1903, 34, p. 772.

[56] Bordet, J.: Adaptation des virus aux organismes vaccinés, Ann. de l'Inst. Pasteur, 1892, 6, p. 328.

[57] Bordet, J.: Sur le mode d'action des sérums preventifs, ibid., 1896, 10, p. 193.

[58] Bordet, J.: Contribution a l'étude du serum antistreptococcique, ibid., 1897, 11, p. 177.

[59] Bordet, J.: Apparition spontanée du pouvoir lysogène dans les cultures pures, Compt. rend. Soc. de biol., 1924, 90, p. 96.

[60] Bordet, J.: Pouvior lysogène actif ou spontané et pouvoir lysogène passif ou provoque, ibid., 1925, 93, p. 1054.

[61] Bordet, J.: Le problème de l'autolyse microbienne transmissible ou de bactériophage, Ann. de l'Inst. Pasteur, 1925, 39, p. 717.

[62] Bordet, J., and Ciuca, M.: Exudats leucocytaires et autolyse microbienne transmissible, ibid., 1920, 83, p. 1293.

[63] Bordet, J., and Ciuca, M.: Le bactériophage de d'Herelle, sa production et son interpretation, ibid., 1920, 83, p. 1296.

[64] Bordet, J., and Sleeswyk: Serodiagnostic et variabilité des microbes suivant le milieu de culture, Ann. de l'Inst. Pasteur, 1910, 24, p. 476.

[65] Braun, H., and Feiler, M.: Ueber Serumfestigkeit des Typhusbacillus, Ztschr. f. Immunitätsforsch. u. exper. Therap., 1914, 21, p. 447.

[66] Braun, H., and Schaeffer: Zur Biologie der Fleckfieber-proteus Bazillen. Ein Beitrag der Wirkungsweise der Desinfektionsmittel, Ztschr. f. Hyg., 1919, 89, p. 339.

[67] Braun, H., and Salomon: Ueber Fleckfieber Proteus-bacillus (Weil-Felix), Centralbl. f. Bakteriol., I, O., 1918, 82, p. 243.

[68] Bredeman, G.: Bacillus amylobacter, A.M. et Bredeman, in morphologischer, physiologischer und systematischer Beziehung, Centralbl. f. Bakteriol., II, 1909, 23, p. 385.

[69] Breinl, Fr.: Variations-erscheinungen in der Dysenterie-gruppe, Ztschr. f. Immunitätsforsch. u. exper. Therap., 1922, 35, p. 176.

[70] Breinl, Fr.: Sur les relations du virus exanthématique et des bacilles proteus "X," Arch. de l'Inst. Pasteur (Tunis), 1924, 13, p. 208.

[71] Breinl, Fr., and Fischer: Variationserscheinungen in der Paratyphus-gruppe, Ztschr. f. Immunitätsforsch. u. exper. Therap., 1922, 35, p. 205.

[72] Breinl, A., and Kinghorn, A.: An experimental study of the parasite of tick fever (Spirochaeta duttoni), Memoir XXI, Liverpool School of Trop. Med., 1906.

[73] Bronfenbrenner, J., and Korb, C.: On variants of B. pestis caviae resistant to lysis by the bacteriophage, Proc. Soc. Exper. Biol. & Med., 1925, 23, p. 3.

[74] Bruckner and Cristéanu: Sur les precipitines du gonocoque et du méningocoque, Compt. rend. Soc. de biol., 1906, 62, p. 1070.

[75] Brutsaert, P.: L'agglutination des microbes resistant, ibid., 1924, 90, p. 645.

[76] Brutsaert, P.: La constitution antigénique des vibrions du cholera, ibid., 1924, 91, p. 1157.

[77] Bruynoghe, R.: Au sujet de la guérison des germes devenus resistants au principe bactériophage, ibid., 1921, 85, p. 20.

➡ Buchanan, R. E., and Traux: Non inheritance of impressed variations in Streptococcus lacticus, J. Infect. Dis., 1910, 7, p. 680.

[79] Bull, C. G.: The fate of typhoid bacilli when injected intraveniously into normal rabbits, J. Exper. Med., 1915, 22, p. 475.

[80] Bull, C. G.: The agglutination of bacteria in vivo, ibid., 1915, 22, p. 484.

[81] Bull, C. G., and McKee, Clara: The biological relationships of the diphtheria group of organisms as shown by complement-fixation tests: I. Bacillus diphtheriae vs. Bacillus hofmanni, Am. J. Hyg., 1924, 4, p. 101.

[82] Bull, C. G., and Pritchett, Ida: The agglutination of blood and agar strains of typhoid bacilli, J. Exper. Med., 1916, 24, p. 3.

[83] Bürgers and Bachmann, W.: Bacteriophagenstudien, Ztschr. f. Hyg., 1924, 101, p. 350.

[84] Burk, A.: Mutation bei einem der Koli-gruppe verwandten Bakterium, Arch. f. Hyg., 1908, 65, p. 235.

[85] Burnet, Et.: Sur la notion de parameletensis, Arch. de l'Inst. Pasteur (Tunis), 1925, 14, p. 247.

[86] Burnet, Et.: Actions d'entrainement entre races et espéces microbienne, ibid., p. 384.

[87] Burnet, F. M.: Observations on the agglutinins in typhoid fever, Brit. J. Exper. Path., 1924, 5, p. 251.

[88] Burri, R.: Ueber scheinbar plötzliche Neuerwerbung eines bestimmten gärungsvermögens durch Bakterien der Coligruppe, Centralbl. f. Bakteriol., II, 1910, 28, p. 321.

[89] Čančik, J.: (Bacteriophagy in pyocyaneus cultures). Čas. Lék. Čes., 1923, 62, p. 25. Abstr., J. Am. M. A., 1923, 80, p. 970.

[90] Cantacuzène, J.: Formation d'une race agglutino-resistante de vibrions au contact des tissus d'un organisme immunisé contre ces vibrions, Compt. rend. Soc. de biol., 1925, 92, p. 1461.

[91] Castellani, Aldo: Biochemical characteristics of certain bacteria, Brit. M. J., 1925, 2, p. 734.

[92] Caublot, P.: Le bactériophage du pneumobacille de Friedländer, Compt. rend Soc. de biol., 1924, 90, p. 622.

[93] Celli, A., and Santori: Ueber ein transitorische Varietät vom Choleravibrio, Centralbl. f. Bakteriol., I. O., 1894, 15, p. 789.

[94] Chamberland, Ch., and Roux, E.: Sur l'attenuation de la virulence de la bactéridie charbonneuse, sous l'influence des substances antiseptiques, Compt. rend. Acad. d. Sc., 1883, 96, p. 1088.

[95] Charrin and Roger: Note sur le dévelopment des microbes pathogènes dans le serum des animaux vaccinés, Compt. rend. Soc. de biol., 1889, 1, p. 667.

[96] Charrin and Roger: Attenuation des virus dans le sang des animaux vaccinés, ibid., 1892, 4, p. 620.

[97] Chauveau, A., and Phisalix, C.: Contribution a l'étude de la variabilité et du transformisme en microbiologie à propos d'une nouvelle varieté de bacille charbonneux (B. anthracis claviformis), Compt. rend. Acad. d. Sc., 1895, 120, p. 801.

[98] Chiari, H., and Loeffler, E.: Ueber ein Uebertrāgbares alkalibildenes Aegens gewisses Coli-stämme, Centralbl. f. Bakteriol., I, O., 1925, 96, p. 95.

[99] Ciuca, M.: Lyse transmissible en absence d'electrolytes libres, Compt. rend. Soc. de biol., 1924, 90, p. 521.

[100] Clark, P. F.: The relation of the pseudodiphtheria and diphtheria bacillus, J. Infect. Dis., 1910, 7, p. 235.

[101] Clark, P. F., and Clark, Alice S.: A "bacteriophage" active against a hemolytic streptococcus, (Abstr.), J. Bact., 1926, 11, p. 89.

[102] Clough, Mildred: A study of the pneumococcus reacting with anti-pneumococcus sera of types I, II and III, with an observation of a mutation of one of the strains, J. Exper. Med., 1919, 30, p. 123.

[103] Cohn, F.: Beitrage zur Biologie der Pflanzen, Breslau, 1875, 1, p. 127.

→ Cole, L. J., and Wright, W. H.: Application of the pure-line concept to bacteria, J. Infect. Dis., 1916, 19, p. 209.

[105] Cole, R.: Ueber Agglutination verschiedener Typhusstämme, Ztschr. f. Hyg., 1904, 46, p. 367.

[106] Collins, G.: Studies on the source of the bacteriophage and on the origin of transmissible bacterial autolysis, 1924, Thesis, Univ. of Michigan.

[107] Colombo, G. L.: Ueber die Komplementbindung als Prufungsmethode der Meningokokken- und Gonokokken-sera und die Specificität ihrer Amboceptoren, Ztschr. f. Immunitätsforsch. u. exper. Therap., 1911, 9, p. 287.

[108] Connal, A.: A study of the cerebro-spinal fluid in the infective diseases of the meninges with special reference to cerebro-spinal fever. Quart. J. Med., 1910, 3, p. 152.

[109] Cooper, M. L.: Capsulated bacteria with special reference to B. typhosus, J. Infect. Dis., 1925, 36, p. 439.

[110] Corbett, L., and Phillips, G. C.: The pseudo-diphtheria bacillus, J. Path. & Bact., 1897, 4, p. 193.

[111] Cowan, Mary: Variation phenomena in streptococci, with special reference to colony form, hemolysin production and virulence, Brit. J. Exper. Path., 1922, 3, p. 187.

[112] Cowan, Mary: Variation phenomena in streptococci: Further studies on virulence and immunity, ibid., 1923, 4, p. 241.

[113] Cowan, Mary: Variation phenomena in streptococci: Further studies on virulence and immunity in mice and rabbits, ibid., 1924, 5, p. 226.

[114] Crowell, M. J.: Morphological and physiological variations in the descendants of a single diphtheria bacillus, J. Bact., 1926, 11, p. 65.

[115] Danysz: Immunization de la actéridee charbonneuse contre l'action du serum du rat, Ann. de l'Inst. Pasteur, 1900, 14, p. 643.

[116] Davison, W. C.: Nature and therapeutic application of bacteriolysants (d'Herelle's phenomenon), Abstr. Bact., 1921, 6, p. 27.

[117] Dawson, A. I.: Bacterial variations induced by changes in the composition of the culture medium, J. Bact., 1919, 4, p. 133.

[118] De Kruif, P.: Dissociates of Microbic species: Mutation in pure-line strain of the bacillus of rabbit septicaemia, Proc. Soc. Exper. Biol. & Med., 1921, 19, p. 34

[119] De Kruif, P.: Dissociation of microbic species. I. Coexistence of individuals of different degrees of virulence in cultures of the bacillus of rabbit septicemia, J. Exper. Med., 1921, 33, p. 773.

[120] De Kruif, P.: Mutation in the bacillus of rabbit septicaemia, ibid., 1922, 35, p. 561.

[121] De Kruif, P.: Virulence and mutation of the bacillus of rabbit septicemia, ibid., p. 621.

[122] De Kruif, P.: Rabbit septicemia types D and G in normal rabbits, ibid., 1922, 36, p. 309.

[123] Delepine, A. S.: The technique of serum diagnosis with special reference to typhoid fever, Brit. M. J., 1897, 1, p. 967.

[124] Denny, F. P.: Observations on the morphology of B. diphtheriae, B. pseudo-diphtheriae and B. xerosis, J. M. Res., 1903, 9, p. 117.

[125] De Martini, L.: Zur Differenzierung der Diphtherie- von dem pseudo-diphtherie Bazillen, Centralbl. f. Bakteriol., I. O., 1897, 21, p. 87.

[126] Deutsch: Die Impfstoff und Sera, 1903. (Quoted from Hess.[250])

[127] Dickson, E. C.: Botulism, Rockefeller Inst. for M. Res., Monograph 8, 1918.

[128] Dineur, E.: (On typhoid agglutination), Bull. de l'Acad. Roy. de Méd. de Belgique, 1898, 4, p. 705. (Quoted from Smith and Reagh.[446])

[129] Dobell, C.: Some recent work on mutation in microorganisms. (Relates to Protozoa), J. Genetics, 1912, 2, p. 201.

[130] Dobell, C.: Some recent work on mutation in microorganisms: II. Mutations in bacteria, J. Genetics, 1912, 2, p. 325.

[131] Doerr: Ueber filtrierbares Virus, Centralbl. f. Bakteriol., I. O., 1911, 50, p. 12. (Quoted after Enderlein.[160])

[132] Donati, A. Ueber die naturliche Immunität gegen Milzbrand, Ztschr. f. Immuntätsforsch. u. exper. Therap., 1910, 5, p. 142.

[133] Dopter, Ch.: Etudes de quelque germes isolés du rhinopharynx voisins du méningocoque (Para-méningocoque), Compt. rend. Soc. de biol., 1909, 67, p. 74.

[134] Dopter, Ch.: Méningite cerebrospinales a para-méningocoque, Bull. de l'Inst. Pasteur, 1911, 9, p. 840. (Abstract.)

[135] Dopter, Ch., and Koch, R.: Sur la coagglutination du méningocoque et due gonocoque, Compt. rend. Soc. de biol., 1908, 65, p. 215.

[136] Dopter, Ch., and Koch, R.: Sur les precipitines du méningocoque et du gonocoque, Compt. rend. Soc. de biol., 1908, 65, p. 285.

137 Dudtschenko, I. C.: (On the morphology and biology of the pest bacillus), J. Microbiol., Petrograd, 1915, 2, p. 79.

138 Durand, H. Pouvoir pathogène du bacille tuberculeux filtré, Compt. rend. Soc. de biol., 1924, 91, p. 11.

139 Durand, H., and Vaudremer, A.: Retour au type classique du Bacille tuberculeux filtré après passage par le peritoine du cobaye, ibid., 1924, 90, p. 916.

140 Durham, H. E.: A theory of relpases in fevers, J. Path. & Bac., 1901, 7, p. 240.

141 Dutton, L. O.: Rôle of the bacteriophage in streptococcus infections: An interpretation of certain cultural characetristics, J. Infect. Dis., 1926, 39, p. 48.

142 Dyar, H. G.: On certain bacteria from the air of New York City, Ann. N. Y. Acad. Med., 1895, 8, p. 322.

143 Eastwood, A.: Report No. 13 on public health and medical subjects, Ministry of Health, 1922. (Quoted by Yoshioka.[494])

144 Eastwood, A.: Bacterial variation and transmissible autolysis, Report No. 18 on public health and medical subjects, Ministry of Health, 1923.

145 Eberson, F.: A bacteriologic study of the diphtheroid organisms with special reference to Hodgkin's disease, J. Infect. Dis., 1918, 23, p. 1.

146 Eijkmann, C.: Ueber thermolabile Stoffwechselprodukte als Ursache der naturlichen Wachstumshemmung der Mikroorganismen, Centralbl. f. Bakteriol., I. O., 1904, 37, p. 436.

147 Eisenberg, Philipp: Ueber die Anpassung der Bakterien an die Abwehrkrafte des infizierten Organismus, Centralbl. f. Bakteriol., I, O., 1903, 34, p. 739.

148 Eisenberg, P.: Ueber sekundäre Bakterienkolonien, ibid., 1906, 40, p. 188.

149 Eisenberg, P.: Untersuchungen über die Variabilität der Bakterien: I. Ueber sporogene und nonsporogene Rassen des Milzbrandbacillus, ibid., 1912, 63, p. 305.

150 Eisenberg, P.: Ueber sogen. Mutationsvorgänge bei Choleravibrionen, ibid., 1912, 66, p. 1.

151 Eisenberg, P.: Weitere Untersuchungen über das Sporenbildungsvermögen bei Milzbrandbacillen, ibid., 1914, 73, p. 81.

152 Eisenberg, P.: Ueber den Variationkreis des B. prodigiosus und B. violaceus, ibid., 1914, 73, p. 448.

153 Eisenberg, P.: Ueber Mutationen in der Gruppe des Bact. fluorescens, Bact. pneumoniae, Sarcina tetragena und Bact. typhi., ibid., 1914, 73, p. 466.

154 Eisenberg, P.: Variabilität in der Typhus-Coli Gruppe, ibid., 1918, 80, p. 385.

155 Eisenberg, P.: Ueber die Variabilität des Schleimbildungsvermögens und die Gramfestigkeit, ibid., 1919, 82, p. 401.

156 Eisler, M., and Silberstern, F.: Beitrage zur Bakterien-agglutination, Ztschr. f. Hyg., 1921, 93, p. 267.

157 Elser, W. J., and Huntoon, F. M.: Studies on meningitis, J. M. Res., 1909, 20, p. 373.

158 Emmerich, R., and Low, O.: Bacteriologische Enzyme als Ursache der erworbenen Immunität und die Heilung von Infektionskrankheiten durch diesselben, Ztschr. f. Hyg., 1899, 31, p. 1.

159 Enderlein, G.: Grundelemente der vergleichenden Morphologie und Biologie der Bakterien (Bakteriologische Studien III), Sitzungsber. Ges. Naturf. Freunde, Berlin, 1916, p. 403.

160 Enderlein, G.: Bakterien-Cyclogenie. Prologomena zu Untersuchungen über Bau, geschlechtliche und ungeschlechtliche Fortpflanzung und Entwicklung der Bakterien, 1925, p. 1.

161 Evans, Alice: Studies on the etiology of epidemic encephalitis: I. The streptococcus, U. S. P. H. A. Pub. Health Rep., 1926, 41, p. 1095.

162 Evans, Alice, and Freeman, W.: Studies on the streptococcus of encephalitis, J. Bact., 1926, 11, p. 110 (abstract).

163 Eyre, J. W.; Leatham, A. N., and Washburn, J. W.: A study of different strains of pneumococci with especial reference to the lesions they produce, J. Path. & Bact., 1906, 11, p. 246.

164 Eyre, J. W., and Washburn, J. W.: Varieties and virulence of the pneumococcus, Lancet, 1899, 1, p. 19.

165 Falk, I. S.; Gussin, H. A., and Jacobson, M. A.: Studies on respiratory diseases: XXI. Electrophoretic potential and virulence of pneumococci (types 1, 2, 3, and 4), J. Infect. Dis., 1925, 37, p. 481.

166 Falk, I. S., and Matsuda, T.: Influence of some salts which change P. D. on the phagocytosis of pneumococci, Proc. Soc. Exper. Biol. & Med., 1926, 23, p. 781.

167 Feiler, M.: Zur Bilogie des Typhus-bazillus: Ein Beitrag zur Wirkungsweise des Desinfektionsmittel und des Hungers auf Bakterien, Ztschr. f. Immunitätsforsch. v. exper. Therap., 1920, 29, p. 303.

168 Fejgin, Bronislawa: Contribution a l'étude des races résistantes du bacille de Shiga-Kruse, Compt. rend. Soc. de biol., 1923, 90, p. 1381.

169 Fejgin, B.: Sur les variations brusqués du Proteus X19 survenues sous l'influence de l'agent lytique anti- HX19, et leur rapport avec les souches isolées des cobayes infectés avec le virus de passage du typhus exanthématique, ibid., 1924, 90, p. 1106.

170 Fejgin, B.: Sur le principe lytique anti-diphtherique, ibid., 1925, 93, p. 365.

[171] Fejgin, B.: Sur la forme filtrant de Bacille d'Eberth, ibid., 1925, 92, p. 1528.

[171a] Fejgin, B.: Sur les cultures secondaires du bacille typhique isolé des organes des cobayes infectés avec le virus de la fièvre typhus, ibid., 1925, 93, p. 1530.

[172] Felix, A.: Ueber Varianten der Proteus X-Stamme, Ztschr. f. Immunitätsforsch. u. exper. Therap., 1922, 35, p. 57.

[173] Felix, A., and Mitzenmacher, F.: Weitere Untersuchungen über den Nachweis der O und H Receptoren bei den Proteusstammen, Wien. klin. Wchnschr., 1918, 31, p. 988.

[174] Felton, L. D., and Daugherty, Katherine: Studies on virulence: II. The increase in virulence in vitro of a strain of pneumococcus, J. Exper. Med., 1924, 39, p. 137.

[175] Ficker, M.: Ueber Wachstumsgeschwindikeit des Bakterium coli comm. auf. Platten, 1895. (Abstract: Baumgarten's Jahresber., 1896, 12, p. 338.)

[176] Fischer, C.: The differentiation of the diphtheria bacillus from organisms morphologically similar, Arch. Ophthalmol., 1909, 38, p. 610.

[177] Fison, E. T.: Widal's serum diagnosis of typhoid fever, Brit. M. J., 1897, 2, p. 266.

[178] Firtsch, G.: Untersuchungen über Variationserscheinungen bei Vibrio proteus (Kommabacillus von Finkler-Prior), Arch. f. Hyg., 1888, 8, p. 369.

[179] Fletcher, W.: Capsulate mucoid forms of paratyphoid and dysentery bacilli, Lancet, 1918, 2, p. 102.

[180] Fletcher, W.: (Title as above), J. Roy. Army Med. Corps, 1920, 34, p. 219.

[181] Flu, P. C.: Ueber Cholerabakteriophagen, Tijdschr. v. Vergilijk. Geneesk., 1924, 10, p. 196. (Quoted from d'Herelle.[248]).

[182] Fontés: Studien über Tuberculose, Ann. de l'Inst. Oswaldo Cruz, 1910, 2, p. 2.

[183] Fraenkel, C.: Einwirkung der Kohlensäure auf die Lebensthatigkeit der Micro-organismen, Ztschr. f. Hyg., 1889, 5, p. 332.

[184] Friedberger, E., and Meissner, Gertrud: Zur Pathogenese der experimentelle Typhus Infektion des Meerschweinchen, Klin. Wchnschr., 1923, 2, p. 450.

[185] Friel, A. R.: Pub. So. African Inst. M. Res., 1915, No. 5. (Quoted from Reimann.[405])

[186] Fuhrmann, F.: Entwicklungszyklen bei Bakterien, Verh. d. Ges. Deutsch. Naturf. u. Arzte, 1906, p. 278. (Quoted from Enderlein.[160]).

[187] Fulmer, E. I.: Acclimatization of yeast to ammonium fluoride and its reversion in wort, J. Physical Chem., 1921, 25, p. 455.

[188] Furth, J.: Variationsversuche mit Paratyphus B (Weil), Ztschr. f. Immunitätsforsch. u. exper. Therap., 1922, 35, p. 155.

[189] Furth, J.: Variationsversuche in der Dysenteriegruppe, ibid., p. 176.

[190] Furth, J.: Rezeptoranalyse und Variationsversuche mit paratyphosus und Aertrycke, ibid., p. 162.

[191] Furth, J.: Variationsversuche bei den Bacillus typhi, ibid., p. 133.

[192] Fusier, M. L., and Meyer, K. F.: Principles in serologic grouping of B. abortus and B. melitensis. Correlation between absorption and agglutination tests, J. Infect. Dis., 1920, 27, p. 185.

[193] Gamaleia: Bakteriolysine-bakterienzerstorenden Fermente, Centralbl. f. Bakteriol., I. O., 1899, 26, p. 661.

[195] Gardiner, A. D.: The growth of branching forms of bacilli ("Three point multiplication"), J. Path. & Bact., 1925, 28, p. 189.

[196] Gardiner, A. D., and Ainley-Walker: An inquiry into the nature of the serological differences exhibited by different cultures of a bacterial species, Brit. J. Hyg., 1921, 20, p. 110.

[197] Gay, F., and Claypole: The "typhoid carrier" state in rabbits as a method of determining the comparative immunizing value of preparations of the typhoid bacillus, Arch. Int. Med., 1913, 12, p. 614.

[198] Gildemeister, E.: Ueber Dauerausscheider vom Paratyphus B, Centralbl. f. Bakteriol., I. O., 1916, 78, p. 129.

[199] Gildermeister, E.: Weitere Mitteilungen über Variabilitätserscheinungen bei Bakterien, die bereits bei ihrer Isolierung aus dem Organismus zu beobachten sind, ibid., 1916, 79, p. 49.

→ Goodman, H. M.: Variability in the diphtheria group of bacilli, J. Infect. Dis., 1908, 5, p. 421.

[201] Gordon, M. H., and Murray, E. G.: Identification of the meningococcus, J. Roy. Army Med. Corps, 1915, 25, p. 411; J. Hyg., 1918, 17, p. 290.

[202] Gorham, F. P.: Morphological varieties of Bacillus diphtheriae, J. Med. Res., 1901, 6, p. 128; 1902, 7, p. 128.

[203] Gory, M.: Transformation muqueuse de Bacillus coli, Compt. rend. Soc. de biol., 1923, 88, p. 49.

[204] Gotschlich, E.: Die Pest-epidemie in Alexandrien im Jahre, 1899, Ztschr. f. Hyg., 1900, 35, p. 195.

[205] Gotschlich, E.: Kolle u. Wassermann: Handbuch der pathogenen Microorganismen, 1913, I, p. 167.

[206] Goyle, A. N.: On bacterial variation with special reference to the alleged convergent phenomenon exhibited by certain distinct pathogenic species (B. typhosus and B. enteritidis, Gaertner), J. Path. & Bact., 1926, 29, p. 149.

[207] Grassberger, R.: Zur Frage zur Scheinfädernbildung in Influenzakulturen, Centralbl. f. Bakteriol., I, O., 1898, 23, p. 353.

[208] Grassberger, R., and Schattenfroh, A.: Sitzungsberichtung der Kaiserl. Akad. der Wissenschaften in Wien. Math. Naturw. Klasse, 1905.

[209] Gratia, A.: Influence de la reaction du milieu sur l'autolyse microbienne transmissible, Compt. rend. Soc. de biol., 1921, 84, p. 275.

[210] Gratia, A.: Autolyse transmissible et variations microbiennes, ibid., 85, p. 251.

[211] Gratia, A.: Studies on the d'Herelle phenomenon, J. Exper. Med., 1921, 34, p. 115; ibid., p. 287.

[211a] Gratia, A.: The Twort-d'Herelle phenomenon, ibid., 1922, 35, p. 287.

[212] Gratia, A.: Variations microbienne et infection charbonneuse, Compt. rend. Soc. de biol., 1924, 90, p. 369.

[213] Gratia, A.: Sur un remarkable example d'antagonisme entre deux souches de coli bacille, ibid., 1925, 93, p. 1040.

[214] Griffith, F.: Second report on the identification of the meningococcus in the nasopharynx, with special reference to serological reactions, Brit. J. Hyg., 1918, 17, p. 124.

[215] Griffith, F.: The influence of immune serum on the biological properties of pneumococci. Report 18 on Public Health and Medical Subjects, Ministry of Health, 1923.

[216] Gruber and Futaki: Ueber die Resistenz gegen Milzbrand und über die Herkunft der Milzbrandfeinlichen Stoffe, München. med. Wchnschr., 1907, 54, p. 249.

[217] Gruschka, Th.: Variationversuche mit den Bac. enteritidis Gaertner, Ztschr. f. Immunitätsforsch. u. exper. Therap., 1922, 35, p. 97.

[218] Günther: Einfuhrung in das Studium der Bakteriologie, 1906.

[219] Gurd, F. B.: A contribution to the bacteriology of the female genital tract with special reference to the detection of gonococcus, J. Med. Res., 1908, 18, p. 291.

[220] Gurney-Dixon, S.: The transmutation of bacteria, 1919.

[221] Hadley, Philip: Toxin production by B. diphtheriae on protein-free media, J. Infect. Dis., 1907, Sup. 3, p. 95.

[222] Hadley, Philip: Studies on fowl cholera: II. Active immunity in rabbits, Centralbl. f. Bakteriol., I, O., 1913, 69, p. 271.

[223] Hadley, Philip: Studies on fowl cholera: IV. The reciprocal relations of virulent and avirulent cultures in active immunization, Rhode Island Agric. Exper. Sta. Bull. 159, 1914, p. 384.

→ Hadley, Philip: Transmissible lysis in Bacillus pyocyaneus, J. Infect. Dis., 1924, 34, p. 260.

[225] Hadley, Philip: The action of the lytic principle on capsulated bacteria, Proc. Soc. Exper. Biol. & Med., 1925, 23, p. 109.

[226] Hadley, Philip: Proliferative reaction to stimuli by the lytic principle (bacteriophage) and its significance, J. Infect. Dis., 1925, 37, p. 35.

[227] Hadley, Philip: Parallelism between serologic and bacteriophagic response in B. typhosus and certain avian paratyphoids, Proc. Soc. Exper. Biol. & Med., 1926, 23, p. 443.

[228] Hadley, Philip, and Dabney, Eugenia: The bacteriophagic relationships between B. coli, Strep. fecalis and Strep. lacticus, ibid., 1926, 24, p. 13.

[229] Hajós, E.: Beiträge zur Frage der wachstumshemmenden Wirkung von Bouillonkulturen, Centralbl. f. Bakteriol., I, O., 1922, 88, p. 583.

[230] Hall, Ivan C.: The reduction of selenium compounds by sporulating anaerobes, J. Bact., 1926, 11, p. 407.

[231] Hamburger, F.: Ueber spefische Virulenzsteigerung in vitro, Wien. klin. Wchnschr., 1903, 4, p. 497.

[232] Hamburger, F., and Czickeli, H.: (Agglutination relationships), ibid., 1924, 37, p. 10.

[233] Hansen, E. Chr.: Recherches sur la physiologie et la morphologie des ferments alcooloques: X. Variation des Saccharomvces, Compt. rend. des Travaux du Laboratoire Carlsb., 1900, 5, p. 47.

[234] Hauduroy, P.: Les culures secondaires après filtration dans le phénomène de d'Herelle, Compt. rend. Soc. de biol., 1924, 91, p. 1209.

[235] Hauduroy, P.: Les cultures secondaires après filtration dans le phénomène de d'Herelle, ibid., p. 1325.

[236] Hauduroy, P.: Présence de formes invisible de microbes visible dans la nature, ibid., 1926, 94, p. 246.

[237] Hauduroy, P., and Peyre, Ed.: Le bactériophage du Bacille pyocyanique, ibid., 1923, 88, p. 689.

[238] Haven, L. C.: Biologic studies of the diphtheria bacillus, J. Infect. Dis., 1920, 26, p. 388.

[239] Heinemann, P. G.: Morphology of a strain of B. diphtheriae, J. Bact., 1917, 2, p. 361.

240 Heller, Hilda: The study of colony formation in deep agar. Studies on pathogenic anaerobes, VI, J. Infect. Dis., 1922, 30, p. 1.

241 Heller, Hilda: Mutations in the genus, Nicolaierillus (B. tetani). Studies on pathogenic anaerobes. VIII. ibid., 1922, 30, p. 33.

242 d'Herelle, F.: Sur un microbes invisible antagoniste des bacilles dysenterique, Compt. rend. Acad. d. Sc., 1917, 165, p. 373.

243 d'Herelle, F.: Sur le microbe bactériophage, Compt. rend. Soc. de biol., 1919, 82, p. 1237.

244 d'Herelle, F.: Sur la resistance de bactéries a l'action du microbe bactériophage, ibid., 1920, 83, p. 97.

245 d'Herelle, F.: Phenomènes coincidant avec l'aquisition de la résistance des bactéries a l'action du bactériophage, ibid., 1921, 84, p. 384.

246 d'Herelle, F.: The bacteriophage: Its rôle in immunity, (Translation), 1922.

247 d'Herelle, F.: Immunity in natural infectious disease. (Translation), 1924.

248 d'Herelle, F.: The bacteriophage' and its behavior. (Translation), 1926.

249 d'Herelle and Hauduroy: Sur les charactères des symbiosis "bacterie-bactériophage," Compt. rend. Soc. de biol., 1925, 34, p. 1288.

250 Hess, H.: Bedeutung der Kapsel für Virulenz des Milzbrandbazillus, Arch. f. Hyg., 1921, 89, p. 237.

251 Hewlett and Knight: The so-called pseudodiphtheria bacillus and its relation to the Loeffler bacillus, Trans. Brit. Inst. Prev. Med., 1897, ser. 1, p. 7.

252 Hirschfeld, E., and Zajdel, J.: Sur la variabilité des bactéries sous l'influence des conditions thermiques défavorables, Compt. rend. Soc. de biol., 1924, 90, p. 1104.

253 Hoder, F.: Ueber Zusammenhange zwischen Bakteriophagen und Bakteriemutation, Ztschr. f. Immunitätsforsch. u. exper. Therap., 1925, 42, p. 197.

254 Hort, E. C.: The life-histories of bacteria, J. Roy. M. Soc., 1916, 11.

255 Hort, E. C.: The meningococcus of Weichselbaum, Brit. M. J., 1917, 2, p. 377.

256 Hort, E. C.: Morphological studies on the life histories of bacteria, Proc. Roy. Soc., B., 1917, 83, p. 468.

257 Hort, E. C.: The life history of bacteria, Brit. M. J., 1917, 1, p. 571.

→ Hort, E. C.: The reproduction of aerobic bacteria, Brit. J. Hyg., 1920, 18, p. 369.

259 Hort, E. C.; Larkin, C. E., and Benians, T. H. C.: Cerebrospinal fever: The place of the meningococcus in its etiology, Brit. M. J., 1915, 1, pp. 541, 715.

260 Hübener: Ueber Paratyphus Bakterien und ihnen ähnliche Bakterien bei gesunden Menchen, Centralbal. f. Bakteriol., I, Ref., 1909, 44, p. 136.

261 Hueppe, F.: Die Formen der Bakterien und ihre Beziehungen zu den Gattung und Arten, 1891.

262 Ishii, O.: A study of spontaneous agglutination in the colon-typhoid group of bacilli, J. Bact., 1922, 7, p. 71.

263 Issaeff, B.: Contribution a l'étude de l'immunité acquisé contre le pneumocoque, Ann. de l'Inst. Pasteur, 1893, 7, p. 260.

264 Izar, G.: Sui considetti batteriofagi, Ac. Gioemia d. sci. nat. in Cattina, 1921. (Quoted from d'Herelle.[248])

265 Jacobsen: Mitteilungen über einen variablen Typhusstamm (B. typhi mutabile) sowie über eine eigentümliche hemmende Wirkung des gewöhnlichen Agar, verursacht durch Autoclavierung, Centralbl. f. Bakteriol., I, O., 1910, 66, p. 208.

266 Jacobson, M. A., and Falk, I. S.: Influence of antiserum and of animal passage upon virulence and electrophoresis of pneumococci, Proc. Soc. Exper. Biol. & Med., 1926, 23, p. 785.

267 Jager, H.: Die Aetologie des infektiösen fieberhaften Icterus (Weil'sche Krankheit), Ztschr. f. Hyg., 1892, 12, p. 525.

268 Jennings, H. S.: Life and death, heredity and evolution in unicellular organisms, 1920.

269 Jollos, V.: Variabilität und Vererbung bei Mikro-organismen, Ztschr. f. Indukt. Abstämmungsl., 1914, 12, p. 14.

270 Jones, D. H.: Morphological and cultural studies of some Azotobacter, Centralbl. f. Bakteriol., II, 1913, 38, p. 14.

271 Jones, D. H.: Further studies on the growth cycle of Azotobacter, J. Bact., 1920, 5, p. 325.

272 Joos, A.: Untersuchungen über die verschiedenen Agglutinine des Typhusserums, Centralbl. f. Bakteriol., I, O., 1903, 33, p. 762.

273 Jordan, E. O.: The interconvertibility of rough and smooth bacterial types, J. Am. M. A., 1926, 86, p. 177.

274 Jordan, E. O.: Further observations on the "rough" and "smooth" strains of bacteria, Proc. Soc. Exper. Biol. & Med., 1926, 23, p. 762.

275 Jötten, K. W.: Ueber das sogenannten d'Herellesche Phänomen., Klin. Wchnschr., 1922, 1, p. 2181.

276 Katzu, S.: Bakteriophagahnliche Erscheinungen bei Milzbrand, Centralbl. f. Bakteriol., I, O., 1925, 96, p. 281.

277 Kiessling, F.: Das Bacterium coli commune, Hyg. Rundschau, 1893, 3, pp. 724, 765.

278 Kimura, S.: Ueber Schleimbildung bei Bakterien unter dem Einfluss von Bakteriophagen, Ztschr. f. Immunitätsforsch. u. exper. Therap., 1925, 42, p. 507.

279 Klein, E.: Report of the Medical Office, Local Gov't. Board, No. 32, 1909, p. 399. (Quoted from Gotschlich.[105])

280 Klöcken, A., and Schiönning, H.: Phénomènes d'accroissement perforant et de formation anormale des conidies chez Dematium pullulans de Bary et autres champignons, Compt. rend, travaux du Laboratoire de Carlsb., 1900, 5, p. 47.

281 Kodama, H.: Die Ursache der näturlichen Immunität gegen Milzbrandbacillen, Centralbal. f. Bakteriol., I, O., 1913, 68, p. 373.

282 Kolle, W.: Ueber den jetzigen Stand der Choleradiagnose, Klin. Jahrb., 1903, 11, p. 357.

283 Kolle, W., and Wassermann, A.: Untersuchungen über Meningokokken, ibid., 1906, 15, p. 507.

284 Koraen, G.: Studien über Biologie und Wuchsformen der Diphtheriebazillen, Ztschr. f. Hyg., 1918, 35, p. 277.

285 Koser, Stewart A.: (Bacteriophage isolated for a true thermophilic bacterial culture), Proc. Soc. Exper. Biol. & Med., 1926, 24, p. 109.

286 Kowalenko, A.: Studien über sogenannte Mutationserscheinungen bei Bakterien unter besonderer Berücksichtiguns der Einzellencultur, Ztschr. f. Hyg., 1910, 66, p. 277.

287 Kraus, R., and Joachim, J.: Ueber Beziehungen der prezipitinogen Substanz zur agglutinogen der Bakterien, Centralbl. f. Bakteriol., I, O., 1904, 36, p. 662.

288 Krumwiede, Ch.; Cooper, G., and Provost, D.: Agglutinin absorption, J. Immunol., 1925, 10, p. 55.

289 Krumwiede, Ch.; Cooper, G., and Provost, D.: Serological duality of paratyphoid cultures, Bact. Abstr., 1924, 8, p. 24.

290 Krumwiede, Ch.; Mishulow, L., and Oldenbusch, C.: The existence of more than one immunological type of B. pertussis, J. Infect. Dis., 1923, 32, p. 22.

291 Kruse, W.: C. Flügge: Die Mikroorganismen, Variabilität der Microorganismen, 1896, p. 475.

292 Kruse, W.: (On bacterial variation), Allgemeine Mikrobiologie, 1910, chapter 18.

293 Kruse, W., and Pansini, S.: Untersuchungen über den Diplococcus pneumoniae und verwandte Bakterien, Ztschr. f. Hyg., 1892, 11, p. 279.

294 Kurth, H.: Ueber die Diagnose des Diphtheriebacillus unter Berücksichtigung abweichsener Culturformen desselben, Ztschr. f. Hyg., 1898, 28, p. 409.

295 Lacy, G. R.: A report of typical and atypical Bacillus dysenteriae Shiga, with special reference to agglutination reactions, Philippine J. Sc., 1925, 28, p. 313.

296 Laubenheimer: Experimentelle Beiträge zur Veranderlichkeit der Agglutination bei Typhus, 1903. (Quoted from Sacharoff.[415])

297 Laurent, Em.: Etude sur variabilité du bacille rouge de Kiel, Ann. de l'Inst. Pasteur, 1890, 4, p. 465.

298 Lawrence, J. B., and Ford, W. W.: Studies on aerobic, spore-forming nonpathogenic bacteria: I. Spore-forming bacteria from milk, J. Bact., 1916, 1, p. 273.

299 Lavrinowicz, A.: Observation sur la morphologie et la biologie des Gonocoques, Compt. rend. Soc. de biol., 1925, 93, p. 789.

300 Ledingham, J. C. B.: A "reversion" phenomenon in bacterial fermentation, Brit. J. Hyg., 1918, 7, p. 409.

301 Lehmann, K. B., and Neumann, R.: Atlas and Grundriss der Bakteriologie, 1896.

302 Lesieur, Ch.: Les bacilles dits "pseudo-diphtheriques," J. physiol. et path. gén., 1901, 3, p. 961.

303 Lieske, R.: Bakterien und Strahlenpilze. In K. Linsbauer, Handbuch der Pflanzenanatomie, Bd. 6, Berlin, 1922. (Quoted from Enderlein.[109])

304 von Lingelsheim: Zur Frage der Variation der Typhus-bazillen und verwändte Gruppe, Centralbl. f. Bakteriol., I, O., 1913, 68, p. 577.

305 Lisbonne and Carrère: Sur l'appararition spontanée du pouvoir lysogène dans les cultures pures, Compt. rend Soc. de biol., 1924, 90, p. 265.

306 Lister, Joseph: A further contribution to the natural history of bacteria and the germ theory of fermentative changes, Quart. J. Microsc. Sc., 1873, 13, p. 380.

307 Loeffler: Experimentelle Beiträge zur Theorie und Praxis der Gruber-Widal'chen Agglutinationsprobe, Centralbl. f. Bakteriol., I, O., 1906, 38, p. 101.

308 v. Loghem: Varibilität und Parasitismus: Eine vergleichende Untersuchung von Bakterien der Typhus-coli Gruppe, ibid., 1919, 83, p. 401.

309 Löhlein, H.: Einiges über Phagocytose von Pest- und Milzbrandbacillus, ibid., I, Ref., 1906, 38, p. 32.

310 Löhnis, F.: Studies upon the life cycles of the bacteria, Mem. Nat. Acad. Sc., 1921, 16, p. 252.

311 Löhnis, E., and Smith, E. R.: Life cycles of the bacteria, J. Agric. Res., 1916, p. 675.

[312] Lombroso and Gerini: Alcune osservazione sulla recente epidemia colerica di Livorno, Riv. crit. di clin. med., 1911, 12, p. 3.

[313] Lourens, F. D. E.: Untersuchungen über der Filtrierbarkeit der Schweinepestbacillen (Bac. suipestifer). Centralbl. f. Bakteriol., I, O., 1907, 44, pp. 420, 504, 630.

[314] Macchiati: Di un carattere per la diagnose delle Batteriacee, Nov. giorn. bot. ital., 1899, 6 (ser. II), p. 1384. (Quoted from Enderlein.[150])

[315] Malfitano: La bactériolyse de la batéridie charbonneuse, Compt. rend. Acad. méd., 1900, 2, p. 295.

[316] Malvoz, E.: Sur l'agglutination du Bacillus typhosus, Ann. de l'Inst. Pasteur, 1897, 11, p. 582.

[317] Manniger, R.: Ueber einer Mutation des Geflügelcholerabazillus, Centralbl. f. Bakteriol., I, O., 1919, 83, p. 520.

[318] Marchoux, E.: Unicité ou pluralité des bacilles lépreux, Bull. de. l'Acad. Med., 1925, 94, p. 1091.

[319] Markoff, W. N.: Studien über die Variabilität der Bakterien, zugleich ein Beitrag zur Morphologie und Biologie des Milzbrandbazillus, Ztschr. f. Infekt. d. Haustiere, 1912, 89, p. 1122.

[320] Massini, R.: Ueber einem in biologischer Beziehung interessanten Koli-stamm (B. coli mutabile), Arch. f. Hyg., 1907, 61, p. 250.

[320a] Meissner, G.: Ueber Bakteriophagen gegen Choleravibrionen, Centralbl. f. Bakteriol., I, O., 1924, 91, p. 149.

[321] Matzuschita: Einwirkung des Kochsalzgehaltes des Nahrbodens auf die Wuchsformen der Mikroorganismen, Ztschr. f. Hyg., 1900, 35, p. 495. (Quoted after Enderlein.[150])

[322] McIntosh, J., and Fildes, P.: The classification and study of the anaerobic bacteria of war wounds, M. Res. Com., Special Rep., 1918, ser. 12, p. 29.

[323] McKinley, E. B.: Notes on d'Herelle's phenomenon, J. Lab. & Clin. Med., 1922-23, 8, p. 311.

[324] McKinley, E. B.: Transformation, sous l'influence du principe lytique faible de la specificite antigénique d'une culture, Compt. rend. Soc. de biol., 1925, 93, p. 1052.

[325] Mellon, R. R.: A study of the diphtheroid group of organisms, J. Bact., 1917, 2, pp. 81, 269.

[325a] Mellon, R. R.: Life cycles of the bacteria and their possible relation to pathology, Am. J. M. Sc., 1920, 159, p. 874.

[326] Mellon, R. R.: The life cycle of the so-called C. hodgkini, and their relation to the mutation changes in the species, J. M. Res., 1920, 42, p. 61.

[327] Mellon, R. R.: Further studies on the diphtheroids, ibid., 1921, 42, p. 111.

[328] Mellon, R. R.: Spontaneous agglutination of bacteria in relation to variability and to action of equilibrated solution of electrolytes, ibid., 1922, 43, p. 345.

[329] Mellon, R. R.: Observations on the origin of biotypes (microbic dissociation) in pure lines of bacteria, Proc. Soc. Exper. Biol. & Med., 1922, 20, p. 191.

[330] Mellon, R. R.: Studies in microbic heredity: I. Observations on a primitive form of sexuality (zygospore formation) in the colon-typhoid group, J. Bact., 1925, 10, p. 481.

[331] Mellon, R. R.: II. The sexual cycle of B. coli in relation to the origin of variants with special reference to Neisser and Massini's B. coli mutabile, ibid., p. 579.

[332] Mellon, R. R. (with E. Yost): III. The hereditary origin of group and specific agglutinogens among the colon-alkaligenes organisms, J. Immunol., 1926, 11, p. 139.

[333] Mellon, R. R. (with E. Grenquist): IV. Observation on group agglutinins in specific sera with the technique of cataphoresis, ibid., p. 161.

[334] Mellon, R. R.: The biogenic low of Haeckel and the origin of heterogeneity within pure lines of bacteria, J. Bact., 1926, 11, p. 203.

[335] Mellon, R. R.: VI. The infective and taxonomic significance of a newly described ascosporic stage for the fungi of Blastomycosis, ibid., p. 229.

[336] Mellon, R. R.: VII. Observations on the genetic origin of the several types of fungi found in the lesions of blastomycosis hominis, ibid., p. 419.

[337] Mellon, R. R.: VIII. The infectivity and virulence of a filtrable stage in the life history of B. fusiformis and rerelated organisms, ibid., 1926, 12, p. 279.

[338] Mellon, R. R., and Jost, E. L.: IX. Observations on the biologic origin and the physicochemical nature of inagglutinability with freshly isolated typhoid bacilli, J. Immunol., 1926, 12, p. 331.

[339] Mellon, R. R.: X. The agglutination absorption reaction as related to the newer biology of bacteria, with special reference to spore formation, ibid.,* p. 355.

[340] Mellon, R. R.: XI. Observations on the genetic origin of Staphylococcus albus and aureus, J. Bact., 1926, 12, p. 409.

* Mellon's papers VIII to XIII in the series have not been available to me for review. Dr. Mellon has, however, furnished me with a list of the papers yet to appear so that they may appear complete in the bibliography.

341 Mellon, R. R.: XII. True bacterial polymorphism in a case of suppurative thyroiditis: its bearing on our so-called secondary infections, ibid.,* 1927.

342 Mellon, R. R.: XIII. The experimental reversal of the aqueous-lipoidal systems in the bacterial cell wall as indicated by the surface tension method, ibid.,* 1927.

343 Mellon, R. R.: Practical aspects of the newer biology of bacteria with special relation to their filtrable phases, J. Am. M. A., 1927, in press.*

344 Mellon, R. R., and Anderson, L.: Immunological disparities of spore and vegetative stages of B. subtilis, J. Immunol., 1919, 4, p. 203.

345 Metchnikoff, E.: Sur l'attenuation des bactérides charbonneuse dans le sang de moutons réfractaires, Ann. de l'Inst. Pasteur, 1887, 1, p. 42.

346 Metchnikoff, E.: L'immunité des cobayes vaccinés contre le Vibrio metchnikovi, Ann. de l'Inst. Pasteur, 1891, 5, p. 465.

347 Metchnikoff, E.: Etudes sur l'immunité: V. Immunité des lapins vaccinés contre le microbe du hog cholera, ibid., 1892, 6, p. 289.

348 Middleton: Heritable variations and results of selection in the fission rate of Stolonychia pustulata, J. Eper. Zool., 1915, 19, p. 451.

349 Miehe: Sind ultramicroscopische Organismen in der Natur verbreitet? 1923, 43, p. 1. (Quoted from Enderlein.[109])

350 Mieszner: Rauschbrand und Pararauschbrand, Centralbl. f. Bakteriol., I, O., 1922, 89, p. 123.

→ Moon, V. H.: Attempt to modify the agglutinability of the typhoid bacillus by selective isolation of individual bacilli, J. Infect. Dis., 1911, 8, p. 463.

352 Morishima, K.-I.: Variations in typhoid bacteria, J. Bact., 1921, 6, p. 275.

353 Migula, H.: System der Bakterien, Bd. 2, 1897-1900.

354 Morin, H., and Valtis, J.: Sur la filtration du bacille de Johne à travers les boughies Chamberland L², Compt. rend Soc. de biol., 1926, 94, p. 39.

355 Müller, Paul Th.: Ueber Immunisierung des Typhusbacillus gegen specifische Agglutinine, München. med. Wchnschr., 1903, 50, p. 56.

356 Müller, Reiner: Ueber Mutationsartige Vorgänge bei Typhus, Paratyphus und verwandten Bakterien, Centralbl. f. Bakteriol., I, O., 1908, 42, p. 57.

357 Müller, Reiner: Kunstliche Erzeugung neuer vererbaren Eigenschaften bei Bakterien, München. med. Wchnschr., 1909, 56, p. 885.

358 Müller, Reiner: Mutationen bei Typhus und Ruhrbakterien (Mutationen als specifisches Kulturmerkmal), Centralbl. f. Bakteriol., I, O., 1911, 58, p. 97.

359 Muto, T.: Ein eigentümlicher Bacillus welche sich schnekenartig bewegende Colonien bildet (B. helicoides), ibid., 1904, 37, p. 321.

360 Nadolczny, M.: Ueber das Verhalten virulenter und avirulenter Kulturen derselben Bakterienspecies gegenüber activen Blute, Arch. f. Hyg., 1900, 37, p. 277.

361 Nägeli, C. v.: Untersuchungen über die niedere Pilze und ihren Beziehung zu den Infektionskrankheiten und der Gesundheitspflege, 1877.

362 Neisser, M.: Ein Fall von Mutation nach de Vries bei Bakterien, Centralbl. f. Bakteriol., I, Ref., 1906, 38, p. 98.

363 Neufeld, F.: Ueber die Agglutination der Pneumokokken und über die Theorieen der Agglutination, Ztschr. f. Hyg., 1902, 40, p. 54.

364 Neufeld, F., and Lindermann, E.: Beitrag zur Kenntnis der Serumfesten Typhusstämme. Centralbl. f. Bakteriol., I, O., 1912, 54, p. 229.

365 Neuman, R.: Studien über die Variabilität der Farbenstoffbildung bei Micrococcus pyogenes, var. aureus und einigen anderen Spalpilzen, Arch. f. Hyg., 1897, 30, p. 1.

366 Nicolle, Ch.: Récherches sur la substance agglutinine, Ann. de l'Inst. Pasteur, 1898, 12, p. 1.

367 Nicolle, Ch.: L'agglutination spontanée des cultures, ses rapports avec l'agglutination par les serums, Compt. rend. Soc. de biol., 1898, 50, p. 1054.

368 Nicolle, Ch.: Sur la nature des virus invisible. Origin microbienne des Inframmicrobes, Arch. de l'Inst. Pasteur (Tunis), 1925, 14, p. 105.

369 Nicolle, Ch., and Blanc, G.: Études sur la fièvre recurrent, ibid., 1914, 9, pp. 81, 82.

370 Nicolle, Ch., and Trenl: Récherches sur le phénomène de l'agglutination, Ann. de l'Inst. Pasteur, 1902, 16, p. 562.

371 Nicolle, Ch., and Debains: Sur les races du pneumocoque, avec remarques générales sur les antigènes, Bull. de l'Acad. de med., 1919, 81, p. 843.

372 Nicolle, Ch.; Debains, and Jouan: Etudes sur les méningocoques et les sérums antimeningococques, Ann. de l'Inst. Pasteur, 1918, 32, p. 150.

373 Noguchi, H.: Etiology of yellow fever: V. Cultivation morphology, virulence and biological properties of Leptospira icteroides, J. Exper. Med., 1919, 30, p. 13.

→ Novy, F. G., and Knapp, R. E.: Studies on Spirillum obermeieri and related organisms, J. Infect. Dis., 1906, 3, p. 291.

[375] Novy, F. G., and Soule, M. H.: Respiration of the tubercle bacillus, ibid., 1925, 36, p. 168.

[376] Orcutt, Marion: Mutation among hog cholera bacteria, J. Exper. Med., 1923, 38, p. 9.

[377] Orcutt, M.: Ibid., 1924, 40, pp. 43, 627.

[378] Orsos, Franz: Die Form der tiefliegenden Bakterien- und Hefen-kolonieen, Centralbl. f. Bakteriol., I, O., 1910, 54, p. 289.

[379] Otto, R., and Munter: Zum d'Herelleschen Phänomen, Deutsche med. Wchnschr., 1921, 47, p. 1579.

[380] Otto, R., and Winkler, W.: Ueber die Natur des d'Herelleschen Bakteriophagen, ibid., 1922, 48, p. 383.

[381] Otto, R.; Winkler, W., and Sukiennikova, N.: Bakteriophages Lysin und Paragglutination, Ztschr. f. Hyg., 1923, 101, p. 119.

[382] Park, W. H., and Williams, A. W.: Pathogenic Microorganisms, 1917, p. 208.

[383] Pasteur, Louis (with the collaboration of Ch. Chamberland and E. Roux): De l'atténuation des virus et de leur retour à la virulence, Compt. rend. Acad. d. Sc., 1881, 92, p. 429.

[384] Pearce, Louise: A comparison of adult and infant types of gonococci, J. Exper. Med., 1915, 21, p. 289.

[385] Penfold, W. J.: Studies on Bacterial Variation, Brit. J. Hyg., 1910, 11, p. 30.

[386] Penfold, W. J.: On the specificity of bacterial mutation, ibid., 1912, 12, p. 195.

[387] Pesch, K. L.: Milzbrand Pseudo-bakteriophagen, Centralbl. f. Bakteriol., I, O., 1925, 93, p. 525.

[388] Petrovanu, G.: L'action de l'eau oxygenée sur quelques groupes de microbes. Catalases microbiennes. Phénomène d'autolyse, Compt. rend. Soc. de biol., 1925, 92, p. 459.

[389] Phisalix, C.: Influence de chaleur sur la proprieté sporogène du B. anthracis, abolution persistante de cette function par heredité des charactères acquis, Arch. de Physiol., 1893, p. 217.

[390] Pico, C. E.: Autolyse transmissible du B. anthracis sans intervention de l'hypothétique virus bactériophage, Compt. rend. Soc. de biol., 1922, 87, p. 836.

[391] Piorkowski, G.: Beitrag zur Streptokokkenfräge. Anwendung des d'Herelleschen Phänomens auf Streptokokken, Med. Klin., 1922, 18, p. 474.

[392] Porges, O., and Prantschoff, A.: Ueber die Agglutinität von Bakterien besonders des B. typhi, Centralbl. f. Bakteriol., I, O., 1906, 41, pp. 466, 546, 658.

[393] Preisz, H.: Studien über Morphologie und Biologie des Milzbrandbacillus, ibid, 1904, 35, p. 280.

[394] Preisz, H.: Studien über das Variieren und das Wesen des Abschwächung des Milzbrandbacillus, ibid., 1911, 53, p. 510.

[395] Preisz, H.: Experimentelle Studien über Virulenz, Empfanglichkeit und Immunität beim Milzbrand, ibid., 1909, 49, p. 341.

[396] Prell, H.: Die Vielgestaltigkeit des Bakterium coli, ibid., 1917, 79, p. 324.

[397] Prigsheim, H.: Die Variabilität niederer Organismen, 1910.

[398] Pryer, R.: The cause of scarlet fever, Am. J. Pub. H., 1925, p. 847.

[399] Quiroga, R.: Bactériophage du Bacille pyocyanique, Compt. rend. Soc. de biol., 1923, 88, p. 363.

[400] Radziewski, A.: Untersuchungen zur Theorie der bacteriellen infection, Ztschr. f. Hyg., 1901, 37, p. 1.

[401] Rahn, O.: Ueber den Einfluss der Stoffwechselprodukte auf das Wachstum der Bakterien, Centralbl. f. Bakteriol., I, O., 1906, 16, pp. 417, 609.

[401a] Ransom and Kitashima: Untersuchungen über die Agglutinationsfähigkeit der Cholera-vibrionen durch Choleraserum, Deutsch. med. Wchnschr., 1897, 19, p. 295.

→ Reddish, G. F.: An investigation into the purity of American strains of Bacillus botulinus, J. Infect. Dis., 1921, 29, p. 120.

[403] Revis, Cecil: The stability of the physiological properties of the coliform organisms, Centralbl. f. Bakteriol., II, 1910, 26, p. 161.

→ Revis, Cecil: Variation in B. coli: The production of two permanent varieties from one original strain by means of brilliant green, Proc. Roy. Soc., 1913, 86, p. 373.

[405] Reimann, H. A.: Variations in specificity and virulence of pneumococcus during growth in vitro, J. Exper. Med., 1925, 41, p. 587.

[406] Ritchie, J.: On meningitis associated with an influenza-like bacillus, Jour. Path. & Bact., 1910, 24, p. 615.

[407] Rodet, J.: Sur l'agglutination du B. coli et du B. d'Eberth: I. Sur les races de B. coli au point de vue de leur aptitude agglutinatif. Variabilité de cette proprieté, J. physiol. et path. gén., 1899, 1, p. 806.

[408] Rodet, J.: Bacilles typhiques cadaverique a charactères speciaux. Variabilité de l'aptitude agglutinatif, ibid., 1900, 2, p. 154.

[409] Roger: Propriétés bactéricides du sérum pour le streptocoque de l'érysipile, Sem. méd., 1890, p. 397.

[410] Roger (On the loss of virulence of pneumococci injected into immune animals): Revue gén. des Sciences, 1891, p. 413. (Quoted from Sanarelli[423].)

➡ Rosenow, E. C.: The transmutations of the streptococcus-pneumococcus group, J. Infect. Dis., 1914, 14, p. 1.

➡ Rosenow, E. C.: Streptococci in relation to the etiology of epidemic encephalitis: Experimental results in 81 cases, J. Infect. Dis., 1924, 34, p. 329.

[413] Rosenthal, W.: Beobachtungen über die Variabilität der Bakterien-verbände und die Colonieformen unter verschiedenen physakalischen Bedingungen, Deutsch. Arch. f. klin. Med., 1895, 55, p. 513.

[414] Roux, E.: Bactéridie charbonneuse asporogène, Ann. de l'Inst. Pasteur, 1890, 4, p. 25.

[415] Roux, E., and Chamberland: Sur l'immunité contre le charbon conferée par des substances chimique, ibid., 1888, 2 p. 405.

[416] Roux, E., and Yersin: Contribution a l'étude de la diphterie, ibid., 1890, 4, p. 385.

[417] Ruzicka: Vergleichende Studien über den Bacillen fluorescens liquefaciens, Arch. f. Hyg., 1899, 34, p. 140.

[418] Sacharoff, H.: Ueber die Gewohnung der Milzbrandbacillen an die bakterizidie Wirkung des Serums, Centralbl. f. Bakteriol., I, O., 1904, 37, p. 411.

[419] Sachs, H.: Zur Serodiagnostic des Fleckfiebers, Deutsch. med. Wchnschr., 1917, 43, p. 964.

[420] Sachs, H., and Schlossberger, H.: Arch. a. d. Königl. Inst. f. exper. Therap. zu Frankfurt-A-M., 1919, 7. (Quoted from Goyle.[200])

[421] Sacquépée, E.: Variabilité de l'aptitude agglutinative du Bacille d'Éberth. Ann. de l'Inst. Pasteur, 1901, 15, p. 249.

[422] Sanarelli, J.: Moyens de défense de l'organisme contre les microbes aprés vaccination et dans la guérison, Ann. de l'Inst., Pasteur, 1893, 7, p. 225.

[423] Sanarelli, J.: Etiologie et pathogénie de la fièvre jaune, Ann. de l'Inst., Pasteur, 1897, 11, p. 433.

[424] Sandberg, G.: Ein Beitrag zur Bakteriologie der milchsäurengährung in Magen mit besonderer Berücksichtigung der "langen" Bazillen, Ztschr., f. klin. Med., 1903, 51, p. 80.

[425] Savage, W. G.: Pseudo-clumping in cultures of the typhoid bacillus, J. Path. & Bact., 1901, 7, p. 388.

[426] Sawtchenko: Contribution a l'étude de l'immunté, Ann. de l'Inst., Pasteur, 1897, 11, p. 865.

[427] Schattenfroh, A., and Grassberger, R.: Ueber Buttersäuregährung, Arch. f. Hyg., 1900, 37, p. 54.

[428] Schattenfroh, A., and Grassberger: Neue Beitrage zur Kenntniss der Buttersäuregährungserreger und ihrer Beziehungen zum Rauschbrand, München. med. Wchnschr., 1901, 48, p. 50.

[429] Scheller, R.: Experimentelle Beiträge zur Theorie der Agglutination, Centralbl. f. Bakteriol., I, O., 1904, 36, pp. 427, 694.

[430] Schick, B., and Ersettig, H.: Zur Frage der Variabilität der Diphtheriebazillen, Wien. klin. Wchnschr., 1903, 16, p. 993.

[431] Schierbeck: Ueber die Variabilität der Milchsäurebakterien mit Bezug auf die Gährungsfähigkeit, Arch. f. Hyg., 1900, 38, p. 294.

[432] Schiller, I.: Sur les produits des microbes en association, Centralbl. f. Bakteriol., I, O., 1914, 73, p. 123.

[433] Schlemmer: Ein Beitrag zur Biologie des Typhusbacillus, Ztschr. f. Immunitätsforsch. u. exper. Therap., 1911, 9, p. 180.

[434] Schmitz, K.: Die Verwandlungsfähigkeit der Bakterien. Experimentelles und Kritisches mit besonderer Berücksichtigung der Diphtheriebazillengruppe, Central. f. Bakteriol., I, O., 1916, 77, p. 369.

[435] Schröter and Gutjahr: Vergleichende Studien der Typhus-Coli-Dysenteriebakterien, ibid., 1911, 58, p. 577.

[436] Schütze, H.: The permanence of the serological paratyphoid B types with observations on the non-specificity of agglutination with "rough" variants, Brit. J. Hyg., 1922, 20, p. 330.

[437] Seligmann, E.: Zur Bakteriologie des fädenziehendes Brotes. Ein Beitrag zur Artenentstehung in Bakterienreiche, Centralbl. f. Bakteriol., I, O., 1919, 83, p. 39.

[438] Shaw, E. A.: On exaltation of bacterial virulence by passage in vitro, Brit. M. J., 1903, 1, p. 1074.

[439] Shibayama, G.: Zweite Mitteilung über Agglutination (Versuche mit Pestbazillen), Centralbl. f. Bakteriol., I, Ref., 1906, 38, p. 229.

[440] Shibley, G. S.: The agglutination of streptococci, J. Exper. Med., 1924, 39, p. 245.

[441] Shippen, L. P.: Toxin formation by a variety of B. botulinus when cultivated aerobically under various conditions. Its possible production in the animal body, Arch. Int. Med., 1919, 23, p. 346.

[442] Sierakowsky, S., and Zajdel, R.: Sur les bactéries provoquant la carie dentaire, Compt. rend. Soc. de biol., 1924, 90, p. 961.

[443] Slawyk and Manicatide: Untersuchungen über 30 verschiedene Diphtheriestämme mit Rücksicht auf die Variabilität derselben, Ztschr. f. Hyg., 1898, 29, p. 181.

[444] Smith, T., and Reagh, A. L.: Agglutination affinities of related bacteria parasitic in different hosts, J. M. Res., 1903, 6, p. 277.

[445] Smith, T., and Reagh, A. L.: The nonidentity of agglutinatins acting on flagella and upon the body of bacteria, ibid., 1903, 10, p. 89.

[446] Sobernheim, G., and Seligmann, E.: Beiträge zur Enteritis-bakterien, Ztschr. f. Immunitätsforsch. u. exper. Therap., 1910, 6, p. 401.

[447] Sobernheim, G., and Seligmann, E.: Beobachtungen über die Umwandlung biologischen Eigenschaften von Bakterien, Deutsch. med. Wchnschr., 1910, 36 (1), p. 351.

[448] Sonnenschien, C.: Zur Kenntnis bakteriophagähnlichen Erscheinungen, Centralbl. f. Bakteriol., I, O., 1925, 95, p. 257.

[449] Sörensen, Ejnar: Einer Untersuchungsreihe über die Veränderung einer Urinbakterien in den menschlichen Harnwegen, ibid., 1912, 62, p. 582.

[450] Soule, M. R.: Microbic dissociation of Bacillus subtilis, J. Infect. Dis., 1927. (In press.)

[451] Stanton, A. T., and Fletcher, W.: Melioidosis, disease of rodents communicable to man, Lancet, 1925, 1, p. 10.

[452] Stearn, Esther: A study of the effect of gentian violet on the viability of some waterborne bacteria, Am. J. Pub. Health, 1923, 13, p. 567.

[453] Steinhardt, Edna: Variation in virulence in organisms acted upon by serum and the occurrance of spontaneous agglutination, J. M. Res., 1904, 13, p. 409.

[454] Stryker, Laura: Variations in the pneumococcus induced by growth in immune serum, J. Exper. Med., 1916, 24, p. 49.

[455] Surmont, H., and Arnould, E.: Recherches sur la production du bacille du charbon asporogène, Ann. de l'Inst., Pasteur, 1894, 8, p. 817.

[456] Tanner, F. W.: The yeasts (translation), 1920.

[457] Takami, T.: Agglutination changes of pneumococcus in animal body, Tohoku J. Eper. Med., 1925, 6, pp. 248, 41.

[458] Teague, O., and McWilliams, Helen: Spontaneous agglutination in typhoid and paratyphoid cultures and its bearing on absorption of agglutinin, J. Immunol., 1917, 2, p. 167.

[459] Teague, O., and Torrey, J. C.: A study of gonococcus by the method of "fixation of the complement," J. M. Res., 1907, 17, p. 223.

[460] Thaysen: Studien ueber functionelle Anpassung bei Bakterien, Centralbl. f. Bakteriol., I, O., 1911, 60, p. 1.

[461] Thjotta, Th., and Eide, O. K.: A mutating, mucoid paratyphoid bacillus isolated from the urine of a carrier, J. Bact., 1920, 5, p. 50.

[462] Toenniessen, E.: Ueber Vererbung und Variabilität bei Bakterien mit besonderer Berücksichtigung der Virulenz, Centralbl. f. Bakteriol., I, O., 1914, 73, p. 241.

[463] Tomaselli, C.: Contributo allo studio del batteriofargo, 1923.

→ Topley, W., and Aryton, Joyce: The excretion of B. enteritidis (Aertrycke) in feces of mice, Brit. J. Hyg., 1924, 22, pp. 234, 305.

[465] Torrey, J. C.: Agglutins and precipitins in antigonococcic serum, J. M. Res., 1907, 16, p. 329.

[466] Torrey, J. C., and Buckell, G. T.: A seriological study of the gonococcus group, J. Immunol., 1922, 7, p. 305.

[467] Trautmann and Dale: Beitrag zum Formenkreis des Diphtheriebazillus, Centralbl. f. Bakteriol., I, Ref., 1910, p. 137.

[468] Twort, F. W.: An investigation into the nature of the ultramicroscopic viruses, Lancet, 1915, 2, p. 124.

[469] Valentine, Eugenia: Differences in peroxide production and methemoglobin formation of green (alpha) streptococci, J. Infect. Dis., 1926, 39, p. 48.

[470] Valtis, J.: Sur la filtrabilité du bacille tuberculeux à travers les bougies Chamberland, Ann. de l'Inst. Pasteur, 1924, 38, p. 452.

[471] Vannod, T.: Ueber Agglutinine und specifische Immunkörper in Gonocokkenserum, Deutsch. med. Wchnschr., 1906, p. 1984.

[472] Vaudremer, A.: Un bacille tuberculeux humain, un bacille tuberculeux bovin acidoresistants facultatifs, Compt. rend. Soc. de biol., 1921, 84, p. 259.

[473] Veber, T.: Sur la filtration du bacille tuberculeux du liquide de pneumothorax artificielle sur bougie Chamberland L^2, ibid., 1926, 94, p. 8.

[474] Villazon, N.: Bactériophage efficacé contre la bacille de la peste, ibid., 1923, 89, p. 754.

[475] Villinger, A.: Ueber die Veränderung einiger Lebenseigenschäften des Bakterium coli commune durch äussere Einfluss, Arch. f. Hyg., 1894, 21, p. 101.

[476] Wagner, G.: Beiträge zur Kenntnis der Milzbrand- and Milzbrandänlichen Bacillen, Centralbl. f. Bakteriol., I, O., 1920, 83, p. 386.

477 Walker, E. W.: Immunization against immune serum, J. Path. & Bact., 1903, 8, p. 34.

478 Walker, E. W., and Murray, W.: The effect of certain dyes upon the cultural characteristics of the Bac. typhosus and some other organisms, Brit. M. J., 1904, 2, p. 16.

479 Walker, E. W.: Studies on bacterial variability, Proc. Roy. Soc. B., 1922, 93, p. 54.

480 Wardsworth, A., and Kirkbride, M.: A study of the changes in the virulence of the pneumococcus at different periods of growth and under different conditions of cultivation in media, J. Exper. Med., 1918, 28, p. 791.

481 Webster, L. T.: Biology of Bacterium lepisepticum: I. Effects of oxygen tension and presence of rabbit blood on growth, dissociation and virulence, J. Exper. Med., 1925, 4, p. 571.

482 Weil, E.: Variationsuntersuchungen, Ztschr. f. Immunitätsforsch. u. exper. Therap., 1922, 35, p. 25.

483 Weil, E., and Felix, R.: Weitere Untersuchungen über das Wesen der Fleckfieber-agglutination, Wien. klin. Wchnschr., 1917, 30, p. 1509.

484 Weil, E., and Felix, R.: Ueber den Doppel-typus der Rezeptoren der Typhus-Paratyphus-gruppe, Ztschr. f. Immunitätsforsch. u. exper. Therap., 1920, 29, p. 24.

485 Wilde: Ueber den Bacillus pneumoniae Friedlander, 1896. (Quoted from Wreschner[492].)

486 Winslow, C-E. A., and Winslow, A. R.: The systematic relationships of the Coccaceae, 1908.

487 White, P. B.: Serological studies with regard to the classification of the Salmonella group. Special Rep. 91. Med. Res. Council, 1925.

488 Wilson, W. J.: Pleomorphism as exhibited by bacteria growing in media containing urea, J. Path. & Bact., 1906, 11, p. 394.

489 Wolbach, S. B.: On the filtrability and biology of the spirochaetes, Am. J. Trop. Dis., 1915, 2, p. 494.

490 Wolf, F.: Ueber Modificationen u. experimentell ausgelöste Mutationen bei B. prodigiosus und anderen Schizophyten, Ztschr. f. indukt. Abstämmungs u. Vererbungsl., 1909, 2, p. 90.

491 Wollstein, M.: Biological relationships of Diplococcus intracellularis and the Gonococcus, J. Exper. Med., 1907, 9, p. 588.

492 Wreschner, H.: Untersuchungen über die biologische Bedeutung der Kapsel beim Micrococcus tetragenus, Ztschr. f. Hyg., 1921, 93, p. 74.

493 Wyschelessky, S.: Bemerkenswerthe Befunde bezüglich des Wachtums des Bazillus der Schweinerotlaufs, Zeitschr. f. Infektionsk., parasitäre Krankh. u. Hyg. d. Haustier., 1912, 12, p. 43.

494 Yoshioka, M.: Untersuchungen über Pneumokokkenimmunität: II. Veränderungen der Agglutination bei Pneumokokken des Typus I, II und III und bei Streptokokken, Ztschr. f. Hyg., 1923, 97, p. 232.

495 Zdansky, E.: Gewinnung specifischer Bakteriophagen und über Bakteriophagentherapeutische Versuche, Med. Klin., 1924, 20, p. 1485.

496 Zinnser, H.: Infection and Resistance, 1920.

497 Zoeller, Chr.: Bacille de Shiga autoagglutinable, Compt. rend. Soc. de biol., 1921, 84, p. 87.

498 Zupnik, L.: Ueber Variabilität de Diphtheriebacillen, Berlin. klin. Wchnschr., 1897, 34, p. 1085.

499 Personal communication to the author.

ADDENDA

500 Bronfenbrenner, J.; Muckenfuss, R. S., and Korb, C.: Studies on the bacteriophage of d'Herelle: VI. On the virulence of the overgrowth in the lysed cultures of Bacillus pestis caviae (M. T. II), J. Exper. Med., 1926, 44, p. 605.

500a Burnet, Et.: Sur la recherche de formes filtrantes des bactéries, Compt. rend. Soc. de biol., 1926, 95, p. 1142.

501 Coleman, G. E., and Meyer, K.: Characteristics of a new strain of Cl. tetani, J. Infect. Dis., 1926, 39, p. 328.

502 Draper, A. A.: The biology of Oidium albicans with special reference to mycelial production, J. Infect. Dis., 1926, 39, p. 261.

503 Goyle, A. N.: A comparison of the pathogenicity for mice of normal and rough forms of Salmonella enteritidis Gärtner, J. Path. & Bact., 1926, 29, p. 365.

504 Holman, W. L., and Meekison, D. M.: Gas production by bacterial synergism, J. Infect. Dis., 1926, 39, p. 145.

505 Julianelle, L. A.: Immunological relationships of encapsulated and capsule-free strains of Encapsulatus pneumoniae (Friedländer's bacillus), J. Exper. Med., 1926, 44, p. 683.

506 Julianelle, L. A.: Immunogical relationships of cell constituents of encapsulatus pneumoniae (Friedländer's bacillus), ibid., p. 735.

507 Karwacki, L.: Pluralité des types bacillaires dans les cultures tuberculeuses, Compt. rend. Soc. de biol., 1926, 95, p. 403.

[608] Karwacki, L.: Mutation du bacille tuberculeux dans le liquide des pleurésies tuberculeuses, ibid., 1926, 95, p. 913.

[609] Kindborg, A.: Die Pneumokokken, Ztschr. f. Hyg., 1905, 51, p. 197.

[610] Lepierre: Subsidio para o estudo do meningococco, Centralbl. f. Bakteriol., I, Ref., 1904, 35, pp. 43, 101.

[611] Lommel, Jeanne: Influence de phenol, du formol et de certaines associations microbiennes sur les propriétés biochemique du colibacille, Compt. rend. Soc. de biol., 1926, 95, p. 711.

[612] Lommel, Jeanne: Modifications des propriétés biochemique du colibacille sous l'influence de milieux additionés des matéries colorantes, ibid., 1926, 95, p. 714.

[613] Ørskov, J., and Larsen, A.: On bacterial variation, J. Bact., 1925, 10, p. 473.

[614] Rosenthal, L.: Sur la production de Bactéridies charbonneuses asporogènes par un procede noveau, Compt. rend. Soc. de biol., 1926., 95, p. 445.

→ Shousha, A. T.: Spontaneous agglutination of the cholera vibrio in relation to variability, J. Hyg., 1924, 22, p. 156.

[616] Sorgente, P.: Weitere Untersuchungen über den Meningokokken, Centralbl. f. Bakteriol., I, O., 1905, 39, p. 1.

[617] J. Immunol., 1924, 9, p. 115.

[618] Ibid., 1926, 11, p. 31.

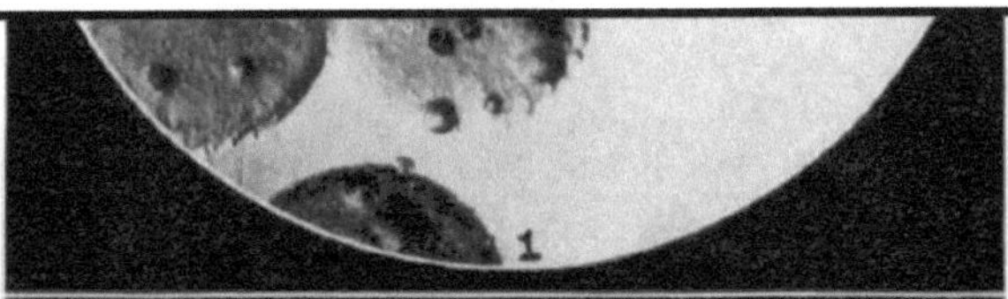

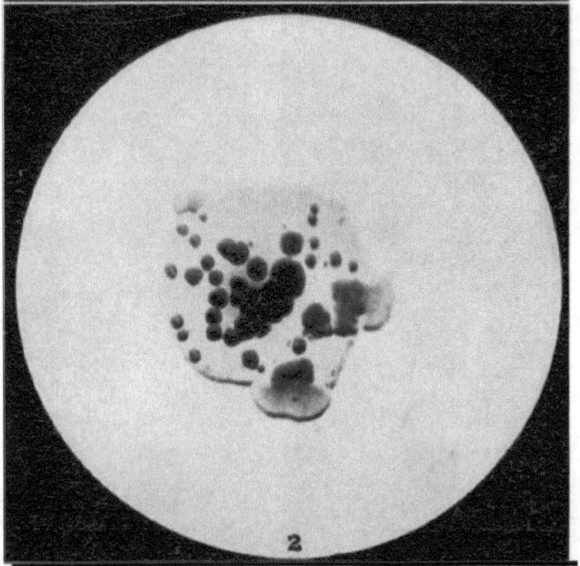

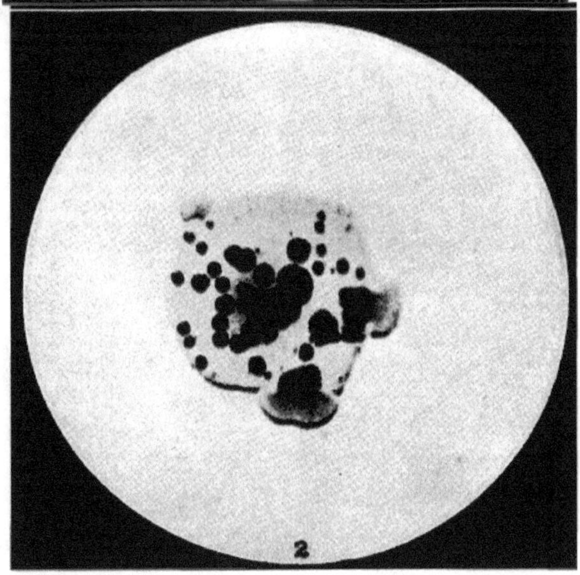

Fig. 1.—Vibro cholerae. Fourteen day old "helle" (S) colonies showing large surface secondary or daughter colonies. From Eisenberg,[150] magnification not given.

Fig. 2.—B. typhosus. Four day old colony on rhamnose agar, showing surface secondary colonies. From Reiner Müller:[858] Taf. II, fig. 3; × 5.

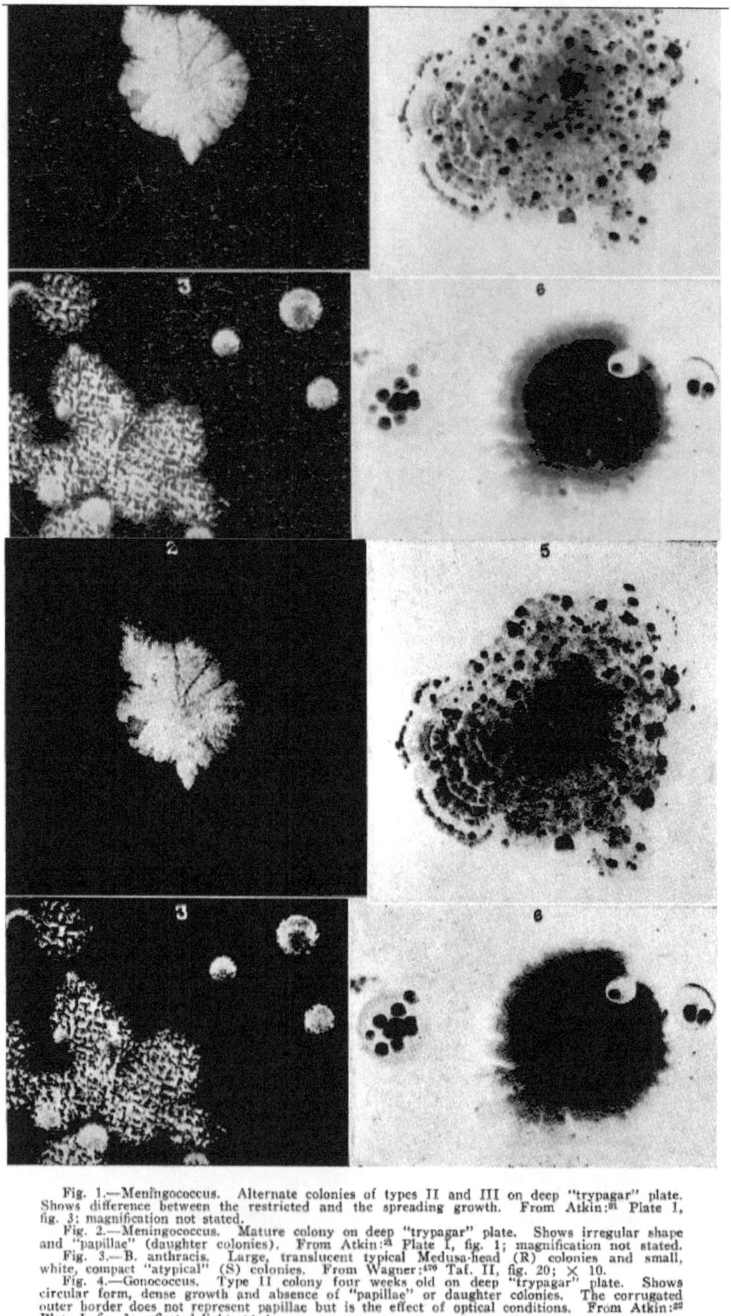

Fig. 1.—Meningococcus. Alternate colonies of types II and III on deep "trypagar" plate. Shows difference between the restricted and the spreading growth. From Atkin:[21] Plate I, fig. 3; magnification not stated.

Fig. 2.—Meningococcus. Mature colony on deep "trypagar" plate. Shows irregular shape and "papillae" (daughter colonies). From Atkin:[21] Plate I, fig. 1; magnification not stated.

Fig. 3.—B. anthracis. Large, translucent typical Medusa-head (R) colonies and small, white, compact "atypical" (S) colonies. From Wagner:[470] Taf. II, fig. 20; × 10.

Fig. 4.—Gonococcus. Type II colony four weeks old on deep "trypagar" plate. Shows circular form, dense growth and absence of "papillae" or daughter colonies. The corrugated outer border does not represent papillae but is the effect of optical conditions. From Atkin:[22] Plate I, fig. 3; reflected light, × 4.

Fig. 5.—Gonococcus. Type I colony four weeks old on deep "trypagar" plate. Shows irregular shape, opaque papillae (daughter colonies) and transparent intervening area. Zonal effect shown on right-hand border. From Atkin:[22] Plate I, fig. 1; transmitted light, × 4.

Fig. 6.—B. typhosus. One "mutated" and three "mutating" colonies on rhamnose agar plate. The latter show daughter colonies. The large colony shows a regeneration fringe. From Reiner Müller:[358] Taf. I, fig. 2; × 5.

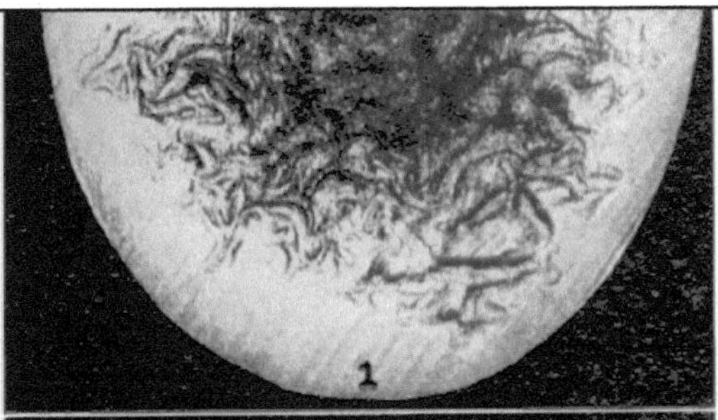

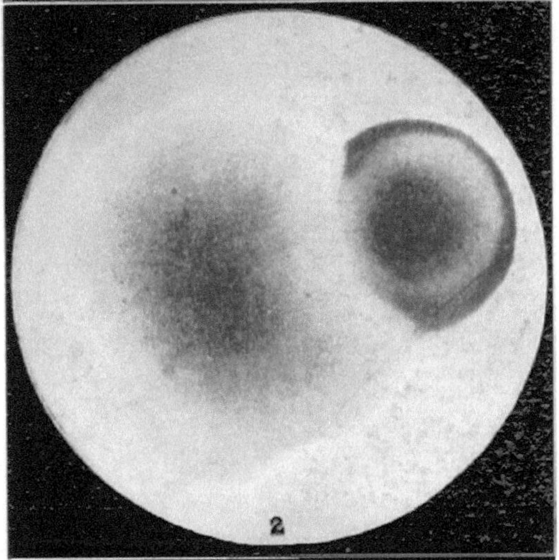

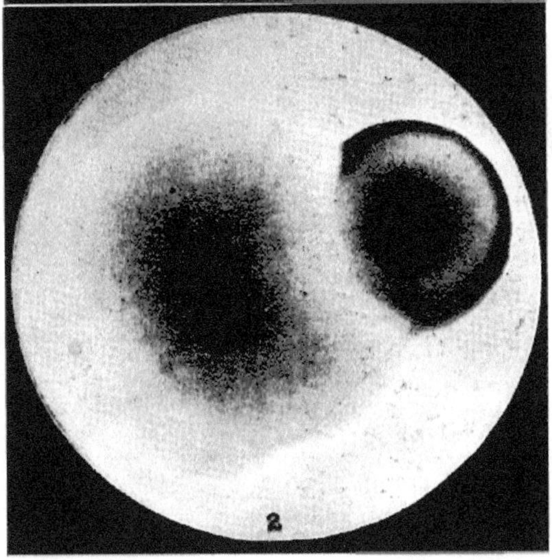

Fig. 1.—B. anthracis. Upper colony: Round, compact, opaque, whitish and "smooth," showing the "Krauskopf" appearance at border. Lower colony: Irregular, spreading, translucent, rough; showing typical anthrax, Medusa-head type. These two colonies are the same as the two colonies shown in the upper left corner of fig. 3 of plate 2. From Wagner:[476] Taf. II, fig. 22; transmitted light, × 50.

Fig. 2.—Vibrio cholerae. "Helle" (S) and dark, granular (O?) colonies, three days old. The dark colony contained many deep secondaries which do not appear in the photograph. From Eisenberg:[150] Taf. I, fig. 5; magnification not stated.

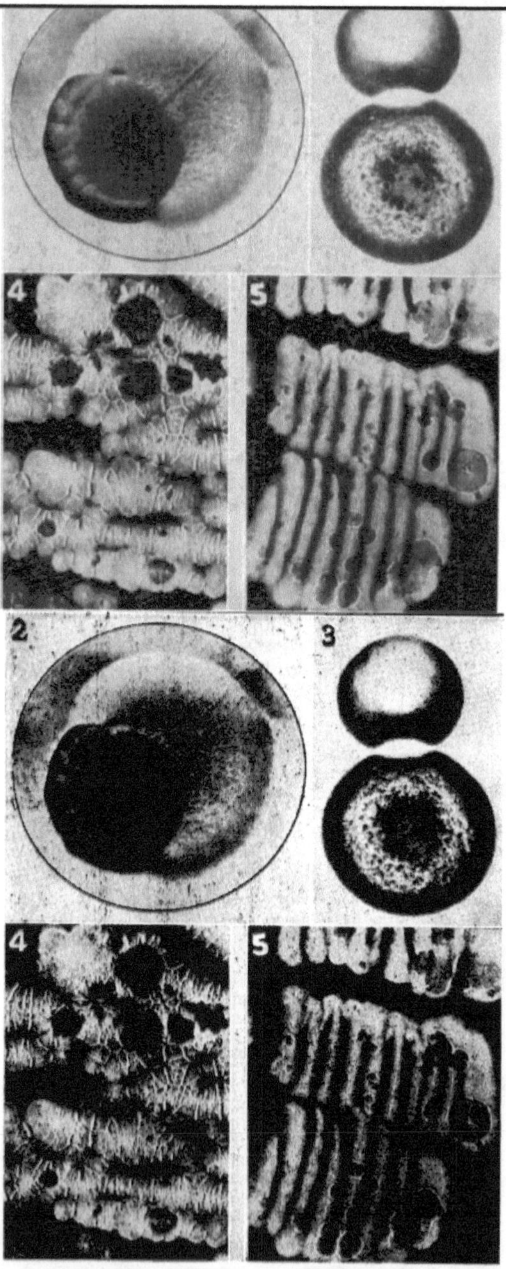

Fig. 1.—B. pyocyaneus. Lytic (and lysogenic) colony seven days old on beef-infusion-agar. Showing mass lysis, lytic pockets and marginal lysis well under way. For earlier stage of same colony see plate 5, fig. 2. From Hadley:[224] Plate II, fig. 7; reflected light, × 1.0.

Fig. 2.—Vibrio cholerae. "Helle" (S) (larger) and "dunkel" (O?) (smaller) colonies five days old on cholera-agar plates. From Eisenberg:[150] Taf. I, fig. 1; magnification not stated.

Fig. 3.—Vibrio cholerae. "Helle" (upper) and transitional (lower) colonies one day old. The latter shows the lysing central area. From Eisenberg:[150] Taf. I, fig. 2; magnification not stated.

Fig. 4.—Monilia culture from throat. Streak of Monilia on agar, showing lysing areas in the background of culture. From Sonnenschien:[448] Taf. I, fig. 5; magnification not stated.

Fig. 5.—B. pyocyaneus. Streaks on agar showing lysing areas against background of culture. From Sonnenschien:[448] Taf. I, fig. 7; magnification not stated.

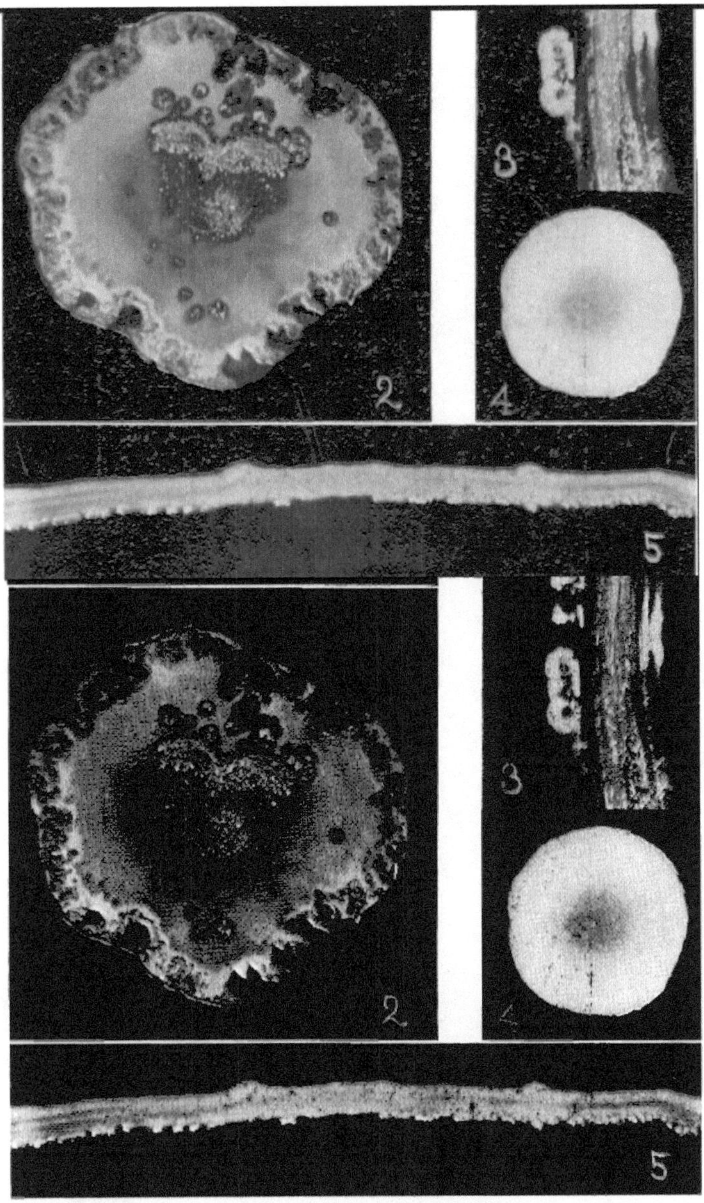

Fig. 1.—B. pyocyaneus. Three days old lytic (and lysogenic) colony on beef-infusion-agar. Shows mass lysis and zonal areas of mass lysis. From Hadley:[224] Plate I, fig. 1; reflected light, × 1.2.

Fig. 2.—B. pyocyaneus. Four days old lytic (and lysogenic) colony on beef-infusion agar. Shows mass lysis, lytic pockets and marginal lysis. (For later stage of same colony see plate 4, fig. 1.) From Hadley;[224] Plate I, fig. 2; reflected light, × 1.0

Fig. 3.—B. pyocyaneus. Streak growth of lytic culture five days old on beef-infusion-agar. Shows marginal lysis and erosions; also the proliferative growth as compared with fig. 5. From Hadley:[224] Plate I, fig. 3; reflected light, × 1.2.

Fig. 4.—B. pyocyaneus. R type colony on beef-infusion-agar, four days old. Shows smooth (unusual) surface. From Hadley:[224] Plate I, fig. 4; reflected light, × 1.0.

Fig. 5.—B. pyocyaneus. Five day old streak of R growth on beef-infusion-agar. Shows even growth free from lytic changes. Compare in extent of growth with fig. 3. From Hadley:[224] Plate I, fig. 5; reflected light, × 1.2.

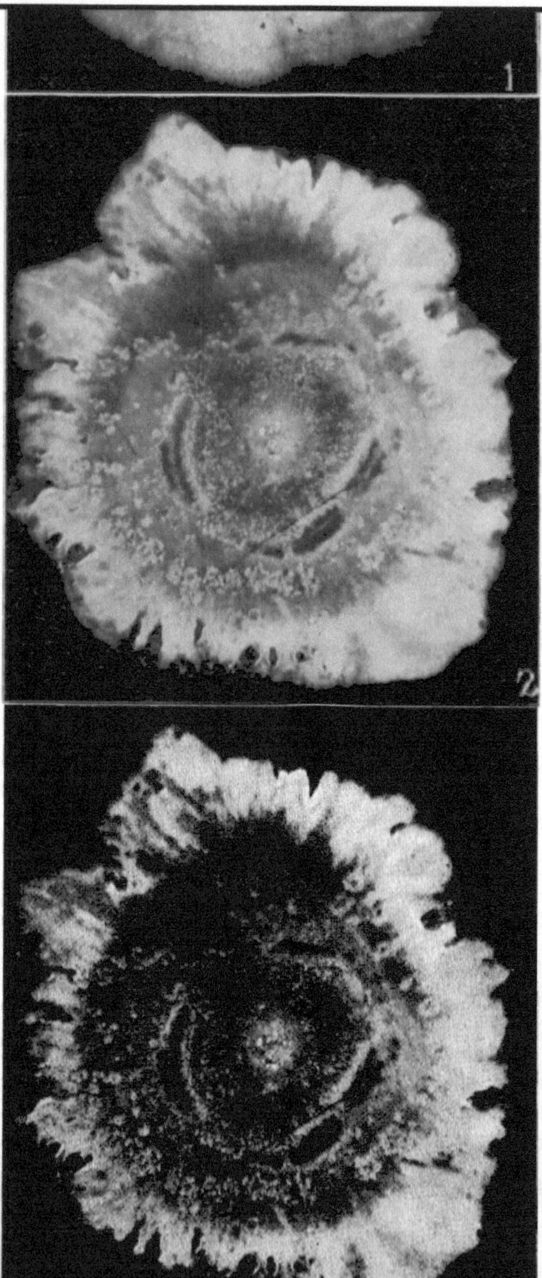

Fig. 1.—B. pyocyaneus. Lytic colony on infusion agar, 8 days old; showing radial and concentric zones of lysis, and the ringed lytic pockets and crystalline deposits. × 1.0. Hadley:[224] Plate 3, fig. 11.

Fig. 2.—B. pyocyaneus. Lytic colony on infusion agar, 8 days old; showing zonal areas of lysis and many lytic pockets. × 1.2. Hadley:[224] Plate 3, fig. 12.

www.ingramcontent.com/pod-product-compliance
Lightning Source LLC
Chambersburg PA
CBHW020727180526
45163CB00001B/134